LINEAR PROGRAMMING
A Modern Integrated Analysis

INTERNATIONAL SERIES IN
OPERATIONS RESEARCH & MANAGEMENT SCIENCE

Frederick S. Hillier, Series Editor
Department of Operations Research
Stanford University
Stanford, California

LINEAR PROGRAMMING
A Modern Integrated Analysis

ROMESH SAIGAL
Department of Industrial and Operations Engineering
The University of Michigan
Ann Arbor, Michigan 48108-2117 USA

Kluwer Academic Publishers
Boston/Dordrecht/London

Distributors for North America:
Kluwer Academic Publishers
101 Philip Drive
Assinippi Park
Norwell, Massachusetts 02061 USA

Distributors for all other countries:
Kluwer Academic Publishers Group
Distribution Centre
Post Office Box 322
3300 AH Dordrecht, THE NETHERLANDS

Library of Congress Cataloging-in-Publication Data

Saigal, Romesh.
    Linear programming : a modern integrated analysis / Romesh Saigal.
        p. cm. -- (International series in operations research & management science ; 1)
    Includes bibliographical references and index.
    ISBN 0-7923-9622-7
    1. Linear programming. I. Title. II. Series.
T57.74.S23   1995
519.7'2--dc20   95-34605
                                                                                       CIP

Copyright © 1995 by Kluwer Academic Publishers

All rights reserved. No part of this publication may be reproduced, stored in a retrieval system or transmitted in any form or by any means, mechanical, photo-copying, recording, or otherwise, without the prior written permission of the publisher, Kluwer Academic Publishers, 101 Philip Drive, Assinippi Park, Norwell, Massachusetts 02061

Printed on acid-free paper.

Printed in the United States of America

Dedicated to

Shailesh, Ashima, David and Veena

Contents

	Preface	ix
1	**Introduction**	1
1.1	The Problem	1
1.2	Prototype Problems	2
1.3	About this Book	4
1.4	Notes	5
2	**Background**	7
2.1	Real Analysis	7
2.2	Linear Algebra and Matrix Analysis	13
2.3	Numerical Linear Algebra	20
2.4	Convexity and Separation Theorems	27
2.5	Linear Equations and Inequalities	31
2.6	Convex Polyhedral Sets	36
2.7	Nonlinear System of Equations	44
2.8	Notes	64
3	**Duality Theory and Optimality Conditions**	67
3.1	The Dual Problem	67
3.2	Duality Theorems	70
3.3	Optimality and Complementary Slackness	73
3.4	Complementary Pair of Variables	78
3.5	Degeneracy and Uniqueness	80
3.6	Notes	83
4	**Boundary Methods**	85
4.1	Introduction	85
4.2	Primal Simplex Method	87
4.3	Bounded Variable Simplex Method	98

	4.4	Dual Simplex Method	103
	4.5	Primal - Dual Method	107
	4.6	Notes	110
5	**Interior Point Methods**		**111**
	5.1	Primal Affine Scaling Method	112
	5.2	Degeneracy Resolution by Step-Size Control	139
	5.3	Accelerated Affine Scaling Method	148
	5.4	Primal Power Affine Scaling Method	166
	5.5	Obtaining an Initial Interior Point	198
	5.6	Bounded Variable Affine Scaling Method	200
	5.7	Affine Scaling and Unrestricted Variables	203
	5.8	Dual Affine Scaling Method	206
	5.9	Primal-Dual Affine Scaling Method	211
	5.10	Path Following or Homotopy Methods	222
	5.11	Projective Transformation Method	244
	5.12	Method and Unrestricted Variables	252
	5.13	Notes	262
6	**Implementation**		**265**
	6.1	Implementation of Boundary Methods	265
	6.2	Implementation of Interior Point Methods	270
	6.3	Notes	304
A	**Tables**		**307**
	Bibliography		**315**
	Index		**338**

Preface

During the past decade, linear programming has gone through a dramatic change. The catalyst for this change is the path breaking work of Narinder K. Karmarkar who, in 1984, presented a polynomial time algorithm generating a sequence of points in the interior of the polyhedron. This method contrasts with the earlier state-of-the art methods that generate a sequence of points on the boundary of the polyhedron. The most notable, and the earliest of these is the simplex method of George B. Dantzig, produced in 1947. The new methods that proceed through the interior of the polyhedron avoid the combinatorial structure of the boundary, which the boundary methods must necessarily contend with. A consequence of this is that even though many of the interior point methods are shown to be polynomial, an example produced in 1971 by Victor L. Klee and George J. Minty demonstrates that in the worst case, simplex method can take an effort which grows exponentially in the data of the problem. Thus, for large problems, the interior point methods may be advantageous. A growing body of computational experience appears to support this conclusion.

Interior point methods trace their roots to the work of I. I. Dikin in 1967. His method is called the (primal) affine scaling method, and was rediscovered by researchers at A. T. & T. Bell Telephone Laboratories while implementing the ideas of Karmarkar in a system called KORBX. Several variants of this method, like the dual and the primal-dual were also generated in the process. Since then, many new methods like the sliding objective function methods, the path-following barrier methods, and potential reduction methods have been generated. Several successful implements of these also exist in the public domain.

This book presents both the boundary and the interior point methods in a unified manner. Approach taken here involves the use of duality theory, in particular the complementary slackness conditions, to derive both the boundary and interior point methods. Most text books on linear programming derive the duality theorems after presenting the computational aspects of the simplex method. And the theorems of the alternative (like Farkas' lemma) are then derived as corollaries to the duality theorems. This approach is unsatisfactory for our purpose. The approach taken here is to prove Farkas' lemma and other theorems of the alternative as convex separation theorems, and then derive the duality theorems directly from this lemma. Thus the tedium of the simplex method is avoided. In addition several concepts like the existence of strict complementary solutions are generally derogated to an exercise at the end of the duality chapter. These results play an important role in interior point methods, and are thus proved explicitly in the duality chapter of this book.

The author has been teaching and developing the methodology for presenting these ideas in several courses taught at the University of Michigan in Ann Arbor, Michigan and the Naval Postgraduate School in Monterey, California and has paid special attention to the mathematical background required for the study of interior point methods. To study the boundary methods, it is sufficient to have a good background in linear algebra, numerical linear algebra and polyhedral convexity. The study of recent methods, in addition, requires sufficient background in real analysis, convexity and separation theorems as well as an understanding of the methodology for solving nonlinear systems of equations, specially Newton's method and homotopy or global Newtons' methods and their implementations. Thus a chapter presenting this background material has been included in this manuscript.

This book is organized into six chapters. Chapter 1 introduces the problem and some standard formulations to orient the reader to linear programming. In the second chapter, careful attention is paid to the necessary mathematical background required for the study of these methods. The third chapter presents the duality theorem and also the important results on complementary slackness conditions. The fourth chapter presents the three important variants of the simplex method, the primal, the dual and the primal dual methods. The fifth chapter presents the interior point methods. It covers the affine scaling methods, the projective transformation methods, the barrier methods as well as the primal-dual homotopy or global Newton methods. Finally the chapter six is devoted to the implementation of both the interior and the boundary methods. Sparsity preserving and stable methods for the boundary and interior methods are also discussed in this chapter.

This book can be used as a text in a one or two semester advanced graduate course in linear programming. A good two semester sequence could cover sections 2.4 - 2.6, terminating with the proof of Farkas' Lemma. Then after proving the duality theorems of Chapter 3, establish the pri-

PREFACE

mal simplex method. Then cover Chapter 5 and establish the primal and dual affine scaling method, and the path following method. During the first semester, cover only the convergence of boundary and interior methods with the non-degeneracy assumption. During the second semester, prove the convergence of these methods under degeneracy, establish the dual simplex method, primal-dual boundary methods; and the power affine scaling method. During this semester, also cover the implementations of Chapter 6.

The writing of this book would never have been completed without encouragement, help and comments from several people. This is gratefully acknowledged from John R. Birge, Steve Chen, S.-C. Fang, Hans Kremers, Masakazu Kojima, Prashant M. Kulkarni, Masakazu Muramatsu, Shinji Mizuno, Sarat Puthenpura, Krishnamachari S. Ramprasad, Lakshman P. Sinha, Manmohan S. Sodhi, Samer Takriti, Takashi Tsuchiya and Terrance C. Wagner. I am also grateful to the editor Gary Folven of Kluwer Academic Publishers, for his kind words and constant encouragement. Some of the research presented in this book was supported by the grant number CCR 9321550 from the National Science Foundation. I thank NSF (and the project director S. Kamal Abdali) for this support.

Romesh Saigal
Ann Arbor, Michigan.
June 1, 1995

LINEAR PROGRAMMING
A Modern Integrated Analysis

Chapter 1

Introduction

1.1 The Problem

This book is concerned with linear programming, that is the problem of minimizing or maximizing a linear function subject to linear equality or inequalities constraints. The canonical form of this problem is

$$\begin{aligned} \text{minimize} \quad & c^T x \\ Ax &= b \\ x &\geq 0 \end{aligned}$$

where A is an $m \times n$ matrix, generally of rank m, b is an m dimensional vector and c is an n dimensional vector. This optimization problem appears in numerous applications in diverse areas as transportation and telecommunications involving problems of crew-scheduling, call routing, facilities planning etc.

History of linear programming starts in 1939, when the first such problem, dealing with organization and planning of production, was posed and solved by L. V. Kantorovich. Starting with his discovery of simplex method in 1949, George B. Dantzig has continued to play a major role in the development of this field. His method has been and continues to be the method of choice for solving this problem.

In 1984, Narinder K. Karmarkar presented a method based on the projective transformation of a simplex which he showed solves any linear program in time that is a polynomial function of data of the problem. This method contrasts with the simplex method for which, in 1971, Victor Klee and George J. Minty produced an example that can be shown to take time that grows exponentially in n, the number of variables in the linear program.

Earlier in 1979, L. G. Khachian had produced a method, called the ellipsoidal method, which was shown to solve any linear program in polynomial time. When implemented, his method was not competitive with simplex method and thus lost favor. On the other hand, implementations of various generalizations of Karmarkar's idea, which go under the name of interior point methods, are very competitive with simplex method and generally outperform it for large problems. The first such method, called the affine scaling method, was presented, by I. I. Dikin, in 1967. Its convergence properties have only recently been understood.

In this chapter we will present some prototype linear programming problems.

1.2 Prototype Problems

Linear programs are very widely formulated and solved. We now present two of the earliest formulations and two prototype problems to motivate the reader to these problems.

1.2.1 Transportation Problem
A single commodity is shipped from m warehouses to n retail stores. The number of available items of this commodity at warehouse i is $s_i \geq 0$, for each $i = 1, 2, \cdots, m$, and the number demanded at each store j is $d_j \geq 0$ for each $j = 1, 2, \cdots, n$. Given that it costs $\$c_{i,j}$ to ship one unit of this commodity from warehouse i to store j, find the least cost quantity to ship from warehouse i to store j, for each $i = 1, 2, \cdots, m$ and $j = 1, 2, \cdots, n$.

When $\sum_{i=1}^{m} s_i = \sum_{j=1}^{n} d_j$, i.e., supply equals demand, this problem is called a balanced transportation problem. In this case this problem has a solution. In case it is not balanced, and supply exceeds demand, a solution can still be found that will meet all demands, but if demand exceeds supply, there is no solution which can meet all demands. In the supply not equal to demand case, an equivalent balanced problem can be formulated and a least cost solution found.

Define $x_{i,j}$ to be the quantity shipped from warehouse i to store j, for each $i = 1, 2, \cdots, m$ and $j = 1, 2, \cdots, n$. Then the transportation linear

1.2 PROTOTYPE PROBLEMS

program is:

$$\begin{aligned}
\text{minimize } & \sum_{i=1}^{m}\sum_{j=1}^{n} x_{i,j} \\
& \sum_{j=1}^{n} x_{i,j} = s_i & \text{for each } i = 1, 2, \cdots, m \\
& \sum_{i=1}^{m} x_{i,j} = d_j & \text{for each } j = 1, 2, \cdots, n \\
& x_{i,j} \geq 0 & \text{for each } i = 1, 2, \cdots, m \\
& & \text{and } j = 1, 2, \cdots, n.
\end{aligned}$$

In this linear program, the size of the matrix A is $(m+n) \times n^2$, of vector b is $m + n$ and of vector c is n^2. This problem has other nice properties. When the data s_i and d_j is integral, even though the variables are required to be reals numbers, there is an optimal solution in which each variable is integer.

1.2.2 Diet Problem

A diet consists of a mixture of several ingredients. This mix is done so that the diet contains a certain level of different nutrients, like vitamin, protein, calcium. fat, carbohydrate, sugar etc. Assume that the diet can contain a mix of n ingredients, $j = 1, 2, \cdots, n$ where the cost of ingredient j is known to be c_j. Also, there are m important nutrients in this diet, $i = 1, 2, \cdots, m$. Further assume that one unit of ingredient j contains $a_{i,j}$ units of nutrient i, for each $i = 1, 2, \cdots, m$ and $j = 1, 2, \cdots, n$. In addition, the acceptable level of nutrient i in the mix is between $l_i \geq 0$ and $u_i \geq 0$ for each $i = 1, 2, \cdots, m$. The diet problem is to find the least cost diet that meets all the required nutritional requirements.

Let x_j be the quantity of ingredient j in the mix. Then the resulting linear program is:

$$\begin{aligned}
\text{minimize } & \sum_{j=1}^{n} c_j x_j \\
l_i \leq & \sum_{j=1}^{n} a_{i,j} x_j \leq u_i & \text{for each } i = 1, 2, \cdots, m \\
& \sum_{j=1}^{n} x_j = 1 \\
& x_j \geq 0 & \text{for each } j = 1, 2, \cdots, n
\end{aligned}$$

This linear program is formulated to produce one unit of the diet.

1.2.3 Profit Maximization Model

A company produces and sells n products, and uses m resources in their production. For each $j = 1, \cdots, n$, the production and sale of one unit of product j yields a profit of \$$c_j$. For each $i = 1, \cdots, m$, the maximum quantity of resource i available for the production of these products is b_i. In addition it is known that producing one unit of product j requires $a_{i,j}$ units of resource i, for each $i = 1, \cdots, m$ and $j = 1, \cdots, n$. Find how many units of each product should this firm produce to maximize its profit.

Let x_j be the quantity of product j produced. Then solving the following linear program gives the profit-maximizing quantities:

$$\begin{array}{rl} \text{maximize} & \sum_{j=1}^{n} c_j x_j \\ & \sum_{j=1}^{n} a_{i,j} x_j \leq b_i \quad \text{for each } i = 1, \cdots, m \\ & x_j \geq 0 \quad \text{for each } j = 1, \cdots, n. \end{array}$$

In this linear program the size of the matrix A is $m \times n$, the vector b is m and the vector c is n.

1.2.4 Cost Minimization Model A company runs n processes j, $j = 1, \cdots, n$, each of which produces m products i, $i = 1, \cdots, m$. The company is required to supply at least b_i units of product i, for each $i = 1, \cdots, m$. For each $j = 1, \cdots, n$, operating processor j at unit level costs \$$c_j$ and produces $a_{i,j}$ units of item i for each $i = 1, \cdots, m$. Find at how many units should the process j be operated?

Let x_j units be the level at which processor j is operated, for $j = 1, \cdots, n$. Then solving the following linear program gives the least cost operating levels:

$$\begin{array}{rl} \text{minimize} & \sum_{j=1}^{n} c_j x_j \\ & \sum_{j=1}^{n} a_{i,j} x_j \geq b_i \quad \text{for each } i = 1, \cdots, m \\ & x_j \geq 0 \quad \text{for each } j = 1, \cdots, n. \end{array}$$

In this linear program the size of the matrix A is $m \times n$, the vector b is m and the vector c is n.

1.3 About this Book

This book presents a unified treatment of both boundary and interior point methods for solving the linear programming problem. Unlike most recent text books on this subject, we follow the approach of D. Gale [75] and prove the duality results before the simplex method is introduced and then derive both types of algorithms form these theorems. This approach has the advantage of giving equal time to both boundary and interior methods, and reduces the tedium of learning the simplex method first. Also, only the revised form of the simplex method is derived, the tableau method need never be introduced. This book also contains many new research results, and presents many results that are only available as technical reports.

This book can be used in a two semester advanced course on linear programming. Selected topics for the first semester include sections 2.4-2.6 on convexity and separation theorems, Farkas' lemma and polyhedral convexity. Then the duality theorems can be proved followed by the development of the revised simplex method, the dual simplex method and the bounded variable version of the simplex method. This can be followed with the development of the primal affine scaling method, the dual affine scaling method, the bounded variable version of the primal affine scaling method, and the path following method of Chapter 5. Proof of convergence under the non-degeneracy assumption can be covered during this semester. In the second semester the primal-dual boundary method, the power primal affine scaling method, superlinear convergent variants, the proof of convergence without the non-degeneracy assumption, and the implementation of these methods can be covered.

1.4 Notes

The linear programming problem is formulated and used extensively in many applications. Here we present two early models and two prototype examples to motivate the reader. Many other formulations can be found in the many text books on this subject. Three notable ones are Bazaraa, Jarvis and Sherali [17], Dantzig [36] and Murty [170].

The first uses of linear programming can be found in the following references: input-output model, Leontief [130], transportation model, Hitchcock [100] and Koopmans [123] and production planning models, Kantorovich [110].

Chapter 2

Background

2.1 Real Analysis

We will deal with, R^n, the n dimensional Euclidean space in this work, and we now give the basic facts from topology that we will need.

2.1.1 Norms Given any vector v in R^n we define the quantity

$$||v||_2 = (\sum_{i=1}^{n} v_i^2)^{\frac{1}{2}}$$

as the Euclidean norm of v. Euclidean norm of a vector represents the shortest distance of the point from the origin. There are many other norms that can be assigned to a given vector, and we will use the norm best suited to the occasion. A general norm can be defined by the following axiomatic system:

A norm is the real valued function

$$||.|| : R^n \to R$$

which satisfies the following properties:

1. $||v|| \geq 0$ for all v; and $||v|| = 0$ if and only if $v = 0$.

2. $||\lambda v|| = |\lambda| ||v||$ for all reals λ.

3. $||u + v|| \leq ||u|| + ||v||$ for all u and v in R^n.

Here are some other norms on $\boldsymbol{R}^n$:

$$||v||_\infty = \max_i |v_i|$$
$$||v||_1 = \sum_{i=1}^{n} |v_i|$$

A fundamental result about norms is that they are all equivalent. This is stated, without proof, in the next theorem.

2.1.2 Theorem Let $||.|| : \boldsymbol{R}^n \to \boldsymbol{R}$ be any norm. Then there exist constants $\alpha > 0$ and $\beta > 0$ such that

$$\alpha ||v||_2 \leq ||v|| \leq \beta ||v||_2$$

for all $v \in \boldsymbol{R}^n$. ∎

2.1.3 Unit Ball Let $x \in \boldsymbol{R}^n$ be arbitrary. Then

$$\bar{B} = \{y : ||x - y|| \leq 1\}$$

is called the closed unit ball containing x, and

$$B = \{y : ||x - y|| < 1\}$$

the open ball containing x. Also, for every $\epsilon > 0$

$$B(x, \epsilon) = \{y : ||x - y|| < \epsilon\}$$

is called the open ball of size ϵ containing x. Similarly a closed ball of size ϵ is defined.

Given a subset $V \subset \boldsymbol{R}^n$, and $x \in V$, we call the open ball $B(x, \epsilon)$ an open neighborhood of x in V if and only if $B(x, \epsilon) \subset V$.

Convergence of Sequences

By $\{x^k\}_{k=1}^{r}$ we represent the finite sequence $x^1, \cdots, x^r$ and by $\{x^k\}_{k=1}^{\infty}$ the infinite sequence $x^1, x^2, \cdots$. By the subsequence K we represent the sequence $\{x^k\}_{k \in K}$ for $K \subseteq \{1, 2, \cdots\}$ which is the infinite sequence $x^{i_1}, x^{i_2}, \cdots$ where $K = \{i_1, i_2, \cdots\}$. In this section we investigate conditions when infinite sequences converge, and, when they have convergent subsequences.

2.1 REAL ANALYSIS

2.1.4 Definition Given an infinite sequence $\{x^k\}_{k=1}^{\infty}$, we say it converges to x^∞ if for every $\epsilon > 0$, there exists an $L \geq 1$ such that for all $k \geq L$, $||x_k - x_\infty|| < \epsilon$; and represent it by

$$\lim_{k \to \infty} x^k = x^\infty.$$

Also, we say that it has a convergent subsequence if there exists a K such that the subsequence $\{x^k\}_{k \in K}$ converges.

We now state several theorems, without proof, about converging sequences. The proofs can be found in any book on real analysis, for example Rudin [190].

2.1.5 Theorem Let $\{x^k\}$ be an infinite sequence in $\mathbf{R}^n$.

1. The sequence $\{x^k\}$ converges to x^∞ if and only if every neighborhood of x^∞ contains all but a finitely many terms of $\{x^k\}$.

2. If $\{x^k\}$ converges to x^∞ and to $x^{\infty'}$ then $x^\infty = x^{\infty'}$.

3. If $\{x^k\}$ converges, then it is bounded. ∎

2.1.6 Theorem Let $\{x^k\}$ and $\{y^k\}$ be sequences that converge respectively to x^∞ and y^∞.

1. $\lim_{k \to \infty}(x^k + y^k) = x^\infty + y^\infty$

2. $\lim_{k \to \infty}(\lambda x^k) = \lambda x^\infty$ for all λ.

3. $\lim_{k \to \infty}(x^k)^T y^k = (x^\infty)^T y^\infty$.

∎

2.1.7 Theorem Every bounded sequence in $\mathbf{R}^n$ contains a convergent subsequence. ∎

Definition An infinite sequence $\{x^k\}$ is called a Cauchy sequence if for every $\epsilon > 0$ there is an integer $L \geq 1$ such that for j and k greater than L

$$||x^j - x^k|| < \epsilon.$$

The importance of Cauchy sequences is the following:

2.1.8 Theorem $\{x^k\}$ converges if and only if it is a Cauchy sequence. ∎

Open, Closed and Compact Sets

2.1.9 Open and Closed Sets A set $V \subset \mathbf{R}^n$ is called open if and only if each $x \in V$ has an open neighborhood in V. A set is called closed if and only if its complement $\mathbf{R}^n \backslash V$ is open. A well known fact about these sets is that an arbitrary union of open sets is open. Also a finite intersection of open sets is open. For closed sets, an arbitrary intersection is closed, and a finite union is closed.

These sets play an important role in analysis. We will state the important properties we need without proof. Most of our properties are related to convergence of sequences.

2.1.10 Theorem A set V is closed if and only if the limits of all convergent sequences of elements of V lie in V. ∎

2.1.11 Closure, Interior and Boundary Let V be an arbitrary set in $\mathbf{R}^n$. Then clo(V) is the set obtained by adding limits of all convergent sequences in V. It is clear that clo(V) is a closed set.

The interior, int(V), of V is the set of all points of V that admit open neighborhoods in V. Thus, int(V) is an open set. The boundary, bdy(V), of V is the set clo(V)\int(V), and is thus a closed set, since it is the intersection of two closed sets.

2.1.12 Compact sets A set V is called bounded if there is a $x \in V$ and an $\epsilon > 0$ such that $V \subset B(x, \epsilon)$. A set V is called compact if and only if it is closed and bounded.

Compactness is usually defined by finite open covers. We have defined it by using the Heine-Borel Theorem. An important property of these sets is the following:

2.1.13 Theorem Every infinite sequence in a compact set has a convergent subsequence. ∎

Continuous Functions

Let $U \subset \mathbf{R}^n$ and $V \subset \mathbf{R}^m$, and let the function

$$f : U \to V$$

be arbitrary. We say it is continuous at $u \in U$ if and only if for every $\epsilon > 0$ there exists a $\delta > 0$ such that $||f(u) - f(v)|| < \epsilon$ for every $v \in V$ with

2.1 REAL ANALYSIS

$||u - v||' < \delta$. A function that is continuous at each point in U is called a continuous function on U. The above definition is equivalent to the fact that a function is continuous on U if and only if the inverse image of every open set in V is open in U.

An equivalent and an important result relating continuity to converging sequences is the following

2.1.14 Theorem f is continuous at u if and only if

$$\lim_{k \to \infty} f(u^k) = f(u).$$

for every sequence $\{u^k\}$ in U that converges to u. ∎

2.1.15 Some Properties of Continuous Functions The following are easy consequences of the definitions

(a) Let f and g be continuous functions from U to V. Then $f + g$ is continuous, and f/g with $g(u) \neq 0$ for all $u \in U$ is continuous.

(b) Let $f : U \to V$ and $g : V \to W$ be continuous functions. Then so is $g \circ f : U \to W$.

The following is a very useful result about continuous functions:

2.1.16 Theorem Let $U \subset \mathbf{R}^n$ be a compact set. Then $f(U)$ is a compact set in $\mathbf{R}^m$. ∎

2.1.17 Lipschitz Continuous We say a function $f : U \to V$ is called *locally* Lipschitz continuous at u if there is a neighborhood N of u and a constant $\alpha(u)$ such that

$$||f(u) - f(v)|| \leq \alpha(u)||u - v||$$

for all $v \in N$. It is Lipschitz continuous if the constant is independent of u and the neighborhood is the whole set U. In this case, $\alpha > 0$ is called the Lipschitz constant.

Differentiable Functions

Let $D \subseteq \mathbf{R}^n$ be an open set and let

$$f : D \subseteq \mathbf{R}^n \to \mathbf{R}^m.$$

We say f is (Fréchet) differentiable at $x \in D$ if there is an $m \times n$ matrix $Df(x)$ such that for every convergent sequence $\{x^k\}$

$$\lim_{x^k \to x} \frac{f(x^k) - f(x) - Df(x)(x^k - x)}{||x^k - x||} = 0.$$

The matrix $Df(x)$ is called the derivative of f at x.

The function f is said to differentiable on D if it is differentiable at every $x \in D$. If the map $Df : D \subseteq \mathbf{R}^n \to \mathbf{R}^{m \times n}$ is continuous, the function f is called continuously differentiable, or C^1 for short.

If Df is differentiable, by $D^2 f$, we represent its derivative, which is an $m \times n \times n$ matrix; and if f is r times differentiable, by $D^r f$ we represent it rth derivative, which is the derivative of $D^{r-1} f$. By C^r we represent all functions that are r times continuously differentiable; i.e., all their derivatives, $Df, D^2 f, \cdots, D^r f$ are continuous functions.

2.1.18 Theorem Let $f : D \subseteq \mathbf{R}^n \to \mathbf{R}^m$ be C^r. Then there exists a C^r mapping S on D such that

$$f(x+h) = f(x) + Df(x)h + \frac{1}{2}D^2 f(x)(h,h) + \cdots + \frac{1}{r!}D^r f(x)(h, h, \cdots, h) + S(h)$$

with $\lim_{h \to 0} \frac{S(h)}{||h||^r} = 0$. ■

2.1.19 Theorem Let $f : D \subseteq \mathbf{R}^n \to \mathbf{R}^m$ be C^r. Then

$$f(x+h) = f(x) + Df(x)h + \frac{1}{2}D^2 f(x)(h,h) + \cdots + \frac{1}{r!}D^r f(\hat{x})(h, h, \cdots, h)$$

where $\hat{x} = x + \lambda h$ for some $0 \leq \lambda \leq 1$. ■

2.1.20 Gradient, Hessian and Jacobian First and second derivatives of a real valued function $f : D \subseteq \mathbf{R}^n \to \mathbf{R}$ are called, respectively, the gradient and Hessian of f. They are represented, respectively, by the n vector ∇f and the $n \times n$ matrix $D^2 f$.

The first derivative of a function $f : D \subseteq \mathbf{R}^n \to \mathbf{R}^n$ is called the Jacobian, and is an $n \times n$ matrix.

2.1.21 Theorem Let $f : D \subseteq \mathbf{R}^n \to \mathbf{R}^m$ be a differentiable function with its derivative Lipschitz continuous on D with constant $\alpha > 0$. There exists a function $e : D \times D \to \mathbf{R}^m$ such that for every x and y in D

$$f(y) = f(x) + Df(x)(y - x) + e(y, x)$$

with $\|e(y,x)\| \leq \frac{1}{2}\alpha\|y-x\|^2$.

Proof: Note that

$$f(y) - f(x) - Df(x)(y-x) = \int_0^1 (Df(x+t(y-x)) - Df(x))(y-x)dt$$

with the understanding that

$$\int_0^1 \begin{pmatrix} g_1(t) \\ \vdots \\ g_m(t) \end{pmatrix} dt = \begin{pmatrix} \int_0^1 g_1(t)dt \\ \vdots \\ \int_0^1 g_m(t)dt \end{pmatrix}.$$

Hence

$$\begin{aligned} \|e(y,x)\| &\leq \int_0^1 \|(Df(x+t(y-x)) - Df(x))(y-x)\|dt \\ &\leq \int_0^1 \alpha t \|y-x\|^2 dt \\ &= \frac{1}{2}\alpha\|y-x\|^2 \end{aligned}$$

and we are done. ∎

2.2 Linear Algebra and Matrix Analysis

In this section we give the necessary background in linear algebra and matrix analysis needed in this work.

2.2.1 Linear Space By a linear space, V, we represent the system that satisfies the following axioms:

1. To every pair of vectors u and v in V there corresponds a vector $u+v$ in V, called their sum, in such a way that:

    (a) the addition is commutative, $u+v = v+u$.

    (b) the addition is associative, $(u+v)+w = u+(v+w)$.

    (c) There exists a null vector, 0 such that $u = u+0$ for every $u \in V$.

    (d) To each vector $u \in V$ there corresponds a unique vector $-u \in V$ such that $u + (-u) = 0$, (called the additive inverse).

2. To every pair $\alpha \in R$ and $v \in V$, there corresponds a vector αu in V, call the product of α and u, in such a way that:

(a) The multiplication is associative, $\alpha(\beta u) = \beta(\alpha u)$.

(b) $1.u = u$ for all u in $\boldsymbol{V}$.

3. Also

   (a) Multiplication by scalars is distributive with respect to vector addition, i.e., $\alpha(u + v) = \alpha u + \alpha v$.

   (b) Multiplication by vectors is distributive with respect to scalar addition, i.e., $(\alpha + \beta)u = \alpha u + \beta u$.

These axioms are not mutually exclusive, but will suffice for our purposes.

2.2.2 Linear Combination Given vectors $v^1, v^2, \cdots, v^l$ in $\boldsymbol{V}$, scalars $\alpha_1, \alpha_2, \cdots, \alpha_l$ in $\boldsymbol{R}$, we call the vector $\alpha_1 v^1 + \alpha_2 v^2 + \cdots + \alpha_l v^l$ a linear combination of the vectors $v^1, v^2, \cdots, v^l$.

2.2.3 Linear Independence A set of vectors $v^1, v^2, \cdots, v^l$ is said to be *linearly dependent* if 0 can be expressed as a linear combination of these vectors, i.e., $0 = \alpha_1 v^1 + \alpha_2 v^2 + \cdots + \alpha_l v^l$ with at least one $\alpha_j \neq 0$; and linearly independent otherwise. Thus, a set of vectors is said to be linearly independent if and only if

$$\alpha_1 v^1 + \alpha_2 v^2 + \cdots + \alpha_l v^l = 0$$

implies that $\alpha_1 = \alpha_2 = \cdots = \alpha_l = 0$.

2.2.4 Basis A set of vectors V in the vector space $\boldsymbol{V}$ is called a basis for $\boldsymbol{V}$ if and only if they are linearly independent and each vector in $\boldsymbol{V}$ can be expressed as a linear combination of the vectors in V. The vector space $\boldsymbol{V}$ is called *finite dimensional*, if the number of vectors in any basis is finite.

2.2.5 Theorem For a finite dimensional space, every independent set of vectors can be extended to a basis, i.e., vectors can be added to each independent set to make it into a basis.

Proof: Let $v^1, v^2, \cdots, v^l$ be a set of linearly independent vectors. As $\boldsymbol{V}$ is finite dimensional, it has a finite basis $u^1, u^2, \cdots, u^r$. Consider the set of vectors $v^1, v^2, \cdots, v^l, u^1, u^2, \cdots, u^r$. If this set is linearly independent, we have a basis for $\boldsymbol{V}$. Otherwise

$$\sum_{i=1}^{l} \lambda_i v^i + \sum_{j=1}^{r} \mu_j u^j = 0$$

2.2 LINEAR ALGEBRA AND MATRIX ANALYSIS

has a non-zero solution. Some $\mu_s \neq 0$, since $\{v^i\}$ are linearly independent. Rewriting

$$u^s = \sum_{i=1}^{l} -\frac{\lambda_i}{\mu_s} v^i + \sum_{j \neq s} -\frac{\mu_j}{\mu_s} u^j$$

we note that u^s is dependent on the other vectors. Dropping u^s from the set, we note that the remaining set is either linearly independent, or the above procedure can be used to drop another u^j; giving us the result. ∎

2.2.6 Theorem Every basis of a finite dimensional linear space contains the same number of vectors.

Proof: Let $U = \{u^1, u^2, \cdots, u^r\}$ be a set of linearly independent vectors, and $V = \{v^1, v^2, \cdots, v^s\}$ be a set of vectors that span the linear space $\boldsymbol{V}$, i.e., each vector of $\boldsymbol{V}$ can be expressed as a linear combination of the vectors of V. By Theorem 2.2.5, U can be extended to a spanning set by adding exactly $s - r$ elements of V. (We actually use the argument of the proof of that theorem here). Hence the number of vectors in the spanning set V, $s \geq r$, the number of vectors in the linearly independent set U. Now, if both U and V are different basis for $\boldsymbol{V}$, then using the linear independence property of U and spanning property of V we get $s \geq r$; we get $r \geq s$ by using the spanning property of U and the linear independence property of V. Thus the theorem follows. ∎

2.2.7 Dimension The number of vectors in any basis is called the dimension of the linear space. This is well defined by Theorem 2.2.6.

2.2.8 Subspace A subset S of a linear space V is called a subspace if it satisfies all the axioms 2.2.1 of the linear space. The dimension of the subspace is the largest number of linearly independent vectors in that subspace.

Euclidean Space

Here we will work with the n dimensional Euclidean space $\boldsymbol{R}^n$, the set of all vectors in the n-dimensional coordinate space. This space has the *standard* basis $u^1, u^2, \cdots, u^n$ where u^j is the jth unit vector $(0, \cdots, 0, 1, 0, \cdots, 0)^T$, with 1 in the jth position and all other components equal to 0.

Matrix Analysis

In this section we will introduce the notation and major results we will need from matrix theory.

2.2.9 Submatrix Given an $m \times n$ matrix A, $I \subseteq \{1, 2, \cdots, m\}$ and $J \subseteq \{1, 2, \cdots, n\}$ we define by A_{IJ} as the $|I| \times |J|$ submatrix whose entries lie in the rows indexed by I and columns indexed by J. By $A_{I.}$ and $A_{.J}$ we represent, respectively, the $|I| \times n$ submatrix whose rows are indexed in I and $m \times |J|$ submatrix whose columns are indexed in J. Whenever there is no chance of confusion, we will drop the period in these definitions.

2.2.10 Rank Given an $m \times n$ matrix A, we define its column rank as the maximum number of linearly independent columns, and its row rank as the maximum number of linearly independent rows. A simple fact is that the row rank and the column rank of any matrix are equal, and that number is called the rank of the matrix A.

2.2.11 Null Space Given an $m \times n$ matrix A, by $\mathcal{N}(A)$ we represent the set $\{x : Ax = 0\}$, called the null space of A.

2.2.12 Column Space Given an $m \times n$ matrix A, by $\mathcal{R}(A)$ we represent the set $\{Au : u \in \mathbf{R}^n\}$, called the column space of A.

2.2.13 Row Space Given an $m \times n$ matrix A, by $\mathcal{R}(A^T)$ we represent the set $\{A^T u : u \in \mathbf{R}^m\}$, called the row space of A.

2.2.14 Left Null Space Given an $m \times n$ matrix A, by $\mathcal{N}(A^T)$ we represent the set $\{x : A^T x = 0\}$, called the left null space of A.

2.2.15 Basic Theorem of Linear Algebra Let A be an $m \times n$ matrix of rank r. Then the following are true:

1. The null space $\mathcal{N}(A)$ is a subspace of $\mathbf{R}^n$ of dimension $n - r$.

2. The row space $\mathcal{R}(A^T)$ is a subspace of $\mathbf{R}^n$ of dimension r.

3. The column space $\mathcal{R}(A)$ is a subspace of $\mathbf{R}^m$ of dimension r.

4. The left null space $\mathcal{N}(A^T)$ is a subspace of $\mathbf{R}^m$ of dimension $m - r$.

2.2 LINEAR ALGEBRA AND MATRIX ANALYSIS

5. The subspaces $\mathcal{N}(A)$ and $\mathcal{R}(A^T)$ are orthogonal complements of each other in $\mathbf{R}^n$. Thus each vector $u \in \mathbf{R}^n$ has a unique decomposition $u = u_1 + u_2$ such that $u_1 \in \mathcal{N}(A)$, $u_2 \in \mathcal{R}(A^T)$ and $u_1^T u_2 = 0$.

6. The subspaces $\mathcal{N}(A^T)$ and $\mathcal{R}(A)$ are orthogonal complements of each other in $\mathbf{R}^m$. Thus each vector $u \in \mathbf{R}^m$ has a unique decomposition $u = u_1 + u_2$ such that $u_1 \in \mathcal{N}(A^T)$, $u_2 \in \mathcal{R}(A)$ and $u_1^T u_2 = 0$.

Proof: See Section 2.4, Strang [217], . ∎

2.2.16 Projection into a Subspace Let $S \subseteq \mathbf{R}^n$ be a subspace, and let $x \in \mathbf{R}^n$ be an arbitrary vector. The projection of x into S is the vector $P(x) \in S$ such that

$$P(x) = \mathrm{argmin}_{u \in S} ||x - u||_2^2 \qquad (2.1)$$

i.e., $P(x)$ is the vector in S that is closest to x (in the Euclidean norm sense). It can be readily seen that the vector $x - P(x)$ must be perpendicular to every vector in S, since such a $P(x)$ will give the smallest Euclidean distance between x and a vector of S.

When the subspace S is one of the four subspaces defined by a matrix, the projection operator P takes on a special form. We explore this in the next theorem.

2.2.17 Theorem Let A be an $m \times n$ matrix of rank m. Then, for all $x \in \mathbf{R}^n$, $P(x) = A^T(AA^T)^{-1}Ax$ is the projection of x into the row space $\mathcal{R}(A^T)$. Also, $x - P(x)$ is the projection of x into the null space $\mathcal{N}(A)$.

Proof: Substituting the row space $\mathcal{R}(A^T)$ for S in 2.1, we obtain

$$P(x) = \mathrm{argmin}_u ||x - A^T u||_2^2.$$

Since the norm square simplifies to $x^T x - 2x^T A^T u + u^T A A^T u$, a convex quadratic form, it can be minimized by setting its derivative equal to zero. This gives the solution $u^* = (AA^T)^{-1}Ax$, with the resulting vector $P(x) = A^T(AA^T)^{-1}Ax$ in $\mathcal{R}(A^T)$, and we have the first part of the Theorem. The second part follows from the Theorem 2.2.15, part 5. ∎

The matrix $P = A^T(AA^T)^{-1}A$ is called a projection matrix. All projection matrices are characterized by the following two properties:

2.2.18 Theorem A matrix P projects into a subspace if and only if

1. It is indempotent; i.e., $P^2 = P$.

2. It is symmetric; i.e., $P^T = P$.

Proof: Let P be a projection matrix projecting into the subspace S. Since $Px \in S$, $P(Px) = Px$ for all x. Thus $P^2 = P$. Also, since $(I - P)x$ is orthogonal to Px for every x, we have $x^T(I-P)^T Px = 0$ and $x^T P^T (I-P)x = 0$ or $x^T Px = x^T P^T Px = x^T PP^T x$ giving the second property.

Now, let P satisfy the properties (1) and (2) of the theorem. Then P is a projection into the column space $\mathcal{R}(P)$. To see this, we need only show that $(I - P)x$ is orthogonal to Px for all x. But $x^T P^T (I - P)x = x^T P^T x - x^T P^T Px = x^T Px - x^T P^2 x = 0$ and we are done. ∎

Matrix Norms

Let A be an $m \times n$ matrix, and let $||.||$ be any norm on $\boldsymbol{R}^m$ and $||.||'$ be any norm on $\boldsymbol{R}^n$. We define the matrix norm, $||A||$ as

$$||A|| = \max{}_{\{x: ||x||'=1\}} ||Ax||.$$

Since, $||.||$ is a continuous function and the set $\{x : ||x|| = 1\}$ is compact, the matrix norm is well defined, and non-negative. We will call the matrix norm $1, 2$ or ∞ when the underlying vector norms are, respectively, $1, 2$, or ∞.

2.2.19 Theorem Matrix norm has the following properties:

1. $||Ax|| \leq ||A||.||x||'$.

2. $||\lambda A|| = |\lambda|.||A||$.

3. $||A + B|| \leq ||A|| + ||B||$.

4. $||AB|| \leq ||A||.||B||$.

Proof: To see (1), note that $||Ax||/||x||' \leq \max_x ||Ax|| = ||A||$.

To see (2), note that $\max_x ||\lambda Ax|| = \max_x |\lambda|.||Ax||$, and we have the result since $|\lambda| > 0$.

To see (3), note that $\max_x ||(A + B)x|| \leq \max_x \{||Ax|| + ||Bx||\} \leq \max_x ||Ax|| + \max_x ||Bx||$.

2.2 LINEAR ALGEBRA AND MATRIX ANALYSIS

To see (4), let A be $m \times n$, B be $n \times p$ and let $||.||, ||.||', ||.||''$ be norms respectively on $\boldsymbol{R}^n, \boldsymbol{R}^m, \boldsymbol{R}^p$. Then $||AB|| = \max_x ||ABx|| \leq ||A|| . \max_x ||Bx||'' \leq ||A|| . ||B||$. ∎

Since this norm satisfies condition (1) of Theorem 2.2.19, it is called the matrix norm *subordinate* to the vector norm. There are matrix norms which are not subordinate to any vector norm. As an example, consider

$$||A|| = (\sum_{i,j} a_{ij}^2)^{\frac{1}{2}},$$

called the the Euclidean norm. For this norm $||I|| = n^{\frac{1}{2}}$, and thus is not subordinate to any vector norm. In this work, we will deal only with matrix norms that are subordinate to some vector norm.

2.2.20 Some Norms All calculations of the norms are for an $m \times n$ matrix A.

(a) For any subordinate norm, norm of the identity matrix, $||I|| = 1$.

(b) $||A||_1 = \max_j \sum_i |a_{ij}|$.

(c) $||A||_\infty = \max_i \sum_j |a_{ij}|$.

(d) $||A||_2 = $ (maximum eigenvalue of $AA^T)^{\frac{1}{2}}$.

(e) Let P be a projection matrix. Then $||P||_2 \leq 1$.

(f) Let $AA^T = I$. Then $||A||_2 \leq 1$ and $||A^T||_2 = 1$.

∎

Eigenvectors and Eigenvalues

Given an $m \times m$ real matrix A, we call a number λ an eigenvalue and a vector q an eigenvector of A if and only if

$$Aq = \lambda q.$$

The eigenvalue can be a complex number. In that case, the eigenvector is also complex. It is evident that eigenvalues of a matrix are the roots of the *characteristic polynomial*

$$\det(A - \lambda I) = 0$$

and each root λ of this polynomial is an eigenvalue of the matrix A. The corresponding eigenvectors lie in the null space $\mathcal{N}(A - \lambda I)$. The dimension of this null space is bounded above by the multiplicity of the root λ of the characteristic polynomial (and also of the eigenvalue λ of A). As is well known, each polynomial of degree m has exactly m roots (counting multiplicities) in the complex plane. In case all eigenvalues of A are real and distinct, it can be diagonalized. We summarize this in the next theorem.

2.2.21 Theorem Let A have all real and distinct eigenvalues, i.e., its eigenvalues are $\lambda_1 < \lambda_2 < \cdots < \lambda_m$. Then there exists an $m \times m$ matrix Q and a diagonal matrix Λ such that $A = Q\Lambda Q^{-1}$. Also, for each i, the ith diagonal entry of Λ is λ_i. ∎

When the matrix A is symmetric, i.e., $A = A^T$, all the eigenvalues are real and the eigenvectors can be chosen to form an orthonormal basis for $\mathbf{R}^m$.

2.2.22 Theorem Let the $m \times m$ matrix A be symmetric. Then

1. All its eigenvalues are real.

2. Its eigenvectors can be chosen to be mutually orthogonal.

3. $A = Q\Lambda Q^T$ for some orthonormal matrix Q with $QQ^T = I$.

In case A is also positive definite, all its eigenvalues are positive. ∎

2.3 Numerical Linear Algebra

One of the fundamental problems of computational linear algebra is that of solving the linear system of equations:

$$\sum_{j=1}^{n} a_{ij} x_j = b_i, \quad i = 1, \cdots, m$$

where the matrix $A = (a_{ij})$ and vector $b = (b_i)$ are given and the vector $x = (x_j)$ of variables must be assigned values so that when these are substituted in the system, all the m linear equations are satisfied.

2.3.1 Solution Set The solution set of the linear equation system is $\{x : Ax = b\}$. This set may be empty.

2.3 NUMERICAL LINEAR ALGEBRA

2.3.2 Non-singular Matrix
Given an $m \times n$ matrix A, we say it is non-singular if $m = n$ and its rank r is m. In case $m = n$ but the rank $r \neq m$, we say the matrix A is *singular*.

Non-singular matrices play a central role in numerical linear algebra. These matrices also have this very important property:

2.3.3 Theorem
Let A be an $m \times m$ matrix. Then the following are equivalent.

1. A is non-singular.
2. $Ax \neq 0$ for every $x \neq 0$.
3. $Ax = b$ has a unique solution for each b in $\mathbf{R}^m$.
4. The inverse, an $m \times m$ matrix A^{-1} exists, is non-singular, and $AA^{-1} = A^{-1}A = I$.

Proof: (1) $\Rightarrow$ (2) . Since A is non-singular it has full column rank, and thus its columns are linearly independent, and we are done.

(2) $\Rightarrow$ (3). Let $Ax^1 = b$ and $Ax^2 = b$. Then $A(x^1 - x^2) = 0$, and thus $x^1 - x^2 = 0$ and we are done.

(3) $\Rightarrow$ (4). Let u^i be the ith unit vector. Then $Ax = u^i$ has the unique solution $x = x^i$. Define the $m \times m$ matrix $X = (x^1, \cdots, x^m)$, and we note that $AX = I$. Let $D = XA$. Then $ADX = AXAX = AX$, and so $A(D - I)X = 0$. Thus, for all $u \neq 0$, $A(D - I)Xu = 0$. Since A is non-singular, $(D - I)Xu = 0$. If $Xu = 0$ then $0 = AXu = u$. But $u \neq 0$, thus $D = I$ and we are done with $A^{-1} = X$.

(4) $\Rightarrow$ (1). If $Au = 0$ then $u = A^{-1}Au = A^{-1}0 = 0$ and we have the non-singularity of A. ∎

We now give some results that connect the matrix norm with the existence of the inverse of a matrix.

2.3.4 Theorem
Let the square matrix C have $||C|| < 1$ for any matrix norm. Then
$$(I + C)^{-1} = I - C + C^2 - C^3 + \cdots.$$

Proof: From the properties of matrix norm, it is clear that $||C^k|| \leq ||C||^k$, and thus $C^k \to 0$ as $k \to \infty$. Now consider the identity
$$(I + C)(I - C + C^2 - \cdots + (-1)^k C^k) = I + (-1)^k C^{k+1}$$
and letting k go to ∞ we have our result. ∎

2.3.5 Theorem Let A be a non-singular matrix, and $C = A^{-1}(B-A)$, with $||C|| < 1$ for any matrix norm. Then $\det(AB) > 0$ and B^{-1} exists with

$$||B^{-1}|| \leq \frac{||A^{-1}||}{1-||C||}.$$

Proof: Consider the matrices $A(t) = I + tC$ for all $0 \leq t \leq 1$. $\det(A(0)) = \det(I) = 1$. Also, if $\det(A(t)) = 0$ then for some $||u|| = 1$

$$Iu + tCu = 0$$
$$Cu = -\frac{1}{t}u.$$

Hence,

$$||C|| = \max_{||x||=1}||Cx|| \geq ||Cu|| \geq \frac{1}{t}||u|| \geq 1.$$

Thus $\det(A(t)) > 0$ (as it is continuous) for each $0 \leq t \leq 1$. Also,

$$A(1) = I + C = I + A^{-1}B - I = A^{-1}B$$

$\det(AB) > 0$ and so B^{-1} exists. Now,

$$(I+C)^{-1} = I - C + C^2 - \cdots$$

and so

$$||(I+C)^{-1}|| \leq 1 + ||C|| + ||C||^2 + \cdots$$
$$= \frac{1}{1-||C||}.$$

But, $||B^{-1}|| = ||B^{-1}AA^{-1}|| \leq ||B^{-1}A||||A^{-1}|| = ||(I+C)^{-1}||||A^{-1}||$ and we are done. ∎

Gaussian Elimination

One of the most widely used techniques for solving linear systems of equations is Gaussian elimination. This technique is based on the following straight forward result:

2.3 NUMERICAL LINEAR ALGEBRA

2.3.6 Theorem Let M be an $m \times m$ non-singular matrix, A be an $m \times n$ matrix, and b be a m vector. Then the solution set of the system

$$Ax = b$$

is the same as the solution set of the system

$$MAx = Mb.$$

Proof: Because of the non-singularity of M, $M(Ax - b) = 0$ if and only if $Ax - b = 0$. ∎

During Gaussian elimination of the system, two special non-singular matrices are employed. These are the *permutation matrix*, which is used to interchange rows/columns of the underlying matrix, and the *elementary matrix*, which is used to scale and add rows of the matrix. Before formally introducing the elimination procedure, we consider these matrices.

2.3.7 Permutation Matrix Given a permutation $\pi = (\pi_1, \cdots, \pi_m)$ of $\{1, \cdots, m\}$, define the matrix P whose ijth entry P_{ij} is given by

$$P_{ij} = \begin{cases} 1 & j = \pi_i \\ 0 & \text{otherwise.} \end{cases}$$

It is readily confirmed that each row and column of P has exactly one non-zero entry, 1; and thus it is non-singular. Also, PA is a matrix whose ith row is occupied by the π_ith row of A, and AP is a matrix whose π_ith column is occupied by the ith column of A.

2.3.8 Principal Rearrangement Given an $m \times m$ matrix A and an $m \times m$ permutation matrix P, the matrix product PAP^T is called the principal rearrangement of A.

2.3.9 Elementary Matrix M is called an elementary matrix if it has exactly *one* column different from the identity matrix I, and its diagonal entries are non-zero. Thus, an elementary matrix, which has its jth column different from the identity matrix, has the form

$$M = I + v(u^j)^T$$

where u^j is the jth unit vector, and v is some vector with its jth entry $v_j \neq -1$. It is readily verified that M is non-singular. Its inverse is

$$M^{-1} = I - \frac{1}{1 + v_j} v(u^j)^T.$$

Given an $m \times n$ matrix A and an elementary matrix M

$$MA = A + v(u^j)^T A = A + vA_{j.}$$

and can be readily seen as adding v_i times the jth row $A_{j.}$ to the ith row $A_{i.}$ of A. It is this property of the elementary matrices that Gaussian elimination uses. It generates a series of m elementary matrices and possibly m permutation matrices (as needed) to transform the given matrix A into an upper triangular matrix U.

2.3.10 Pivot Operation
The above operation with an elementary matrix is referred to as a pivot on the *pivot row* j of the matrix A. The jth element v_j of v is called the *pivot element*. Also, the column of the pivot element is called the *pivot column*.

In Gaussian elimination, the pivot element is chosen so that, it is non-zero; and, after the pivot, all entries in the pivot column below the pivot element are reduced to zero. We will use the notation $M^{(j)}$ to represent the elementary matrix that pivots on the row j of A, for $j = 1, \cdots, m$.

In the process of elimination, if some pivot element encountered is zero, row interchanges are done to make a row with a non-zero element in the pivot position the pivot row.

2.3.11 Triangular Matrix
Given an $m \times n$ matrix U, we say it is a *upper* triangular matrix if its columns can be partitioned into sets $S_1, \cdots, S_r$, $r \leq m$, such that each column v in S_i has the property that $v_j = 0$ for all $j \geq i + 1$; and *lower* triangular if the partition of the columns is such that each column v in S_i has $v_j = 0$ for all $j \leq i - 1$.

The aim of Gaussian elimination is to find elementary matrices which reduce a given $m \times m$ non-singular matrix into an upper triangular form. This is done by eliminating all entries in the pivot column below the diagonal to zero; and rows are permuted from *below* the diagonal element.

Systems of equations with non-singular triangular matrices are readily solvable. Systems with lower triangular matrices can be solved by *forward solve*, i.e., since the first row contains only one variable, its value can be determined easily. Having determined the values of the first i variables, the $i + 1$st equation has only the $i + 1$st variable unknown. It can then be readily determined. Systems with upper triangular matrices can be solved by *backward solve*. Here the last equation has only one variable, the last one. After determining its value, one works backwards to get, sequentially, the values of the other variables.

2.3 NUMERICAL LINEAR ALGEBRA

2.3.12 LU Factorization Theorem For each $m \times n$ matrix A of rank r, there exists a permutation matrix P, elementary matrices $M^{(1)}, \cdots, M^{(r)}$ and an upper triangular matrix U such that

$$PA = M^{(1)} M^{(2)} \cdots M^{(r)} U.$$

In addition,
$$L = M^{(1)} M^{(2)} \cdots M^{(r)}$$

is a lower triangular matrix.

In case A is a non-singular $m \times m$ matrix,

$$PA = LU$$

where L and U are, respectively, non-singular $m \times m$ lower and upper triangular matrices. ■

2.3.13 Solving Factored System Let A be a non-singular matrix, and be factored as $PA = LU$. The system $Ax = b$ is equivalent to (see Theorem 2.3.6)

$$LUx = Pb.$$

This system can thus be readily solved by the following steps:

(a) (Forward Solve)
$$Lx' = Pb$$

(b) (Backward Solve)
$$Ux = x'.$$

2.3.14 Cholesky Factor A very important system of equations that arises in practice involves a positive definite and symmetric matrix. Such systems are generally solved by Cholesky Factorization. We now establish the main result regarding this factorization.

2.3.15 Theorem Let A be an $m \times m$ symmetric and positive definite matrix. There exists a unique lower triangular matrix L with positive diagonal elements, such that

$$A = LL^T.$$

The matrix L is called the *Cholesky* factor of A.

Proof: We prove the existence by an induction argument. The result is clearly true for $m = 1$, since then $L = A^{\frac{1}{2}}$. Now assume the theorem is true when $m = k-1$ and consider a matrix A with $m = k$. Since A is symmetric, we can write

$$A = \begin{pmatrix} B & a \\ a^T & \alpha \end{pmatrix}$$

for some $(k-1) \times (k-1)$ symmetric matrix B, and a $k-1$ vector a. Also, A is positive definite if and only if B is. Thus, by the induction hypothesis, $B = LL^T$ for some lower triangular matrix L.

Let $Lb = a$. As A is positive definite, $\alpha - b^T b > 0$. To see this, consider the vector $x^T = (-a^T B^{-1}, 1)$ and note that

$$\begin{aligned} 0 &< x^T A x \\ &= \alpha - a^T B^{-1} a \\ &= \alpha - b^T b. \end{aligned}$$

Define $\beta^2 = \alpha - b^T b$ and note that

$$A = \begin{pmatrix} L & 0 \\ b^T & \beta \end{pmatrix} \begin{pmatrix} L^T & b \\ 0 & \beta \end{pmatrix}$$

and we are done.

To see the uniqueness, assume that $A = L_1 L_1^T = L_2 L_2^T$ with both L_1 and L_2 lower triangular matrices with positive diagonal elements. Then $D = L_2^{-1} L_1 = L_2^T L_1^{-T}$. Since $L_2^{-1} L_1$ is lower triangular and $L_2^T L_1^{-T}$ is upper triangular, D must be diagonal. Also, $L_1 = L_2 D$, $L_2^T = D L_1^T$. Thus $D = D^{-1}$. D has positive diagonal elements as both L_1 and L_2 have. Thus $D = I$ and we are done. ∎

2.3.16 LQ and QR Factors Another factorization scheme used extensively for the solution of least squares problems as well as in the stable implementations of the simplex method is the QR factorization of a matrix A, where R is an upper triangular matrix and the columns of Q are orthonormal, i.e., $Q^T Q = I$. Here, we also introduce the LQ factorization, whose factors are the transpose of the factors of the QR scheme.

2.3.17 Theorem

1. Let A be an $m \times n$ matrix of rank m. Then there exist a lower triangular matrix L and an $m \times n$ matrix Q with $QQ^T = I$ such that

$$A = LQ.$$

2. Let A be an $m \times n$ matrix of rank n. Then there exist a upper triangular matrix R and an $m \times n$ matrix Q with $Q^T Q = I$ such that

$$A = QR.$$

Proof: We will only prove the first part of this theorem. It is readily seen that the result of the second part can be obtained by using the first result on A^T.

Since A has full row rank, the matrix AA^T is symmetric and positive definite. Thus, from Theorem 2.3.15, there is a lower triangular matrix L such that $AA^T = LL^T$. Defining $Q = L^{-1}A$, we have our result. ∎

2.4 Convexity and Separation Theorems

Convex sets play a central role in optimization theory. We now develop some of their important properties.

2.4.1 Convex Set Given a set C in $\boldsymbol{R}^n$, we say C is convex if and only if, for every pair of points u and v in C and $0 \leq \lambda \leq 1$, $(1 - \lambda)u + \lambda v \in C$; that is, the straight line joining any two points of C lies in C.

The following result is a straight forward consequence of the above definition.

2.4.2 Theorem Let $\{C_t\}_{t \in T}$ be an arbitrary collection of convex subsets of $\boldsymbol{R}^n$. $\cap_{t \in T} C_t$ is convex.

Proof: This is readily verified. ∎

2.4.3 Examples Some examples of convex sets follow:

(a) The empty set, $\emptyset$.

(b) **Line.** Let u and v be arbitrary points in $\boldsymbol{R}^n$. Then a line between u and v is the convex set $\{(1 - \lambda)u + \lambda v : 0 \leq \lambda \leq 1\}$.

(c) **Ray.** Let u and v be arbitrary points in $\boldsymbol{R}^n$; Then a ray at u in the direction v is the convex set $\{u + \lambda v : \lambda \geq 0\}$.

(d) **Hyperplane.** Let a be an arbitrary non-zero vector in $\boldsymbol{R}^n$, and γ be a scalar. Then a hyperplane H is the convex set $H = \{x : a^T x = \gamma\}$.

(e) **Halfspace.** Given a hyperplane $H = \{x : a^T x = \gamma\}$, the convex sets $\{x : a^T x \leq \gamma\}$ and $\{x : a^T x \geq \gamma\}$ are called the halfspaces generated by the hyperplane H.

(f) Let A be an arbitrary $m \times n$ matrix. The set $C = \{x : Ax = 0\}$ is convex.

(g) Let A be an arbitrary $m \times n$ matrix, and b be a m vector. The set $C = \{x : Ax = b\}$ is convex.

(h) A and b be as before. The set $C = \{x : Ax = b, x \geq 0\}$ is convex.

(i) Let $\{a_i\}_{i=1}^m \subset \mathbf{R}^n$, and $\gamma_i \in \mathbf{R}$, $i = 1, \cdots, m$. The set $C = \{x : a_i^T x \leq \gamma_i, i = 1, \cdots, m\}$ is convex.

2.4.4 Definition Given a convex set C and a scalar λ, we define their product as the set $\lambda C = \{\lambda v : v \in C\}$; and given two convex sets C_1 and C_2, their sum as the set $C_1 + C_2 = \{u + v : u \in C_1, v \in C_2\}$. The following theorem establishes that convexity is preserved under these operations.

2.4.5 Theorem Let C_1 and C_2 be convex subsets of $\mathbf{R}^n$. Then λC_1 and $C_1 + C_2$ are convex for all $\lambda \in \mathbf{R}$.

Proof: To see that λC_1 is convex, let $u, v \in \lambda C_1$, $\lambda \neq 0$. Since C_1 is convex and $\frac{1}{\lambda} u$ and $\frac{1}{\lambda} v$ are in C_1 for each $0 \leq \mu \leq 1$, $\frac{\mu}{\lambda} u + \frac{1-\mu}{\lambda} v \in C_1$. Thus, $\lambda(\frac{\mu}{\lambda} u + \frac{1-\mu}{\lambda} v) = \mu u + (1 - \mu)v \in \lambda C_1$.

To see that $C = C_1 + C_2$ is convex, let $u, v \in C$. Then for some $u^1, v^1 \in C_1$ and $u^2, v^2 \in C_2$, $u = u^1 + u^2$, $v = v^1 + v^2$. Now, consider

$$\begin{aligned} \mu u + (1-\mu)v &= \mu(u^1 + u^2) + (1-\mu)(v^1 + v^2) \\ &= \mu u^1 + (1-\mu)v^1 + \mu u^2 + (1-\mu)v^2 \\ &= \hat{u} + \hat{v} \end{aligned}$$

is in C since by the convexity of C_1 and C_2, $\hat{u}$ and $\hat{v}$ are respectively in C_1 and C_2. ∎

2.4.6 Convex Combination Given set of vectors $\{u^i\}_{i=1}^m$, scalars $\{\lambda_i\}_{i=1}^m$, we call the sum $\lambda_1 u^1 + \lambda_2 u^2 + \cdots + \lambda_m u^m$ a *linear combination* of the vectors u^i, $i = 1, \cdots m$; and, an *affine combination* if $\lambda_1 + \lambda_2 + \cdots \lambda_m = 1$. It is called a *convex combination* if it is an affine combination and $\lambda_i \geq 0$ for each $i = 1, \cdots m$.

2.4 CONVEXITY AND SEPARATION THEOREMS

2.4.7 Convex Hull Let $S \subset \mathbf{R}^n$ be an arbitrary set. The intersection of all convex sets containing S is called the convex hull of S. We represent this set by hull(S). From Theorem 2.4.5 it is a convex set; and, is the *smallest* convex set containing S.

2.4.8 Theorem The convex hull of any set S, hull(S) is the set of all convex combinations of arbitrary collections of elements of S.

Proof: It is clear that $S \subseteq \text{hull}(S)$. Also, if T is the collection of all convex combinations of collections of elements of S, as hull(S) is convex, $T \subseteq \text{hull}(S)$. Now let u and v be two elements in T. Then for some collections $\{u^i\}_{i=1}^k$ and $\{v^j\}_{j=1}^l$, $u = \sum_{i=1}^k \lambda_i u^i$ and $v = \sum_{j=1}^l \mu_j v^j$; $(1-\lambda)u + \lambda v = \sum_{i=1}^k (1-\lambda)\lambda_i u^i + \sum_{j=1}^l \lambda \mu_j v^j$ is a convex combination of $\{u^i\} \cup \{v^j\} \subset S$. Thus T is convex and thus hull(S) $\subseteq T$ and we are done. ∎

Separation Theorems

One of the more important properties of convex sets is that when they do not intersect, they can be separated by hyperplanes; that is, there exists a hyperplane such that the first set lies in one of the halfspaces generated by the hyperplane while the second in the other. These theorems have many applications, and will play a key role in the our development of the duality theory of linear programming.

2.4.9 Definition Given a hyperplane $H = \{x : a^T x = \gamma\}$ and two sets C_1 and C_2 in $\mathbf{R}^n$, we say the hyperplane H separates them if C_1 lies in one of the halfspaces defined by H, with C_2 in the other. We say the separation is *strict* if the hyperplane does not intersect C_1 or C_2.

Our goal here is to obtain conditions on the two sets under which the existence of the separating hyperplane can be established. We now prove our first theorem.

2.4.10 Weak Separation Theorem Let C be an *non-empty closed* and *convex* subset of $\mathbf{R}^n$, such that $0 \notin C$. Then, there is a $z \in C$ such that $z^T x \geq z^T z$ for all $x \in C$.

Proof: Define $\alpha > 0$ so large that the set $\hat{C} = C \cap \{x : x^T x \leq \alpha\} \neq \emptyset$. Such an α clearly exists. Define the problem:

$$\min_{x \in \hat{C}} x^T x.$$

Since $x^T x$ is continuous in x, and $\hat{C}$ is closed and bounded, this problem has a solution. Let z be a solution. Then $z^T z \leq x^T x$ for all $x \in \hat{C}$. But, by the definition of α, for $x \in C \setminus \hat{C}$, $x^T x > \alpha$ we have

$$z^T z \leq x^T x \text{ for all } x \in C.$$

Now, let $x \in C$ be arbitrary. Since C is convex, for every $0 \leq \lambda \leq 1$, $\lambda x + (1 - \lambda) z \in C$, or $z + \lambda(x - z) \in C$. Hence

$$\begin{aligned} z^T z &\leq (z + \lambda(x - z))^T (z + \lambda(x - z)) \\ &= z^T z + 2\lambda z^T (x - z) + \lambda^2 (x - z)^T (x - z). \end{aligned}$$

Assume that $z^T(x - z) < -\epsilon < 0$. Define $\hat{\lambda} > 0$ so small that $\hat{\lambda}(x - z)^T(x - z) < 2\epsilon$. Then

$$0 \leq 2z^T(x - z) + \hat{\lambda}(x - z)^T(x - z) < -2\epsilon + 2\epsilon = 0$$

leading to a contradiction. Thus $z^T(x - z) \geq 0$ and we are done. ∎

2.4.11 Comment This theorem is a separation theorem, since the hyperplane $H = \{x : z^T x = z^T z\}$ separates the convex sets C and $\{0\}$. We call it the weak separation theorem since the hyperplane does intersect C (the element z belongs to H and C).

2.4.12 Strict Separation Theorem Let C_1 and C_2 be two disjoint non-empty closed convex subsets of $\mathbf{R}^n$, and let C_2 be bounded. Then, there exists a hyperplane H which strictly separates them; i.e., there is a vector z in $\mathbf{R}^n$ and a scalar γ such that

$$\begin{aligned} z^T x &< \gamma \text{ for all } x \in C_1 \\ z^T x &> \gamma \text{ for all } x \in C_2 \end{aligned}$$

with $H = \{x : z^T x = \gamma\}$.

Proof: From Theorem 2.4.5 , $-C_2 = -1 \times C_2$ is convex, and so is $C = C_1 + (-C_2)$. Since C_2 is closed and bounded, C is closed. To see this, let $\{x^k\}_{k=1}^{\infty} \subset C$ be such that $x^k \longrightarrow x$. We now show that $x \in C$. Since, for each k, $x^k \in C$ there exist $y^k \in C_1$ and $z^k \in C_2$ such that $x^k = y^k - z^k$. Since C_2 is compact, on some subsequence K, $z^k \longrightarrow z$ for $k \in K$. Hence, for $k \in K$, $x^k + z^k \longrightarrow x + z$, or $y^k = x^k + z^k \longrightarrow x + z$. Since C_1 is closed, $x + z \in C_1$. Hence $x = (x + z) - z \in C$, so C is closed.

Now as $C_1 \cap C_2 = \emptyset$, $0 \notin C$, and from Theorem 2.4.10, there is a $z^0 \in C$ with
$$(z^0)^T x \geq (z^0)^T z^0 \text{ for all } x \in C.$$
Let $x \in C$ be arbitrary, and for some $u \in C_1$ $v \in C_2$ $x = u - v$,
$$(z^0)^T (u - v) \geq (z^0)^T z^0 > 0.$$
Hence for all $u \in C_1$ $v \in C_2$, $z_0^T u \geq z_0^T v + z_0^T z_0$. Thus
$$\begin{aligned}
\infty &> \alpha \\
&= \inf_{u \in C_1} (z^0)^T u \\
&\geq \sup_{v \in C_2} (z^0)^T v + (z^0)^T z_0 \\
&> \sup_{v \in C_2} (z^0)^T v \\
&= \beta \\
&> -\infty
\end{aligned}$$

Let $\gamma = \frac{1}{2}(\alpha + \beta)$, and define the hyperplane $H = \{x : (z^0)^T x = \gamma\}$. Note that for $u \in C_1$, $(z^0)^T u \geq \alpha > \gamma$, and for $v \in C_2$, $(z^0)^T u \leq \beta < \gamma$. Thus H strictly separates C_1 and C_2, and we are done. ∎

2.5 Linear Equations and Inequalities

Given an $m \times n$ matrix A of rank r and a vector $b \in \mathbf{R}^m$, we consider now the system of m equations in n variables

$$Ax = b. \tag{2.2}$$

We also consider this system as

$$\sum_{j=1}^n A_j x_j = b \tag{2.3}$$

where A_j is the jth column of the matrix A, and as

$$\sum_{j=1}^n a_{ij} x_j = b_i \text{ for } i = 1, \cdots, m \tag{2.4}$$

where a_{ij} is the ijth element of the matrix A.

2.5.1 Solution We say a vector $\bar{x}$ is a solution of the linear equation (2.2) if $A\bar{x} = b$. We define S as the set of all solutions of equation (2.2).

2.5.2 Theorem Let A be an $m \times n$ matrix of rank r, and $b = 0$. Then S has dimension $n - r$. If $r = n$, $S = \{0\}$.

Proof: Since, in this case, S is the null space of the matrix A, the result follows from Theorem 2.2.15. ∎

2.5.3 Theorem Let A be an $m \times n$ matrix of rank r and $b \neq 0$. If $S \neq \emptyset$, we can write $S = x^0 + \mathcal{N}(A)$ where x^0 is some vector in S and $\mathcal{N}(A)$ is the null space of the matrix A.

Proof: Since $A(x^0+u) = Ax^0+Au = b$ for all $u \in \mathcal{N}(A)$, $x^0+\mathcal{N}(A) \subseteq S$. To see the contrary, let $x \in S$, and consider $x-x^0$. As $A(x-x^0) = Ax-Ax^0 = 0$ $x - x^0 \in \mathcal{N}(A)$. As $x = x^0 + (x - x^0) \in x^0 + \mathcal{N}(A)$ for every $x \in S$, $S \subseteq x^0 + \mathcal{N}(A)$ and we have our result. ∎

2.5.4 Comment A consequence of Theorem 2.5.3 is that the set S is an affine space, and thus is a copy of $\boldsymbol{R}^{n-r}$.

2.5.5 Theorem Let A be an $m \times n$ matrix of rank r, and $b \neq 0$.

1. If $\operatorname{rank}(A) = \operatorname{rank}(A, b)$, dimension of S is $n - r$.

2. If $\operatorname{rank}(A) < \operatorname{rank}(A, b)$, $S = \emptyset$.

Proof: To see (1), since $\operatorname{rank}(A) = \operatorname{rank}(A, b)$, b lies in the column space of A, thus the system (2.2) has a solution. Thus, the result follows from the from Theorems 2.5.2 and 2.5.3. To see (2), note that the condition implies b does not lie in the column space of A; thus the system (2.2) has no solution. ∎

2.5.6 Theorem $S = \emptyset$ if and only if for some $\gamma \neq 0$ the system $y^T A = 0$, $y^T b = \gamma$ has a solution.

Proof: If $S \neq \emptyset$ there can be no y since the contrary would imply that $\gamma = 0$. Now assume that $S = \emptyset$. Then b cannot be expressed as a linear combination of the columns of A. Let the rank of A be r, and let $A_1, \cdots, A_r$ be the r columns of A that are linearly independent. Consider the system

$$y^T A_j = 0 \text{ for } i = 1, \cdots, r$$
$$y^T b = \gamma.$$

This system is $(r+1) \times m$ and has rank $(r+1)$. Hence, from Theorem 2.5.5 this system has a solution for every $\gamma \neq 0$. Also, since any column of A can be expressed as a linear combination of the columns $A_1, \cdots, A_r$, $y^T A = 0$, and we are done. ∎

Linear Inequalities

The structure of the solution to a system of linear equations is readily identified as an affine space, and was determined by the tools of linear algebra. The structure of linear inequalities is very rich, and we will need the tools of convexity theory to unravel it. We now explore this structure. Consider the system

$$Ax = b \qquad x \geq 0. \tag{2.5}$$

where A is an $m \times n$ matrix of rank r, b is a m vector and x is an n vector.

2.5.7 Feasible Solution We say a vector $\bar{x}$ is a feasible solution of system (2.5) if it satisfies all the equality constraints as well as the inequality constraints of the system; and a *solution* if it only satisfies the equality constraints. Let P be the set of all feasible solutions of (2.5).

2.5.8 Comment The set P of feasible solutions is very structured. As can be readily seen, the solution set of system (2.5) is the intersection of an affine space with the non-negative orthant in $\mathbf{R}^n$, thus has feasible solutions that belong to the boundary of P.

2.5.9 Interior Solution A feasible solution $\bar{x} \in P$ is called an interior solution if $\bar{x}_i > 0$ for all $i = 1, \cdots, n$; and a *boundary* solution otherwise.

Given a boundary solution x, define $I(x) = \{j : x_j > 0\}$ and $N(x) = \{j : x_j = 0\}$.

Basic Feasible Solution

2.5.10 Basis A subset $B \subset \{1, \cdots, n\}$ is called a basis if and only if the vectors $\{A_j : j \in B\}$ form a basis for the column space $\mathcal{R}(A)$ of A. The columns $A_j : j \in B$ are called the basic columns, and the matrix B (represented by the same symbol as for the set) is called the basis matrix.

2.5.11 Basic Feasible Solution A feasible solution $\bar{x}$ is called a basic feasible solution if and only if $I(\bar{x}) \subseteq B$ for some basis B; i.e., the columns $A_j, j \in I(\bar{x})$ are linearly independent. We call B the defining basis for $\bar{x}$. Thus, if $N = \{1, \cdots, n\} \backslash B$, we see that the basic feasible solution satisfies

$$\bar{x}_B = B^{-1}b$$
$$\bar{x}_N = 0$$

with $B^{-1}b \geq 0$.

2.5.12 Non-degenerate basic feasible solutions A basic feasible solution $\bar{x}$ is called non-degenerate if and only if $I(\bar{x}) = B$; for some basis B (equivalently, $\bar{x}_B > 0$); and degenerate otherwise.

A degenerate basic feasible solution, can have many defining basis. This creates special problems for algorithms that solve the linear programming problem. We will investigate this further later.

An important fact about inequality systems is that if they have feasible solutions, they also have basic feasible solutions.

2.5.13 Existence of Basic Feasible solutions Assume that the set of feasible solutions $P \neq \emptyset$. Then P contains a basic feasible solution.

Proof: Since $P \neq \emptyset$, let $\bar{x} \in P$, and $I(\bar{x}) = l \leq n$. If $A_j, j \in I(\bar{x})$ are linearly independent, then $\bar{x}$ is a basic feasible solution, and we are done. Otherwise, using Theorem 2.4.2, as the rank of $A_{I(\bar{x})}$ is less than l, the system $\sum_{j \in I(\bar{x})}^{l} A_j y_j = 0$ has a solution with at least one $j \in I(\bar{x}), y_j \neq 0$. Defining $y_j = 0$ for $j \notin I(\bar{x})$ we note that $Ay = 0$. Now let

$$\theta_1 = \begin{cases} \min\{\bar{x}_j/y_j : y_j > 0\} \\ +\infty & \text{if } y \leq 0 \end{cases}$$

$$\theta_2 = \begin{cases} \max\{-\bar{x}_j/y_j : y_j < 0\} \\ +\infty & \text{if } y \geq 0 \end{cases}$$

$$\theta = \min\{\theta_1, \theta_2\}$$

Let $x = \bar{x} - \theta y$. It is readily confirmed $x \in P$ and $I(x) < I(\bar{x})$. Either x is a basic feasible solution, or we can continue the above construction and reduce the number of non zeros in x, and we are done. ■

Convex separation theorems play a central role in establishing the existence of a solution to linear inequalities, and these theorems generally go by the name of *Theorems of the Alternative*. The most celebrated of these results is

2.5 LINEAR EQUATIONS AND INEQUALITIES

2.5.14 Farkas' Lemma Let A be an $m \times n$ matrix and b a m vector. Then, exactly one of the following systems of inequalities has a solution:

1. $Ax = b$, $x \geq 0$.

2. $A^T y \leq 0$, $b^T y > 0$.

Proof: Assume that both systems (1) and (2) have a solution, and let the solution be $\bar{x}$ and $\bar{y}$ respectively. Then the contradiction

$$0 < b^T \bar{y} = \bar{x} A^T \bar{y} \leq 0$$

can be obtained by substituting (1) to obtain the equality; and (2) to obtain the inequalities.

Now assume (1) has no solution. We generate a solution for (2), thus obtaining the lemma. Define $C_1 = \{Ax : x \geq 0\}$ and $C_2 = \{b\}$. C_1 and C_2 are closed convex sets with C_2 bounded. As (1) has no solution, $C_1 \cap C_2 = \emptyset$, and thus from Theorem 2.4.12, there is a $z \in \mathbf{R}^m$ and $\gamma \in \mathbf{R}$ such that

$$z^T u < \gamma \quad \text{for all} \quad u \in C_1$$
$$z^T u > \gamma \quad \text{for all} \quad u \in C_2.$$

Since $0 \in C_1$, $\gamma > 0$. Also, for each $u \in C_1$, $\rho u \in C_1$ for all $\rho > 0$. Thus $\rho z^T u < 0$ for all $\rho \geq 0$, and so $z^T u \leq 0$ for all $u \in C_1$. As $A_j \in C_1$ for each $j = 1, \cdots n$, we obtain a solution $y = z$ for the system (2) and we are done.∎

The following are a simple consequence of Farka's Lemma:

2.5.15 Theorem Let A be an $m \times n$ matrix and b a m vector. Then, exactly one of the following systems of inequalities has a solution:

1. $Ax \geq b$, $x \geq 0$.

2. $A^T y \leq 0$, $b^T y > 0$ and $y \geq 0$.

Proof: By introducing slack variables, the first system can be readily seen equivalent to

$$Ax - s = b, \ x \geq 0, \ s \geq 0.$$

and the result follows from Theorem 2.5.14. ∎

2.5.16 Motzkin's Theorem of the Alternative Let A, B, C be $m \times n_1, m \times n_2$ and $m \times n_3$ matrices with $n_1 \geq 1$. Exactly one of the following two systems has a solution:

1. $Ax + By + Cz = 0$, $0 \neq x \geq 0$, $y \geq 0$

2. $A^T u < 0$, $B^T u \leq 0$, $C^T u = 0$.

Proof: by a change of variables, the unrestricted variables z in the system (1) can be replaced by the difference of restricted variables $z = z_1 - z_2$, $z_1 \geq 0$, $z_2 \geq 0$; and since the right hand side of the system is 0, the constraint $0 \neq x \geq 0$ can be replaced by the equivalent constraint $e^T x = \sum_{j=1}^{m_1} x_j = 1$ $x \geq 0$ to generate the following system equivalent to (1):

$$\begin{array}{rcl} Ax + By + Cz_1 - Cz_2 & = & 0 \\ e^T x & = & 1 \\ x \geq 0 \quad y \geq 0 \quad z_1 \geq 0 \quad z_2 \geq 0 & & \end{array}$$

Applying the Farkas' Lemma 2.5.14 to the above system we obtain

$$\begin{array}{rcl} A^T u + e\theta & \leq & 0 \\ B^T u & \leq & 0 \\ C^T u & \leq & 0 \\ -C^T u & \leq & 0 \\ \theta & > & 0 \end{array}$$

and the above is readily seen equivalent to system (2) of this theorem. ∎

2.6 Convex Polyhedral Sets

The geometry of convex polyhedral sets plays a critical role in linear programming theory. In this section we introduce the main geometric ideas of these convex sets. Some definitions and preliminary results follow:

2.6.1 Convex Polyhedral Set A convex set P in $\mathbf{R}^n$ is called polyhedral if it is an intersection of a finite number of half spaces. Since each half space can be represented by a linear inequality, $\{x : a^T x \geq \alpha\}$ for some n-vector a; a convex set is polyhedral if and only if there is an $m \times n$ matrix A and a m vector b such that $P = \{x : Ax \geq b\}$.

2.6 CONVEX POLYHEDRAL SETS

2.6.2 Theorem Intersection of a finite collection of convex polyhedral sets is a convex polyhedral set.

Proof: Follows from the fact that each set is an intersection of a finite collection of half spaces, and as finite sum of finite numbers is finite, the intersection of all the sets is a finite intersection of half spaces. ∎

2.6.3 Affine Space Any translation of a subspace is called an affine space, i.e., if V is a subspace of R^n, $V + \{a\}$, for every $a \in R^n$, is called an affine space. The dimension of an affine space is the same as the dimension of the translated subspace, and thus is the largest number k of the vectors $\{v^i\}_{i=1}^k$ such that the set of vectors $v^1 - a, \cdots, v^k - a$ are linearly independent.

2.6.4 Affine Combination Given a collection of vectors $\{v^i\}_{i=1}^k$ and scalars $\{\lambda_i\}_{i=1}^k$ with $\sum_{i=1}^k \lambda_i = 1$, we call the vector $v = \sum_{i=1}^k \lambda_i v^i$ an affine combination of the vectors $\{v^i\}_{i=1}^k$.

2.6.5 Affine Hull Given an arbitrary set S in R^n, the affine hull of S, aff(S), is the intersection of all affine subspaces containing S. It is readily seen that aff(S) is an affine space. Thus, if S is an affine space, $S = $ aff(S).

2.6.6 Theorem The affine hull, aff(S), of S is the set of all affine combinations of arbitrary collections of elements of S. ∎

2.6.7 Dimension of a Set Let $S \subseteq R^n$ be arbitrary. We define the dimension of S, dim(S), as the dimension of the affine space, aff(S) containing it. Thus, it is the largest number k of vectors $\{v^i\}_{i=1}^k$ in S such that for some $a \in S$, the vectors $v^1 - a, \cdots, v^k - a$ are linearly independent. We leave the reader to verify that such a k gives the dimension of the smallest translated subspace in which the set lies.

Polyhedral Sets in Affine Space

Systems encountered in many applications consist of both equations and inequalities. An example being the system

$$Ax = b \quad x \geq 0.$$

Geometrically, the solution to this system is a convex polyhedral set. To see this, regard the solution set of the equality as an affine space, and of the inequalities, a convex polyhedral set in R^n. The set of interest is the intersection of the affine space and the polyhedral set which, we show later, is again a polyhedral set, but lies in the affine space. Since our goal is to investigate the geometric properties of these sets, we will obtain results independent of the underlying space. Translating these results to non-geometric properties will require careful consideration of the underlying spaces. We now consider polyhedral sets in affine spaces.

2.6.8 Half Affine Space We define the intersection of a half space with an affine space as a half affine space.

2.6.9 Polyhedral Set in an Affine Space An intersection of a finite number of half affine spaces of an affine space is referred to as a polyhedral set in that affine space.

2.6.10 Theorem The intersection of a convex polyhedral set with an affine space is a convex polyhedral set that lies in that space.

Proof: Follows readily from the above definitions. ∎

Geometric Properties of Polyhedral sets

We now investigate the geometric properties of the convex polyhedral sets. These properties relate mostly to the boundary structure of these sets. The boundary has many combinatorial properties as well, but we are only interested in certain geometric properties useful in linear programming. These relate mostly to the faces of these sets.

2.6.11 Face A face F of a convex polyhedral set P is a collection of all vectors in S such that any two vectors are in F if and only if the center of the line joining these two vectors also belongs to F; i.e, if $v_1, v_2 \in P$ then $v_1, v_2 \in F$ if and only if $\frac{1}{2}(v_1 + v_2) \in F$. Dimension, $\dim(F)$, of the face is the dimension of $\text{aff}(F)$.

2.6.12 Theorem The face of a polyhedral set is also a polyhedral set.

Proof: Let F be a face of P. Then $F = \text{aff}(F) \cap P$, and the intersection of an affine space with a polyhedral set is also a polyhedral set, see Theorem 2.6.10. ∎

A consequence of this theorem is that the intersection of two faces of P is either empty or their common face, as well as a face of P.

2.6 CONVEX POLYHEDRAL SETS

2.6.13 Special Faces Of special interest are the following faces of P.

(a) **Facet**: A face of dimension $\dim(P) - 1$ is called a facet of P.

(b) **Vertex**: A zero dimensional face is called a vertex of P.

(c) **Edge**: A one dimensional face is called an edge of P.

The structure of a vertex is very simple. The structure of an edge is also simple, since it is a one dimensional polyhedral set.

2.6.14 Theorem One dimension convex polyhedral set is either a line or a ray. Thus it has either 0, 1 or 2 vertices.

Proof: There are only three possible one dimensional convex sets, a line, a ray and a one dimensional affine space, and we are done.

Convex Polyhedral sets in Linear Programming

Convex polyhedral sets encountered in linear programming are generally defined by systems of inequalities in one of the following forms:

(a) $\{x : Ax = b, x \geq 0\}$.

(b) $\{x : Ax \geq b, x \geq 0\}$.

(c) $\{x : Ax \geq b\}$.

By simple transformations, involving adding more variables, the forms (b) and (c) can be converted into a form equivalent to (a). We illustrate this for (c).

Note that the system (c) is equivalent to

$$Ax - s = b, \quad s \geq 0$$

which is equivalent to

$$Ax^1 - Ax^2 - s = b, \quad x^1 \geq 0, x^2 \geq 0, s \geq 0$$

and is the required form. We now establish some important results relating to polyhedral sets defined by a system of inequalities of the form (a).

2.6.15 Polytopes A bounded convex polyhedral set is called a polytope. The following theorem gives a characterization.

2.6.16 Theorem Let $P = \{x : Ax = b, x \geq 0\} \neq \emptyset$. P is bounded if and only if $Ax = 0, 0 \neq x \geq 0$ has no solution.

Proof: Assume that there is a u such that $Au = 0, 0 \neq u \geq 0$ and let $v \in P$. Then $\{v + \rho u : \rho \geq 0\} \subseteq P$, and thus P is unbounded. To see the other half of the theorem, assume that $P \neq \emptyset$ and it is unbounded. Then there exist $\{\rho_k v^k\} \in P$ with $||v^k|| = 1$ and $v^k \to v^*$, $\rho_k \to \infty$. Also, $\rho_k > 0$ and $v^k \geq 0$ for every k. Thus $Av^k = (1/\rho_k)b$, and letting k go to ∞ we see that $Av^* = 0, 0 \neq v^* \geq 0$. ∎

2.6.17 Homogeneous Solution Every solution to the system of inequalities $x \geq 0, Ax = 0$ is called a homogeneous solution of the system (a).

For any such solution $u \neq 0$, for all $x \in P$, the ray $\{x + \rho u : \rho \geq 0\} \subset P$. In this case we say the ray lies in the convex polyhedral set.

Characterization of Vertices and Edges

We now establish an important relationship between the vertices of P and basic feasible solutions of $Ax = b$, $x \geq 0$.

2.6.18 Theorem $u \in P = \{x : Ax = b, x \geq 0\}$ is a vertex if and only if u is a basic feasible solution of the inequality system.

Proof: u is a basic feasible solution if and only if the set of vectors A_j, $j \in I(u)$ are linearly independent (since any set of linearly independent vectors can be extended to a basis for an arbitrary set of vectors). Let F be the smallest face of P that contains u. If for some $v, w \in P$, $u = \frac{1}{2}(v + w)$ then $v_i = w_i = u_i = 0$ for every $i \notin I(u)$, and $\sum_{j \in I(v)} A_j v_j = \sum_{j \in I(w)} A_j w_j = b$. But as $I(v) = I(w) = I(u)$, it is easily seen that $\{A_j\}_{j \in I(u)}$ are linearly independent if and only if $v_j = w_j$ for every $j \in I(u)$ establishing the result. ∎

2.6.19 Theorem $v \in P = \{x : Ax = b, x \geq 0\}$ is an edge if and only if $|I(v)| - \text{rank}(A_{I(v)})$ is 1.

Proof: It is easily seen that $Av = \sum_{j \in I(v)} A_j v_j = b$. since $v_j = 0$ for all $j \notin I(v)$. Hence v belongs to the face $F = \{y : \sum_{j \in I(v)} A_j y_j = b, y_j \geq 0, j \in I(v)\}$; and F is an edge if an only if its dimension is 1 and we have our result. ∎

Since an edge has at most two vertices, we can define:

2.6 CONVEX POLYHEDRAL SETS

2.6.20 Adjacent Vertices We say two vertices are adjacent if and only if they belong to the same edge.

2.6.21 Theorem Two vertices v^1 and v^2 are adjacent if and only if

1. $|I(v^1) \setminus I(v^2)| = 1$
2. $|I(v^1) \cup I(v^2)| - \text{rank}(A_{I(v^2) \cup I(v^2)}) = 1$

■

An important result about convex polyhedral sets is the following:

2.6.22 Theorem Every convex polyhedral set has a finite number of vertices and edges.

Proof: There are at most $(n!)/(m!(n-m)!)$ basis for the matrix A, and each generates at most one vertex. Also each pair of vertices generates at most one edge, and we are done. ■

Degeneracy of Convex Polyhedral Sets

Though there is a relationship between the vertices of P and the basic feasible solutions of the inequality system, it is not one to one. In general it is many to one. This creates special problems for algorithms that solve the linear programming problem. We now investigate this relationship.

2.6.23 Non-degenerate Faces A vertex v of P is non-degenerate if the $m \times |I(v)|$ matrix $A_{I(v)}$ has rank m; and degenerate otherwise. Thus, for a non-degenerate vertex, $A_{I(v)}$ is an $m \times m$ matrix of rank m.

Likewise, we say a face F is non-degenerate if for all $v \in \text{int}(F)$, the rank of $A_{I(v)}$ is m; and degenerate otherwise. Thus, for a non-degenerate edge, $A_{I(v)}$ is an $m \times (m+1)$ matrix of rank m. Since P is a face of P, this also defines the non-degeneracy of P.

The following is the implication of non-degeneracy.

2.6.24 Theorem Let P be non-degenerate. Then there is a one to one correspondence between the vertices of P and the basic feasible solutions of the system of inequalities.

Proof: Let v be a vertex of P. It must be non-degenerate, thus $A_{I(v)}$ is an $m \times m$ matrix of rank m, and thus defines a basis for A. Thus $I(v)$ is a subset of only one basis $B = I(v)$ and we are done. ■

2.6.25 Theorem A non-degenerate vertex has exactly $n - m$ edges incident on it, and thus has at most $n - m$ adjacent vertices.

Proof: If v is a non-degenerate vertex, then $|I(v)| = m$. Define the $n - m$ sets $I(v) \cup \{j\}$ for each $j \in N(v)$. Each of these sets generates an edge containing v, and at most one adjacent vertex. ∎

Characterization of Convex Polyhedral Sets by Vertices

Polytopes are completely characterized by their vertices. In case of unboundedness, the structure is also determined by the vertices. We now prove this result.

2.6.26 Resolution Theorem Let P be convex and polyhedral. Then $x \in P$ if and only if x can be expressed as a sum of a convex combination of vertices of P and a homogeneous solution of P.

Proof: Let $x \in P$, and be such that the vectors $A_j, j \in I(x)$ are linearly independent. Then we are done as x is a vertex.

Define $n(x) = |I(x)| - \text{rank}(A_{I(x)})$ called the nullity of $A_{I(x)}$. We have shown that the theorem holds when the nullity is zero. We prove this theorem by an induction argument. For this purpose assume that the theorem is true for all x such that $n(x) \leq k - 1$ for some $k \geq 1$. Now let $x \in P$ be such that $n(x) = k$. For this case, $A_{I(x)}$ is an $m \times |I(x)|$ matrix of rank $|I(x)| - k < |I(x)|$ and thus its null space is non-empty. Hence there exists a vector $y \neq 0$ such that $y_j = 0$, $j \notin I(x)$ and

$$\sum_{j \in I(x)} A_j y_j = 0.$$

Now, there are three cases, depending on the sign of y.

Case 1. $y \leq 0$. Define a $\theta > 0$ such that $z = x + \theta y \geq 0$ with at least one component of z equal to zero. Then $|I(z)| \leq k - 1$, and from the induction hypothesis we can express z as a sum of a convex combination of vertices, and a homogeneous solution. Since $x = z - \theta y$, $-\theta y \geq 0$ and sum of homogeneous solutions is homogeneous, we are done.

Case 2. $y \geq 0$. We define a $\theta > 0$ such that $z = x - \theta y \geq 0$ with at least one component of z equal to zero. Use the same argument as in case (i) to complete this case.

2.6 CONVEX POLYHEDRAL SETS

Case 3. $y \not\geq 0$, $y \not\leq 0$. We can define two vectors $z^1 = x + \theta_1 y \geq 0$, $\theta_1 > 0$ and the vector $z^2 = x - \theta_2 y \geq 0$, $\theta_2 > 0$ such that they both have at least one component zero. Then

$$x = \frac{\theta_2}{\theta_1 + \theta_2} z^1 + \frac{\theta_1}{\theta_1 + \theta_2} z^2$$

and can be readily seen as a convex combination of z^1 and z^2. By induction hypothesis, both z^1 and z^2 can be expressed as the required sum. Since a convex combination of two convex combinations of vertices is again a convex combination of vertices; and, a convex combination of two homogeneous solutions is a homogeneous solution, we are done. ∎

2.6.27 Corollary Let P be a polytope. Then $x \in P$ if and only if it can be expressed as a convex combination of the vertices of P.

Proof: Follows from Theorem 2.6.26 since the only homogeneous solution in this case is the zero solution. ∎

Polyhedral Sets with Integral Data

We consider some properties of polyhedral sets when the data A, b is integer. In particular we will consider the polyhedron $\{x : Ax = b, x \geq 0\}$ when the data is integer, A is $m \times n$ and b is in $\mathbf{R}^m$. We will show that any extreme point of this polyhedron cannot have arbitrarily small or arbitrarily large components. For this purpose, define

$$L = \lceil \sum_{i=1}^{m} \sum_{j=1}^{m} \log_2(1 + |A_{i,j}|) + \sum_{i=1}^{m} \log_2(1 + |b_i|) \rceil$$

where $\lceil \alpha \rceil$ is the smallest integer greater than and equal to α. We now prove a simple lemma.

2.6.28 Lemma Let B be any nonsingular submatrix of (A, b). Then $1 \leq |Det(B)| \leq 2^L$.

Proof: From the definition of determinant, if Π is the set of all permutations of $\{1, 2, \cdots, m\}$,

$$|Det(B)| = |\sum_{\sigma \in \Pi} \prod_{i=1}^{m} B_{i, \sigma_i}|$$

$$\leq \sum_{\sigma \in \Pi} \prod_{i=1}^{m} |B_{i,\sigma_i}|$$

$$\leq \prod_{i=1}^{m} (\sum_{i=1}^{m} |B_{i,j}|)$$

$$\leq \prod_{i=1}^{m} \prod_{j=1}^{m} (1 + |B_{i,j}|)$$

$$\leq \prod_{i=1}^{m} \prod_{j=1}^{n} (1 + |A_{i,j}|) \prod_{i=1}^{m} (1 + |b_i|)$$

and the upper bound follows from the definition or L. Lower bound follows since $Det(B)$ is a non-zero integer. ∎

We can then show that

2.6.29 Bounds of Basic Feasible Solutions Let x be a vertex of the polyhedron. Then, for each j, $x_j \leq 2^L$, and for each j such that $x_j > 0$, $x_j > 2^{-L}$.

Proof: Since x is a vertex, for some basis B we note that $x_j \geq 0$ for each $j \in B$ and $x_j = 0$ for each $j \notin B$. Let j be an index such that $x_j > 0$. Using Cramer's rule, we see that

$$x_j = \frac{|Det(B^j)|}{|Det(B)|}$$

where B^j is the matrix obtained when the jth column of B is replaced by b. The result now follows form Lemma 2.6.28. ∎

In case the polyhedron is bounded, the following corollary can be established.

2.6.30 Corollary Let the polyhedron be bounded. Then each vector x in the polyhedron is such that if $x_j > 0$ then $2^{-L} \leq x_j \leq 2^L$.

Proof: Follows from Corollary 2.6.27 and 2.6.29. ∎

2.7 Nonlinear System of Equations

Let $D \subset \mathbf{R}^n$ be an open set, and let

$$f : D \to \mathbf{R}^n$$

be such that

2.7 NONLINEAR SYSTEM OF EQUATIONS

(a) it is continuously differentiable,

(b) its derivative, Df, is Lipschitz continuous, with Lipschitz constant $\alpha > 0$, i.e., for all x and y in D

$$||Df(y) - Df(x)|| \leq \alpha ||y - x||.$$

We consider here the problem of finding an x such that

$$f(x) = 0.$$

In this section, we introduce Newton's method for solving this equation, and prove most of its important properties. We also introduce some of its more important variants.

2.7.1 Newton's Method Newton's method is the most widely used procedure for solving this problem. We now present this method.

Step 1 Let x^0 and $\epsilon > 0$ be arbitrary. Set $k = 0$.

Step 2 $x^{k+1} = x^k - Df(x^k)^{-1} f(x^k)$

Step 3 If $||f(x^{k+1})|| < \epsilon$ or $Df(x^{k+1})$ is singular, stop. Otherwise set $k = k+1$ and go to step 2.

Even though the method can be started at any initial vector x^0, its convergence, and thus success, is critically related to this choice. The most celebrated of the results that establish the convergence as a function of this initial point is the following theorem:

2.7.2 Kantorovich's Theorem Let x^0 be such that

1. $||Df(x^0)^{-1}|| \leq \beta$

2. $||Df(x^0)^{-1} f(x^0)|| \leq \gamma$

3. $\alpha\beta\gamma \leq \frac{1}{2}$.

where $\alpha > 0$ is the Lipschitz constant. Then the sequence $\{x^k\}_{k=1}^{\infty}$ is well defined, and converges to a zero of f.

Proof: For each k

$$||Df(x^k)^{-1}|| \leq 2^k \beta \tag{2.6}$$

$$||Df(x^k)^{-1}f(x^k)|| \leq 2^{-k}\gamma. \qquad (2.7)$$

We now prove (2.6) and (2.7) by an induction argument. By the hypothesis of the theorem, (2.6) and (2.7) are true for $k = 0$. Now assume that they are true for all $i \leq k$ for some $k \geq 0$, and consider $i = k$. Then,

$$\begin{aligned} ||Df(x^i)^{-1}(Df(x^{i+1}) - Df(x^i))|| &\leq 2^i\beta\alpha||x^{i+1} - x^i|| \\ &\leq \alpha 2^i\beta 2^{-i}\gamma \\ &= \alpha\beta\gamma \\ &\leq \frac{1}{2}. \end{aligned}$$

Hence, from Theorem 2.3.5 $Df(x^{i+1})^{-1}$ exists and

$$||Df(x^{i+1})^{-1}|| \leq \frac{2^i\beta}{1 - \frac{1}{2}} = 2^{i+1}\beta.$$

and we have (2.6) by the induction hypothesis. Also, from the Theorem 2.1.21

$$\begin{aligned} f(x^{i+1}) &= f(x^i) + Df(x^i)(x^{i+1} - x^i) + e(x^{i+1}, x^i) \\ &= e(x^{i+1}, x^i) \end{aligned}$$

with $||e(x^{i+1}, x^i)|| \leq \frac{1}{2}\alpha||x^{i+1} - x^i||^2$.

Thus, from (2.6),

$$\begin{aligned} ||f(x^{i+1})|| &\leq \frac{1}{2}\alpha||x^{i+1} - x^i||^2 \\ &\leq \frac{1}{2}\alpha 4^{-i}\gamma^2. \end{aligned} \qquad (2.8)$$

Hence:

$$\begin{aligned} ||Df(x^{i+1})^{-1}f(x^{i+1})|| &\leq 2^{i+1}\beta\frac{1}{2}\alpha 4^{-i}\gamma^2 \\ &= 2^{-i}\alpha\beta\gamma^2 \\ &\leq 2^{-(i+1)}\gamma \end{aligned}$$

and we have (2.7) by the induction hypothesis.

We are now ready to prove the theorem. Note that

$$||x^{i+1} - x^i|| = ||Df(x^i)^{-1}f(x^i)|| \leq 2^{-i}\gamma.$$

2.7 NONLINEAR SYSTEM OF EQUATIONS

Thus, for any $l \geq 1$ we have

$$\begin{aligned}
||x^{l+i} - x^i|| &\leq ||x^{l+i} - x^{l+i-1}|| + \cdots + ||x^{i+1} - x^i|| \\
&\leq (2^{-(l+i-1)} + 2^{-(l+i-2)} \cdots 2^{-i})\gamma \\
&\leq 2^{-(i-1)}\gamma.
\end{aligned}$$

which goes to zero as i goes to ∞. Hence, $\{x^k\}$ is a Cauchy sequence, and thus converges, say to x^∞. As f is continuous, from equation (2.8), $f(x^\infty) = 0$ and we are done. ∎

One of the most important properties of Newton's method is its rapid convergence to the solution. We now establish this result.

2.7.3 Theorem Assume that the sequence $\{x^k\}$ generated by Newton's method is well defined and converges to x^∞. Also, assume that $Df(x^\infty)$ is non-singular. Then,

1. $f(x^\infty) = 0$

2. There is a $L \geq 1$, and a $\delta > 0$ such that for all $k \geq L$,

$$||x^{k+1} - x^\infty|| \leq \delta ||x^k - x^\infty||^2$$

i.e., the convergence rate is quadratic.

Proof: Since $Df(x^\infty)^{-1}$ exists, let $\bar{\beta} = ||Df(x^\infty)^{-1}||$. Also, as $||x^k - x^\infty|| \to 0$, there exists a $L \geq 1$ such that for all $k \geq L$

$$\begin{aligned}
||Df(x^\infty)^{-1}(Df(x^k) - Df(x^\infty))|| &\leq ||Df(x^\infty)^{-1}|| \, ||(Df(x^k) - Df(x^\infty))|| \\
&\leq \alpha\bar{\beta}||x^k - x^\infty|| \\
&\leq \frac{1}{2}.
\end{aligned}$$

For $k \geq L$, from Theorem 2.3.5, $Df(x^k)^{-1}$ exists and $||Df(x^k)^{-1}|| \leq 2\bar{\beta}$. Also, from Theorem 2.1.21 with $||e(x^{k+1}, x^k)|| \leq \frac{1}{2}\alpha ||x^{k+1} - x^k||^2$ we have

$$\begin{aligned}
f(x^{k+1}) &= f(x^k) + Df(x^k)(x^{k+1} - x^k) + e(x^{k+1}, x^k) \\
&= e(x^{k+1}, x^k)
\end{aligned}$$

and thus $||f(x^{k+1})|| \leq \frac{1}{2}\alpha ||x^{k+1} - x^k||^2$. Thus, as f is continuous, $f(x^{k+1}) \to f(x^\infty)$, we have $f(x^\infty) = 0$.

Now

$$x^{k+1} - x^\infty = x^k - x^\infty - Df(x^k)^{-1}f(x^k)$$

and
$$f(x^\infty) = f(x^k) + Df(x^k)(x^\infty - x^k) + e(x^\infty, x^k)$$
or
$$Df(x^k)^{-1}f(x^k) = (x^k - x^\infty) - Df(x^k)^{-1}e(x^\infty, x^k).$$
Thus
$$x^{k+1} - x^\infty = -Df(x^k)^{-1}e(x^\infty, x^k).$$
Hence, for $k \geq L$
$$\begin{aligned}\|x^{k+1} - x^\infty\| &\leq \|Df(x^k)^{-1}\|\|e(x^\infty, x^k)\| \\ &\leq 2\bar{\beta}.\frac{1}{2}\alpha\|x^k - x^\infty\|^2\end{aligned}$$
and we have the theorem with $\delta = \alpha\bar{\beta}$. ∎

We will now prove a result relating the limiting Newton direction with the distance to the solution.

2.7.4 Theorem Let x^* be such that $f(x^*) = 0$, and let $\|Df(x^*)^{-1}\| \leq \beta$. There exists a $\theta > 0$ such that for all $\|x - x^*\| < \theta$
$$\frac{1}{2}\|x - x^*\| \leq \|x' - x\| \leq \frac{3}{2}\|x - x^*\|.$$
where $x' = x - Df(x)^{-1}f(x)$, the produce of a Newton iterate.

Proof: Let $\theta = \frac{1}{3\alpha\beta}$, and let x be such that $\|x - x^*\| \leq \theta$. Then
$$\begin{aligned}\|Df(x^*)^{-1}(Df(x) - Df(x^*))\| &\leq \alpha\beta\|x - x^*\| \\ &\leq \frac{1}{3}.\end{aligned}$$
Thus from Theorem 2.3.5 $Df(x)^{-1}$ exists, and $\|Df(x)^{-1}\| \leq \frac{3}{2}\beta$. Consider
$$0 = f(x^*) = f(x) + Df(x)(x - x^*) + e(x, x^*)$$
and thus
$$x' - x = -Df(x)^{-1}f(x) = (x - x^*) + Df(x)^{-1}e(x, x^*)$$
and our result follows from the fact that $\alpha\beta\theta \leq \frac{1}{3}$, and Theorem 2.1.21. ∎

2.7 NONLINEAR SYSTEM OF EQUATIONS

Finite Difference Newton's Method

One of the major disadvantage of Newton's method is that it requires the explicit use of the derivative, which is an $n \times n$ matrix. For large n, obtaining the derivative without error may be difficult and sometimes impossible. One way around this is to use the finite difference approximation to this derivative matrix. This is done in the finite difference Newton's method which follows:

Step 1 Let x^0 be arbitrary, ϵ and ϵ_0 be positive (and small) numbers, and $\theta > 0$. Set k=0.

Step 2 For each $j = 1, \cdots, n$ define

$$A_k(j) = \frac{f(x^k + \epsilon_k u^j) - f(x^k)}{\epsilon_k}$$

where u^j is the jth unit vector and $A_k(j)$ is the jth column of the matrix A_k.

Step 3

$$x^{k+1} = x^k - A_k^{-1} f(x^k)$$

Step 4 Set

$$\epsilon_{k+1} = \min\{\frac{\epsilon_k}{2}, \theta ||A_k^{-1} f(x^k)||\}$$

Step 5 If $\epsilon_{k+1} \leq \epsilon$ or $||f(x^{k+1})|| \leq \epsilon$ stop. Otherwise $k = k+1$, and go to Step 2.

In Step 2 of this method, a finite difference approximation to the derivative, A_k is computed. The work involved is exactly the same as that for the Newton's method, $O(n^2)$ real valued function evaluations. The step size ϵ_k for this calculation is chosen in Step 3, in such a way that it decreases. This guarantees that the method will attain quadratic convergence. We now prove a local convergence theorem and establish the quadratic convergence for this method.

The following lemma establishes that the matrix computed at Step 2 is an approximate derivative.

2.7.5 Lemma For each k, $||A_k - Df(x^k)|| \leq \frac{1}{2}\alpha n \epsilon_k$, where α is the Lipschitz constant of the derivative.

Proof: From Theorem 2.1.21, for each j and k

$$f(x^k + \epsilon_k u^j) = f(x^k) + \epsilon_k Df(x^k)u^j + e^j$$

where $e^j = e(x^k + \epsilon_k u^j, x^k)$ with $||e^j|| \leq \frac{1}{2}\alpha\epsilon_k^2$. Thus

$$A_k = Df(x^k) + E$$

where $\frac{1}{\epsilon_k}e^j$ is the jth column of E for each $j = 1, \cdots, n$. Since $||E|| \leq \frac{1}{2}\alpha n \epsilon_k$, we are done. ∎

We now prove a local convergence theorem that parallels the one proved for Newton's method.

2.7.6 Theorem Let ϵ_0 and A_0 be chosen such that

1. $||A_0^{-1}|| \leq \beta$
2. $||A_0^{-1} f(x^0)|| \leq \gamma$
3. $\epsilon_0 \leq \theta\gamma$

and let

$$(2n\theta + 1)\alpha\beta\gamma \leq \frac{1}{2},$$

where α is the Lipschitz constant of the derivative. Then the sequence $\{x^k\}_{k=1}^\infty$ is well defined, and converges to a limit x^∞. Also $f(x^\infty) = 0$.

Proof: We use an induction argument to show that

$$||A_k^{-1}|| \leq 2^k \beta \tag{2.9}$$

$$||A_k^{-1} f(x^k)|| \leq 2^{-k}\gamma. \tag{2.10}$$

By the hypothesis of the theorem, these results are true for $k = 0$. Thus we assume that these results are true for all $i \leq k$, and let $i = k$. Now, using the result of Lemma 2.7.5

$$\begin{aligned}
||A_i^{-1}(A_{i+1} - A_i)|| &\leq ||A_i^{-1}(A_{i+1} - Df(x^{i+1}))|| + \\
&\quad ||A_i^{-1}(Df(x^{i+1}) - Df(x^i))|| + ||A_i^{-1}(Df(x^i) - A_i)|| \\
&\leq ||A_i^{-1}||(\frac{1}{2}\alpha n \epsilon_{i+1} + \alpha ||x^{i+1} - x^i|| + \frac{1}{2}\alpha n \epsilon_i)
\end{aligned}$$

2.7 NONLINEAR SYSTEM OF EQUATIONS

$$\begin{aligned}
&\leq 2^{i-1}\alpha\beta((n\theta+2)\|A_i^{-1}f(x^i)\| + n\theta\|A_{i-1}^{-1}f(x^{i-1})\|) \\
&\leq \alpha\beta\gamma(\frac{3}{2}\theta n + 1) \\
&\leq \frac{1}{2}.
\end{aligned}$$

We used the facts that $\epsilon_{i+1} \leq \theta\|A_i^{-1}f(x^i)\|$, $\|x_{i+1}-x_i\| = \|A_i^{-1}f(x^i)\|$ and the induction hypothesis to obtain the above. Thus, from Theorem 2.3.5 $\|A_{i+1}^{-1}\| \leq 2\|A_i^{-1}\| \leq 2^{i+1}\beta$ and, by induction, we have the first result.

Now, using Theorem 2.1.21, we can write

$$\begin{aligned}
f(x^{i+1}) &= f(x^i) + Df(x^i)(x^{i+1}-x^i) + e(x^{i+1},x^i) \\
&= f(x^i) + A_i(x^{i+1}-x^i) + (Df(x^i)-A_i)(x^{i+1}-x^i) + e(x^{i+1},x^i).
\end{aligned}$$

Hence

$$\|f(x^{i+1})\| \leq \frac{1}{2}\alpha\|A_i^{-1}f(x^i)\|(n\epsilon_i + \|A_i^{-1}f(x^i)\|) \tag{2.11}$$

and thus

$$\begin{aligned}
\|A_{i+1}^{-1}f(x^{i+1})\| &\leq \|\|A_{i+1}^{-1}\|\|\|f(x^{i+1})\| \\
&\leq \alpha\beta\gamma(2n\theta+1)\|A_i^{-1}f(x^i)\| \\
&= 2^{-(i+1)}\gamma
\end{aligned}$$

and, by induction, we have the second result.

We now use equations (2.9) and (2.10) to prove the theorem. Consider

$$\begin{aligned}
\|x^{l+k}-x^k\| &\leq \|x^{l+k}-x^{l+k-1}\| + \cdots + \|x^{k+1}-x^k\| \\
&= \|A_{l+k-1}^{-1}f(x^{l+k-1})\| + \cdots + \|A_k^{-1}f(x^k)\| \\
&\leq \gamma(2^{-(l+k-1)} + \cdots + 2^{-k}) \\
&\leq 2^{-(k-1)}\gamma
\end{aligned}$$

and for all $l \geq 1$ as $k \to \infty$, $\|x^{l+k}-x^k\| \to 0$. Hence $\{x^k\}$ is a Cauchy sequence and thus converges, say to x^∞. It is readily seen, from (2.11), that $f(x^\infty) = 0$. ∎

We make a comment on the choice of θ. If $\theta = \frac{1}{2n}$ this result requires twice as close a starting point than does Newton's method.

We now establish that the convergence rate of the finite difference Newton's method is also quadratic.

2.7.7 Theorem Let the finite difference Newton's method generate the sequence $\{x^k\}_{k=1}^{\infty}$ which converges to x^∞ with $||Df(x^\infty)^{-1}|| \leq \beta$. Then

1. $f(x^\infty) = 0$

2. There is a $L \geq 1$ and a $\rho > 0$, such that for all $k \geq L$
$$||x^{k+1} - x^\infty|| \leq \rho ||x^k - x^\infty||^2;$$

i.e., the convergence rate is quadratic.

Proof: Consider

$$\begin{aligned} ||A_k - Df(x^\infty)|| &\leq ||A_k - Df(x^k)|| + ||Df(x^k) - Df(x^\infty)|| \\ &\leq \frac{1}{2}\alpha\epsilon_k + \alpha||x^k - x^\infty||. \end{aligned}$$

Thus $||Df(x^\infty)^{-1}(A_k - Df(x^\infty))|| \leq \frac{1}{2}\alpha\beta n \epsilon_k + \alpha\beta ||x^k - x^\infty||$. As $x^k \to x^\infty$ and $\epsilon_k \leq \theta ||x^{k+1} - x^k||$, there exists a $L \geq 1$ such that for all $k \geq L$,

$$\frac{1}{2}\alpha\beta n \epsilon_k + \alpha\beta ||x^k - x^\infty|| < \frac{1}{2}.$$

Thus, from Theorem 2.3.5, for all $k \geq L$, A_k^{-1} exists and $||A_k^{-1}|| \leq 2\beta$.
From Theorem 2.1.21,

$$\begin{aligned} f(x^\infty) &= f(x^k) + Df(x^k)(x^\infty - x^k) + e(x^\infty, x^k) \\ &= -A_k(x^{k+1} - x^\infty) + (Df(x^k) - A_k)(x^\infty - x^k) + e(x^\infty, x^k). \end{aligned}$$

Therefore

$$||f(x^\infty)|| \leq ||A_k|| ||x^{k+1} - x^\infty|| + \frac{1}{2}\alpha n \epsilon_k ||x^\infty - x^i|| + \frac{1}{2}\alpha ||x^\infty - x^i||^2$$

and as k goes to ∞, we see that the right hand side approaches 0, and we have the first part of the theorem.
Thus, from the earlier expression,

$$A_k(x^{k+1} - x^\infty) = (Df(x^k) - A_k)(x^\infty - x^k) + e(x^\infty, x^k).$$

Therefore

$$||x^{k+1} - x^\infty|| \leq ||A_k^{-1}|| \cdot ||Df(x^k) - A_k|| \cdot ||x^k - x^\infty|| + \frac{1}{2}\alpha ||x^k - x^\infty||^2.$$

2.7 NONLINEAR SYSTEM OF EQUATIONS

Let $\theta_k = ||A_k^{-1}||.||Df(x^k) - A_k|| + \frac{1}{2}\alpha||x^k - x^\infty||$. Thus, $\theta_k \leq \alpha\beta n\epsilon_k + \frac{1}{2}\alpha||x^k - x^\infty||$. Using Theorem 2.1.21, we can write

$$f(x^k) = f(x^\infty) + Df(x^\infty)(x^k - x^\infty) + e(x^k, x^\infty).$$

Thus

$$\begin{aligned}
\epsilon_k &= \theta||A_k^{-1}f(x^k)|| \\
&\leq \theta||Df(x^\infty)||||x^k - x^\infty|| + \frac{1}{2}\alpha\theta||x^k - x^\infty||^2 \\
&< \theta'||x^k - x^\infty||.
\end{aligned}$$

Thus $\theta_k \leq (n\alpha\beta\theta' + \frac{1}{2}\alpha)||x^k - x^\infty||$ and we have our result with $\rho = \frac{1}{2}\alpha + n\alpha\beta\theta'$. ∎

Parallel Chord Method

The parallel chord method is a useful but often overlooked variant of Newton's method. The iterations for this method are simple, yet they exhibit linear convergence rate.

We now present this method:

Step 1 Let x^0 be arbitrary, A be an approximation to $Df(x^0)$, $\epsilon > 0$. Set k=0.

Step 2 $x^{k+1} = x^k - A^{-1}f(x^k)$

Step 3 If $||f(x^{k+1})|| \leq \epsilon$, stop. Otherwise set $k = k + 1$ and go to Step 2.

This method requires no calculations of the derivative of f at x, but uses an approximation. Thus each iteration is *cheap*. The price is paid in the rate of convergence, which is linear. A very effective hybrid can be generated by integrating Newton's method and the parallel chord method. This method can also be very useful in the corrector step of the predictor-corrector methods.

We now show the convergence rate of this method.

2.7.8 Theorem Let the parallel chord method generate the sequence $\{x^k\}_{k=1}^\infty$ which converges to x^∞. Then

1. $f(x^\infty) = 0$

2. If
$$||A^{-1}(A - Df(x^\infty))|| \leq \frac{1}{2},$$
then there is a $L \geq 1$ and a $0 < \rho < 1$ such that for all $k \geq L$
$$||x^{k+1} - x^\infty|| \leq \rho ||x^k - x^\infty||$$
i.e., the convergence rate is linear.

Proof: Using Theorem 2.1.21 with $||e(x^{k+1}, x^k)|| \leq \frac{1}{2}\alpha ||x^{k+1} - x^k||^2$, we can write
$$\begin{aligned} f(x^{k+1}) &= f(x^k) + Df(x^k)(x^{k+1} - x^k) + e(x^{k+1}, x^k) \\ &= (Df(x^k) - A)(x^{k+1} - x^k) + e(x^{k+1}, x^k). \end{aligned}$$

As the sequence $\{x^k\}$ converges, letting k go to infinity, it is readily seen that the right hand side of the above expression goes to zero, while the left hand side to $f(x^\infty)$. Thus we have the first part of the theorem.

Again using Theorem 2.1.21 we see that
$$f(x^k) = f(x^\infty) + Df(x^\infty)(x^k - x^\infty) + e(x^k, x^\infty).$$

Adding and rearranging terms we can write the above as
$$\begin{aligned} 0 &= f(x^k) + A(x^{k+1} - x^k) \\ &= f(x^\infty) + (Df(x^\infty) - A)(x^k - x^\infty) + A(x^{k+1} - x^\infty) + e(x^k, x^\infty). \end{aligned}$$
Thus, as $f(x^\infty) = 0$,
$$0 = (A - Df(x^\infty))(x^k - x^\infty) - A(x^{k+1} - x^\infty) - e(x^k, x^\infty).$$
So
$$x^{k+1} - x^\infty = A^{-1}(A - Df(x^\infty))(x^k - x^\infty) - A^{-1}e(x^{k+1}, x^k).$$
. Now define $L \geq 1$ such that for every $k \geq L$
$$\alpha ||A^{-1}|| ||(x^k - x^\infty)|| < 1$$
and we have our result with $\rho = ||A^{-1}(A - Df(x^\infty))|| + \frac{1}{2}\alpha ||A^{-1}|| ||x^k - x^\infty||$.
∎

Global Newton Method

As seen in Theorem 2.7.2, convergence of Newton's method is critically related to the choice of the 'starting point' x^0. This is undesirable, and we call any method that has this property a *local* method. A method whose convergence can be established independent of the starting point, but is based on some overall properties of the functions involved, is called a *global* method. On the other hand, a very desirable feature of Newton's method is its quadratic convergence rate and its basic simplicity and ease of implementation. Because of this there have been many attempts to globalize its convergence.

We now investigate the more popular of these attempts, which go by various names, including embedding, continuation or homotopy methods. The idea here is to imbed the problem of solving a system of n nonlinear equations in n variables into $\boldsymbol{R}^{n+1}$ by the following methodology:

Define a continuously differentiable function

$$F : \boldsymbol{R}^n \times [0,1] \to \boldsymbol{R}^n$$

such that

$$F(x,0) = r(x)$$
$$F(x,1) = f(x)$$

where $r : \boldsymbol{R}^n \to \boldsymbol{R}^n$ is a *simple* function with at least one known zero x_0. The idea then is to solve the seemingly more difficult problem,

$$F(x,t) = 0$$

whose solution is generally a union of smooth curves, one of whose connected components contains the starting point $(x_0, 0)$. The globalization idea now involves tracing this connected component, starting at $(x_0, 0)$. Success is attained if this connected component intersects $\boldsymbol{R}^n \times \{1\}$, i.e., $(x_1\ 1)$ for some computed x_1. Then, it is readily seen that, x_1 is a zero of f.

Some popular forms for the mapping F that have been proposed are

$$F(x,t) = (1-t)r(x) + tf(x)$$

called the *convex* homotopy, and

$$F(x,t) = f(x) - (1-t)f(x_0)$$

called the *global* homotopy.

2.7.9 Regular and Critical Values We say $y \in \mathbf{R}^n$ is a regular value of F if and only if the $n \times (n+1)$ derivative matrix $DF(x)$ has full rank n for every $x \in F^{-1}(y)$; and a critical value otherwise.

The famous theorem of Sard states that if F is at least twice continuously differentiable, the set of regular values of F is dense; thus almost all values of F are regular. A consequence of the fact that y is a regular value of F is that the set $F^{-1}(y)$ is at least once differentiable 1-manifold, and is thus a union of at least once differentiable curves. If zero is a regular value of F, the existence of the curve starting at $(x_0, 0)$ is thus established by this fact. But zero may be a critical value of F. In that case a transversality argument can be applied to a perturbed function F_a to assure that zero becomes a regular value of F_a. We demonstrate this perturbation for the convex homotopy

$$F_a(x,t) = (1-t)(x-a) + tf(x) \qquad (2.12)$$

where we have chosen $r(x) = x - a$. The idea now is to consider a as a parameter which can be chosen freely from an open set $A \subset \mathbf{R}^n$. A transversality argument applied to the function $H(x,t,a) = F_a(x,t)$, with

$$H : \mathbf{R}^n \times [0,1] \times A \to \mathbf{R}^n$$

shows that for almost all $a \in A$, zero is a regular value of F_a. Within our framework, this result is easy to implement, since if x_0 was the desired starting point, zero can be made a regular value by choosing a starting point randomly in an open set A containing x_0. For this to hold, f has to be at least thrice continuously differentiable.

We summarize, without proof, the above discussion in the theorem that follows.

2.7.10 Theorem Let F be twice continuously differentiable. If zero is a regular value of F, $F^{-1}(0)$ is a union of twice continuously differentiable curves. ∎

We will henceforth make the following assumption:

2.7.11 Assumption Zero is a regular value of F, and $F^{-1}(0)$ is a union of at least once continuously differentiable curves.

A Global Existence Theorem

To demonstrate the global nature of the embedding, we now prove a global existence theorem using the convex homotopy represented by the equation (2.12).

2.7 NONLINEAR SYSTEM OF EQUATIONS

2.7.12 Leray Schauder Theorem Let there exist an open bounded set $D \subset \mathbf{R}^n$, containing a, such that for all $x \in \text{boundary}(D)$, and $\lambda < 0$

$$f(x) \neq \lambda(x - a).$$

Then f has a zero in closure(D), i.e., $f(x) = 0$ has a solution $x^* \in$ closure(D).

Proof: Let $(c, t) : [0, \beta) \to \mathbf{R}^n \times [0, 1]$ be the connected component of $F^{-1}(0)$ containing $(a, 0)$, i.e.,

$$\begin{aligned} F(c(r), t(r)) &= 0 \text{ for all } r \in [0, \beta) \\ c(0) &= a \\ t(0) &= 0. \end{aligned}$$

Assume that for some $r \in [0, \beta)$, $c(r)$ lies on the boundary of D. Then

$$0 = F(c(r), t(r)) = (1 - t(r))(c(r) - a) + t(r)f(c(r))$$

and thus $f(c(r)) = \frac{1-t(r)}{-t(r)}(c(r) - a)$. As $c(r)$ lies in the boundary of D, $\frac{1-t(r)}{-t(r)} \geq 0$ or $t(r) \geq 1$. If $t(r) = 1$ then $c(1)$ on the boundary of D is a solution, otherwise the curve c does not hit the boundary for $t(r) \in [0, 1]$, and, as $a \in D$ which is bounded, we are done. ∎

A consequence of this theorem is that if f satisfies the *global* condition of the Leray-Schauder theorem, tracing the curve (path) starting at $(a, 0)$ will lead to the solution x_1 of f. We now show how Newton's method can be used to implement this path tracing, and thus generate a globalized version of Newton's method.

Predictor-Corrector Method

We now consider the problem of tracing the path $(c(r), t(r))$ for $r \in [0, \beta)$ where

$$F(c(r), t(r)) = 0 \text{ for all } r \in [0, \beta).$$

For this purpose, a convenient variable to parameterize the path with is the *arc* length, s. To do this, make the change of variable $r = h(s)$ with $0 = h(0)$, and consider the path

$$p(s) = (c(h(s)), t(h(s)))$$

with the additional requirement that

$$ds = ((\frac{dt(r)}{dr})^2 + \sum_{j=1}^{n}(\frac{d_j c_j(r)}{dr})^2)^{\frac{1}{2}} dr.$$

Then $p(s)$ satisfies

$$F(p(s)) = 0 \tag{2.13}$$
$$\|\dot{p}(s)\|_2 = 1$$

where $\dot{p}(s)$ is the $(n+1)$ dimensional vector whose jth entry, for each $j = 1, \cdots, (n+1)$, is

$$\dot{p}_j(s) = \frac{dp_j(s)}{ds}$$

and the path has been parameterized by the arc length s.

In case it can be shown that the behavior of t is monotone along the path, the homotopy parameter itself can be used to parameterize the path, in which case a desired solution is found when $t = 1$. If the path is reparameterized, one must reach an s such that $p_{n+1}(s) = 1$; and the desired solution is $x_j = p_j(s)$ for all $j = 1, \cdots, n$.

2.7.13 Initial Value Problem Differentiating equation (2.13) we obtain the following initial value problem

$$DF(p(s))\dot{p}(s) = 0 \tag{2.14}$$
$$\|\dot{p}(s)\| = 1$$
$$p(0) = (a, 0)^T$$

and the predictor-corrector methods are designed to integrate this problem and thus follow the path p.

The strategy of a generic predictor-corrector method is the following:

Predictor Step Choose a direction d, generally a tangent to the path p at $p(s)$; i.e., a solution to equation (2.14). Move a certain distance $\theta > 0$, called the step length, along this direction, such that the arc length s increases. Higher order predictors can also be used during the predictor step.

This step will generally lead to a vector not on the path, and there is a possibility of diverging away from the path if only these steps are performed. Thus there arises the need for the corrector step.

2.7 NONLINEAR SYSTEM OF EQUATIONS

Corrector Step

corrector step Since the function F takes the value zero on the path, by taking several Newton, or Newton type iterations the corrector step reduces the distance to the path. These corrector steps are usually taken in a subspace orthogonal to the predictor direction d and thus involve finding a zero of a system of n nonlinear equations in n variables.

Implementing a predictor-corrector method involves finding a good predictor direction, an adequate step length so that the corrector step will be successful, and a good corrector method. We now discuss each of these aspects in some detail.

2.7.14 Predictor Direction Since the path is generally nonlinear, the tangent to the path at $p(s)$ can be a good predictor direction. This can be computed by solving

$$DF(p(s))d = 0$$
$$||d|| = 1.$$

Since we have assumed that zero is a regular value of F, the rank of DF is full (i.e., its null space has dimension 1) and thus the above system has exactly two solutions, d and $-d$ say. Along one of these directions the path approaches the solution, while along the other it moves away. We now orient the path, so that by preserving this orientation we guarantee that we move towards the solution.

Initially, for $s = 0$, let the solution $\bar{d}$ be such that $\bar{d}_{n+1} > 0$; i.e., the homotopy parameter increases along this direction. Define

$$\gamma = \text{sign det} \begin{pmatrix} DF(p(0)) \\ \bar{d}^T \end{pmatrix}.$$

We can preserve the orientation of the path at an arbitrary s by choosing that solution d such that

$$\gamma \det \begin{pmatrix} DF(p(s)) \\ d^T \end{pmatrix} > 0.$$

2.7.15 Corrector Step The corrector step is taken in the subspace orthogonal to the null space of the derivative matrix, DF, i.e., in the row space $\mathcal{R}(DF^T)$ whose dimension is n. We now show how this is done.

Let y be the vector obtained after the predictor step, which needs to be corrected. One way is to attempt a move to the *nearest* point z on the path to y. This can be done by solving

$$\min ||y - z||^2$$
$$F(z) = 0.$$

Using the Lagrangean theory, a necessary condition that z be the nearest point on the path to y is that there exist a vector u such that

$$(y - z) - DF(z)^T u = 0$$
$$F(z) = 0.$$

This problem is hard to solve, so to get a version of Newton's method that is applicable here, we modify this necessary condition by substituting $DF(y)$ in place of $DF(z)$ in the first condition, and replacing the nonlinear equation of the second by its linear approximation at y. We thus obtain the system

$$(y - z) - DF(y)^T u = 0 \qquad (2.15)$$
$$F(y) + DF(y)(z - y) = 0. \qquad (2.16)$$

It is verified by direct substitution that the system of equations (2.15) and (2.16) has the solution

$$u = (DF(y)DF(y)^T)^{-1} F(y)$$
$$(y - z) = DF(y)^T (DF(y)DF(y)^T)^{-1} F(y).$$

This solution is also unique, since $(y - z)$ is orthogonal to the null space of $DF(y)$, and thus for some $d \in \mathcal{N}(DF(y))$, equations (2.15) and (2.16) are equivalent to the system $d^T(y - z) = 0$ and $DF(y)(y - z) = -F(y)$. This is a linear system with the underlying matrix nonsingular.

If we define the right inverse or the Moore-Penrose inverse $DF(y)^+ = DF(y)^T (DF(y)DF(y)^T)^{-1}$, the Step 2 of Newton's method during the correction phase is modified to

$$z = y - DF(y)^+ F(y).$$

It is seen that the *correction* vector $\Delta = z - y$ lies in the row space $\mathcal{R}(DF(y)^T)$, and thus the correction step moves in the space orthogonal to the null space $\mathcal{N}(DF(y))$.

2.7 NONLINEAR SYSTEM OF EQUATIONS

2.7.16 Step Length Analysis Important parameter of predictor-corrector method is the step length, $\theta > 0$. If it is too large, the corrector step will fail (i.e., the vector to be corrected will be outside the domain of convergence of Newton's method); and, if it is too small, too little progress towards the solution will be made. In a good method, this parameter is constantly adjusted, increased if the corrector step converges very rapidly, and decreased if the corrector step diverges. We now develop a methodology to achieve this.

Assume that $x = p(s)$ is on the path (thus $F(x) = 0$) and determines the predictor direction d with $DF(x)d = 0$. Also, assume that, at x, the curvature of the path $D^2 F(x)(d,d) \neq 0$. Using the direction d and a step size θ, so far undetermined, define the predicted point

$$y = x + \theta d.$$

What value should be assigned to θ? We now present a popular method for finding this value.

Assuming θ is known, consider the following corrector steps:

$$z_1 = y - DF(y)^+ F(y)$$
$$z_2 = z_1 - DF(y)^+ F(z_1).$$

The ratio

$$c(x,\theta) = \frac{\|z_1 - z_2\|}{\|y - z_1\|} = \frac{\|DF(y)^+ F(z_1)\|}{\|DF(y)^+ F(y)\|}$$

is a function of x and θ, and is a measure of how fast the corrector iterates are converging. If this ratio is small, the iterates are converging fast. On the other hand if the ratio is large, the iterates are converging slowly. In the former case, the step size chosen to calculate the ratio should be increased for greater efficiency; and, in the later case, it should be decreased.

By the use of Taylor's formula 2.1.18, one can readily establish that there is a constant $\eta(x)$ such that

$$c(x,\theta) = \eta(x)\theta^2 + O(\theta^3) \tag{2.17}$$

where $\lim_{\theta \to 0} \frac{O(\theta^3)}{\theta^2} = 0$. The goal now is to find a $\hat{\theta}$ such that

$$c(x,\hat{\theta}) = \hat{\eta} \tag{2.18}$$

where $\hat{\eta}$ is some prescribed convergence factor determined by the function F. The smaller the constant $\hat{\eta}$, the closer the iterates will stay to the path.

It readily follows that the value

$$\hat{\theta} = \theta \sqrt{\frac{\hat{\eta}}{c(x,\theta)}} \tag{2.19}$$

is obtained by ignoring the higher order terms of θ in equation (2.17) and equation (2.18).

2.7.17 Step Size Adjustment We now present a step size adjustment procedure that is based on the calculated value $\hat{\theta}$.

Step 1 Let $\hat{\eta}$, $\frac{1}{2} \leq \alpha_1 < \alpha_2 \leq 2$ be determined by F; and let θ be the existing step size.

Step 2 Calculate $c(x,\theta)$, and $\hat{\theta}$ by the equation (2.19).

Step 3 Set

$$s = \begin{cases} 1 & \alpha_1 \leq \hat{\theta}/\theta \leq \alpha_2 \\ 2 & \hat{\theta}/\theta < \alpha_1 \\ \frac{1}{2} & \alpha_2 < \hat{\theta}/\theta \end{cases}$$

Step 4 If $s \neq 1$, set

$$\theta = s\theta$$

and re-do the predictor step. Otherwise continue the corrector step until an acceptable point is reached.

The value of s in the above procedure determines the increasing (doubling) or decreasing (halfing) of the existing step size, or accepting it. The increase of the step size is referred to as *acceleration* and the decrease as *de-acceleration*.

2.7.18 A Prototype Predictor Corrector Method The following procedure puts together all the elements of the predictor- corrector method and is a prototype.

Step 1 (a) x^0, the initial starting point.

  (b) $\epsilon > 0$, the stopping accuracy.

  (c) $\frac{1}{2} \leq \alpha_1 < \alpha_2 \leq 2$, the acceleration-deceleration parameters.

  (d) $\hat{\eta} > 0$ the contraction parameter.

  (e) $\theta_0 > 0$ the initial step size.

2.7 NONLINEAR SYSTEM OF EQUATIONS

(f) The initial predictor direction d^0 given by
$$DF(x^0)d^0 = 0.$$

(g) The path orienting parameter, γ, given by
$$\gamma = \text{sign det} \begin{pmatrix} DF(x^0) \\ (d^0)^T \end{pmatrix}.$$

(h) k=0.

(i) Go to Step 3

Step 2 Predictor Direction: Find d^k such that

(a)
$$DF(x^k)d^k = 0$$

(b)
$$\gamma \det \begin{pmatrix} DF(x^k) \\ (d^k)^T \end{pmatrix} > 0$$

Step 3 $\theta = \theta_k$.

Step 4 (a) Predictor Step:
$$y^k = x^k + \theta d^k.$$

(b) Two Corrector Steps:
$$\begin{aligned} z^{k,1} &= y^k - DF(y^k)^+ F(y^k) \\ z^{k,2} &= z^{k,1} - DF(y^k)^+ F(z^{k,1}) \end{aligned}$$

(c) Effective Contraction: Compute
$$c = \frac{\|z^{k,1} - z^{k,2}\|}{\|y^k - z^{k,1}\|}$$
$$\alpha = \sqrt{\frac{\hat{\eta}}{c}}$$
$$s = \begin{cases} 1 & \alpha_1 \leq \alpha \leq \alpha_2 \\ 2 & \alpha < \alpha_1 \\ \frac{1}{2} & \alpha > \alpha_2 \end{cases}$$

(d) If $s = 1$ then go to step 5, otherwise set

$$\theta = s\theta$$

and go to step 4.

Step 5 $j = 2$.

Step 6 If $\|F(z^{k,j})\| < \epsilon$ then

(a) $x^{k+1} = z^{k,j}$.

(b) If $x_{k+1}^{k+1} > \epsilon$. Stop. Solution found.

(c) $k = k + 1, \theta_k = \theta$. Go to step 2

otherwise

(a) $z^{k,j+1} = z^{k,j} - DF(z^{k,j})^+ F(z^{k,j})$

(b) Go to step 6.

2.8 Notes

This chapter covers mathematical background needed to study the many methods for solving a linear programming problem. It also covers several concepts that are not traditionally considered part of the background needed to study linear programming.

Section 2.1 gives a quick review of the necessary concepts from real analysis. These include sequences and their convergence; sets: open, closed, compact and closure; functions: continuous and differentiable and their properties. A good reference for this material is Rudin [190].

Section 2.2 gives a quick review of the necessary concepts in linear algebra and matrix analysis. It covers the definition a vector space, and concepts of linear combination; linear independence; basis; dimension and subspace. It also covers the concept of a submatrix; rank; subspaces null space, column space, row space, left null space; projection into a subspace; matrix norms; and, eigenvalues and eigenvectors. Good references for this material are Strang [217] and Wilkinson [245].

Section 2.3 covers the concepts of numerical linear algebra. It covers the solution of linear system of equations; Gaussian elimination; LU- factorization; Cholesky factorization; LQ-factorization; and, QR-factorization. Good references for this section are Strang [217] and Wilkinson [245].

2.8 NOTES

Section 2.4 covers the topics of convex sets and separation theorems. These concepts were developed by Minkowski [147] and are covered in the following books: Eggleston [61], Rockafellar [184], Stoer and Witzgall [218] and Valentine [240].

Section 2.5 covers the topics of linear equations and inequalities, and the results of Farkas [65] and Motzkin [165] are proved. Our approach to duality is in variance with the now popular method where the duality theorems are proved using the simplex method. We follow the approach of Gale [74] where the duality theorems are proved as a corollary to these results. The lemma of Farkas is proved as a corollary to the strict separation theorem.

Section 2.6 covers the topics of convex polyhedral sets, and lays the geometric foundation for the study of linear programming. More comprehensive details can be found in Eggleston [61] and Valentine [240]. Topics covered include affine space; affine hull; face; vertex; edge; homogeneous solution; degeneracy; characterization of polyhedral sets by vertices; and polyhedral sets with integer data.

Section 2.7 covers the study of Newton's method and its globalization by embedding or continuation as well as several of its variants. This method and its globalization has become indispensable for the study of interior point methods for linear programming. A new and inductive proof (due to Saigal [199]) of the convergence theorem of Kantorovich [111] is given. This theorem is also generalized to the finite difference Newton's method. In addition, the asymptotic quadratic convergence of Newton's method and its finite difference variant is established. Globalization of Newton's method via embedding or continuation is also studied. Roots of these techniques can be traced to the work of Poincarè [178], Leray and Schauder [131] among others, and an extensive study of these can be found in Allgower and Georg [6].

Chapter 3

Duality Theory and Optimality Conditions

3.1 The Dual Problem

A fundamental problem in optimization is the characterization of the solution of the optimization problem. In linear programming theory, this is achieved by Duality theory. This theory establishes a relationship between optimal solutions of two closely related linear programs, called the pair of dual linear programs. We now introduce this theory.

3.1.1 Dual of the Standard Form

Given a linear program in standard form:

$$\begin{aligned} \text{minimize} \quad & c^T x \\ Ax & \geq b \\ x & \geq 0 \end{aligned}$$

where A is an $m \times n$ matrix, b is a m vector and c is a n vector, we define its dual as the linear program:

$$\begin{aligned} \text{maximize} \quad & b^T y \\ A^T y & \leq c \\ y & \geq 0. \end{aligned}$$

Both these linear programs are defined by the same data A, b, and c; and are very closely related in terms of their feasibility and optimality properties. We call this pair of dual linear programs the *standard* form of the dual. It is the general practice in linear programming theory to refer to the first of these problems as the *primal* problem, and the second as the *dual* problem.

However a simple fact about this pair is that either of the two can be called the primal, and the other its dual. This is so because of the next theorem.

3.1.2 Theorem The dual of the dual is the primal.

Proof: We shall prove this result for the case where the dual is generated by the data A, b and c; i.e., the second problem. It is readily seen that an equivalent linear program is

$$\begin{aligned} \text{minimize} \quad & -b^T y \\ -A^T y \geq & -c \\ y \geq & 0 \end{aligned}$$

which has the same form as the primal, but is defined by the data $-A$, $-b$ and $-c$. Its dual is

$$\begin{aligned} \text{maximize} \quad & -c^T x \\ -(A^T)^T x \leq & -(b^T) \\ x \geq & 0 \end{aligned}$$

which is readily seen equivalent to the primal problem with the data A, b and c and we are done. ∎

Duals of the Other Forms

Since the pair of duals is defined on the same data set, their form will depend on the form of the given linear program. We now explore some of these pairs.

3.1.3 Dual of the Canonical Form Given a linear program in canonical form:

$$\begin{aligned} \text{minimize} \quad & c^T x \\ Ax = & b \\ x \geq & 0 \end{aligned}$$

we define its dual as

$$\begin{aligned} \text{maximize} \quad & b^T y \\ A^T y \leq & c. \end{aligned}$$

This form is consistent with the standard form of dual pairs, since the canonical primal is equivalent to the following standard primal

$$\begin{aligned} \text{minimize} \quad & c^T x \\ Ax \geq & b \\ -Ax \geq & -b \\ x \geq & 0 \end{aligned}$$

3.1 DUAL PROBLEM

defined on the data A, $-A$, b, $-b$ and c. Its dual in the standard form is

$$\begin{aligned}
\text{maximize} \quad & b^T y_1 & - & \quad b^T y_2 & \\
& A^T y_1 & - & \quad A^T y_2 & \leq \quad c \\
& y_1 \geq 0 & & \quad y_2 \geq 0 &
\end{aligned}$$

by the transformation $y = y_1 - y_2$ is readily seen equivalent to the canonical dual with the data A, b and c. This transformation results in a variable y which is unrestricted in sign. It can be readily established that for this form the dual of the dual is also the primal; and thus either of the problems can be called the primal, and its related problem the dual. It is again standard practice to call the first problem the primal, and the second the dual.

3.1.4 Dual of the Mixed Form

Given a linear program in mixed form:

$$\begin{aligned}
\text{minimize} \quad & c_1^T x_1 & + & \ c_2^T x_2 & + & \ c_3^T x_3 & \\
& A_{11} x_1 & + & \ A_{12} x_2 & + & \ A_{13} x_3 & \geq b_1 \\
& A_{21} x_1 & + & \ A_{22} x_2 & + & \ A_{23} x_3 & = b_2 \\
& A_{31} x_1 & + & \ A_{32} x_2 & + & \ A_{33} x_3 & \leq b_3 \\
& x_1 \geq 0 & & & & \ x_3 \leq 0 &
\end{aligned}$$

its dual is

$$\begin{aligned}
\text{maximize} \quad & b_1^T y_1 & + & \ b_2^T y_2 & + & \ b_3^T y_3 & \\
& A_{11}^T y_1 & + & \ A_{21}^T y_2 & + & \ A_{31}^T y_3 & \leq c_1 \\
& A_{12}^T y_1 & + & \ A_{22}^T y_2 & + & \ A_{32}^T y_3 & = c_2 \\
& A_{13}^T y_1 & + & \ A_{21}^T y_2 & + & \ A_{33}^T y_3 & \geq c_3 \\
& y_1 \geq 0 & & & & \ y_3 \leq 0. &
\end{aligned}$$

By simple transformations it can also be confirmed that this pair of duals is consistent with the standard and the canonical pair of duals; and that the dual of the dual is the primal.

It is clear that a variable of one problem of the dual pair is related to a specific constraint of the other, and vice versa. The following table summarizes the relationship between the signs of variables of one problem with the type of the corresponding inequality of the other.

Minimum Problem	Maximum Problem
Sign of Variable	Type of constraint
$\geq$	$\leq$
unrestricted	$=$
$\leq$	$\geq$
Type of constraint	Sign of Variable
$\geq$	$\geq$
$=$	unrestricted
$\leq$	$\leq$

To see why the table makes sense, we now illustrate how the type of the constraint of the minimization problem effects the sign of the corresponding variable of the maximization problem. The same illustration explains the converse. For this purpose consider the first set of constraints of the minimization problem of the mixed pair. If the right hand side, b_1 is increased, the set of all feasible solutions of this problem decreases (i.e., some feasible solutions may be lost after this change). Hence the minimum value of the objective cannot decrease, and this is reflected in the non-negativity of the corresponding dual variable y_1. On the other hand, if the right hand side of an equality constraint is increased, the effect of this on the minimum value of the objective is uncertain, which is reflected in the unrestricted sign of the corresponding dual variable. Similarly for the third type of constraint. Thus, it is evident that the signs of the variables of the dual behave like the signs of the Lagrange multipliers of the Lagrangian theory. As we shall subsequently see, the variables of the dual are precisely the multipliers of the Karush-Kuhn-Tucker theory of constrained optimization.

3.2 Duality Theorems

There is an intimate relationship between the pairs of dual problems, and each gives considerable information about the feasibility and optimality of the other. These relationships are explored by the duality theorems.

3.2.1 Weak Duality Theorem Let $\bar{x}$ be a feasible solution of the primal, and $\bar{y}$ be a feasible solution of the dual. Then

$$c^T \bar{x} \geq b^T \bar{y}.$$

Proof: Since $\bar{x} \geq 0$ and $A^T \bar{y} \leq c$ we have

3.2 DUALITY THEOREMS

$$c^T \bar{x} \geq \bar{y}^T A \bar{x}$$

and as $\bar{y} \geq 0$ and $A\bar{x} \geq b$, we have

$$\bar{y} A \bar{x} \geq \bar{y}^T b$$

and we are done. ∎

This theorem immediately shows the following relationship between the feasibility properties of one and the optimality properties of the other.

3.2.2 Corollary

1. If both problems have feasible solutions then they both have optimum solutions.

2. If the primal problem has a sequence of feasible solutions $\{x_k\}$ such that $c^T x_k \to -\infty$, then the dual problem cannot have a feasible solution.

3. If the dual problem has a sequence of feasible solutions $\{y_k\}$ such that $b^T y_k \to \infty$, then the primal problem cannot have a feasible solution.

Proof: If one problem has a feasible solution, by Theorem 3.2.1, the objective value of the other problem becomes bounded, and thus if both have feasible solutions, both objective values are bounded. Thus both problems have optimum solutions, and the first part follows. The remaining parts follow by contradiction. ∎

We now establish the converse of the above corollary as well as some other important properties of the pair.

3.2.3 Strong Duality Theorem

1. If both problems of a pair of duals have feasible solutions, then they both have optimum solutions, and the optimum objective values are the same.

2. Let the dual have a feasible solution. Then the primal problem has no feasible solution if and only if the objective function of the dual problem is unbounded above.

3. Let the primal have a feasible solution. Then the dual problem has no feasible solution if and only if the objective function of the primal problem is unbounded below.

4. Both problems of the dual pair may have no feasible solutions.

Proof: One part of the theorem follows from the Corollary 3.2.2. To prove part (1), we still need to show that the optimum objective values are the same. To do this, consider the system

$$\begin{array}{rcl} Ax & \geq & b \\ -A^T y & \geq & -c \\ -c^T x + b^T y & \geq & 0 \\ x \geq 0 \quad y \geq 0 & & \end{array}$$

If this system has a solution, say $(\bar{x}, \bar{y})$ then feasible solutions to the dual pair have been found with

$$c^T \bar{x} \leq b^T \bar{y}$$

and the result follows from the Weak Duality Theorem 3.2.1. Thus assume this system has no solution. Then, from the Theorem of the Alternative 2.5.15, the following system has a solution:

$$\begin{array}{rcl} A^T u - \lambda c & \leq & 0 \\ -Av + \lambda b & \leq & 0 \\ b^T u - c^T v & > & 0 \\ u \geq 0 \quad v \geq 0 \quad \lambda \geq 0 & & \end{array}$$

First, assume $\lambda \neq 0$. As the right hand side of the system is zero, we can assume without loss of generality that it has a solution with $\lambda = 1$. Then u is readily seen as a solution to the dual and v as a solution to the primal with

$$b^T u - c^T v > 0$$

which contradicts the Weak Duality Theorem 3.2.1. Now assume $\lambda = 0$. If $c^T v \geq 0$ then $b^T u > 0$ and using the Theorem of the Alternative 2.5.15, we see that the primal has no feasible solution, a contradiction. On the other hand, if $c^T v < 0$, then for a feasible solution $\bar{x}$ of the primal and every $\theta > 0$, the vector $x(\theta) = \bar{x} + \theta v$ is a feasible solution of the primal with $c^T x(\theta) = c^T \bar{x} + \theta c^T v \longrightarrow -\infty$ as $\theta \longrightarrow \infty$. Thus from the Weal Duality Theorem 3.2.1, the dual has no feasible solution, a contradiction. Thus part (1) follows.

We now complete the proof of part (2). The proof of part (3) is completed in the same way. Since the primal problem has no feasible solution, using the Theorem of the Alternative 2.5.15 the system below has a solution:

$$\begin{array}{rcl} A^T y & \leq & 0 \\ b^T y & > & 0 \\ y & \geq & 0. \end{array}$$

3.3. OPTIMALITY AND COMPLEMENTARY SLACKNESS

Let this solution be $\bar{y}$. Also, let y^* be a feasible solution of the dual. Consider, for each $k \geq 0$
$$y_k = y^* + k\bar{y} \geq 0$$
and note that $A^T y_k = A^T y^* + k A^T \bar{y} \leq c$ and is thus feasible for each $k \geq 0$. Also, $b^T y_k = b^T y^* + k b^T \bar{y} \to \infty$ as $k \to \infty$, and we are done.

To demonstrate part (4), we present the following example:

$$\begin{array}{rrcrl}
\text{minimize} & -x_1 & - & x_2 & \\
& x_1 & - & x_2 & \geq 1 \\
& -x_1 & + & x_2 & \geq 1 \\
& x_1 \geq 0 & & x_2 \geq 0 &
\end{array}$$

and its dual

$$\begin{array}{rrcrl}
\text{maximize} & y_1 & + & y_2 & \\
& y_1 & - & y_2 & \leq -1 \\
& -y_1 & + & y_2 & \leq -1 \\
& y_1 \geq 0 & & y_2 \geq 0 &
\end{array}$$

and we have demonstrated part (4). ∎

We have proved the strong duality theorem by using a corollary to Farkas' lemma. This lemma was established as a convex separation theorem. In the traditional approach, strong duality theorem is proved by the simplex method, and then Farkas' lemma is shown by using this theorem. In our approach, we will derive the simplex method, and other algorithms from these duality results.

3.3 Optimality and Complementary Slackness

In optimization theory, optimality conditions are not stated as duality type theorems. But duality theorems we have proved in the previous section can be used to derive such conditions for the linear programming problem. These are generally used to derive the simplex method and other algorithms for finding an optimal solution to the linear program. Before we do this we introduce a new class of variables.

3.3.1 Slacks Slacks are nonnegative variables added to convert inequality constraints into equality constraints. For example, it is readily seen that the systems
$$Ax \geq b$$
and
$$Ax - s = b, \ s \geq 0$$

are equivalent in the sense that any solution to the first system generates a unique solution to the second system, and conversely. In linear programming theory, this equivalence plays a central role.

Complementary Slackness Conditions

As was seen in the duality theorems, there is a close relationship between the pair of dual problems. There is also a natural connection between the variables of one problem and the constraints of the other. This relationship is made precise by the complementary slackness conditions, and exists between the variables of one problem and the slacks of the other. Such a relationship exists for every pair of duals, but we will study in detail the standard pair of duals. Investigating other pairs is identical to this study.

Consider a pair of standard dual linear programs, and convert the inequalities into equalities by adding slack variables to the primal and the dual. Thus the primal system is equivalent to

$$\begin{aligned} \text{minimize} \quad & c^T x \\ & Ax \quad - \quad s \quad = b \\ & x \geq 0 \qquad s \geq 0 \end{aligned}$$

and the dual is equivalent to

$$\begin{aligned} \text{maximize} \quad & b^T y \\ & A^T y \quad + \quad t \quad = c \\ & y \geq 0 \qquad t \geq 0. \end{aligned}$$

Variable s is called the primal slack, and variable t the dual slack. We note that all variables in this pair of linear programs are non-negative. The complementary slackness theorem establishes a relationship between the optimum value of the variable of one problem with the optimum value of the slack variable associated with the corresponding constraint of the other.

3.3.2 Complementarity Conditions Given any pair of duals, each variable of one is associated with a specific constraint of the other. Complementarity conditions specify the special relationship between the optimal values of a variable in one problem and the slack on the corresponding constraint of the other. It essentially states that at least one of these two variables must be zero. We state the complementarity conditions for the pair of standard duals, with the understanding that such conditions are readily generated for any pair of duals.

3.3 OPTIMALITY CONDITIONS

Let x, s, y and t be any feasible solutions of a standard pair of duals. We say they satisfy the complementarity conditions if and only if

1. $x_j > 0$ implies $t_j = 0$ for each $j = 1, \cdots, n$.
2. $t_j > 0$ implies $x_j = 0$ for each $j = 1, \cdots, n$.
3. $y_i > 0$ implies $s_i = 0$ for each $i = 1, \cdots, m$.
4. $s_i > 0$ implies $y_i = 0$ for each $i = 1, \cdots, m$.

3.3.3 Complementary Slackness Theorem A pair of primal and dual feasible solutions $\bar{x}, \bar{s}$ and $\bar{y}, \bar{t}$ are optimal for their respective problems if and only if they satisfy the complementarity conditions.

Proof: We will prove this theorem for the pair of standard duals. The proof for the other pairs is identical. From the weak duality theorem, $\bar{x}$ and $\bar{y}$ are optimal for the primal and the dual respectively if and only if

$$c^T \bar{x} - b^T \bar{y} = 0.$$

But

$$\begin{aligned} c^T \bar{x} &= \bar{y}^T A \bar{x} + \bar{t}^T \bar{x} \\ &= \bar{y}^T (b + \bar{s}) + \bar{t}^T \bar{x} \\ &= \bar{y}^T b + \bar{y}^T \bar{s} + \bar{t}^T \bar{x} \end{aligned}$$

and thus $c^T \bar{x} = b^T \bar{y}$ if and only if

$$\bar{y}^T \bar{s} + \bar{t}^T \bar{x} = 0.$$

Since all the variables are non-negative, the above holds if and only if $x_j t_j = 0$ for all j and $y_i s_i = 0$ for all i, and thus we are done. ∎

Although we have proved the Complementary Slackness Theorem for a pair of standard duals, such a theorem holds between the variables and slacks of any pair of duals. Since equality constraints have no slacks and their corresponding dual variables are unrestricted in sign, they generate no complementary conditions.

Complementary Slackness Theorem is an optimality condition for linear programs, and most algorithms designed to solve this problem are based on this condition. This theorem also establishes that the solution of a linear program lies on the boundary of the feasible region. There are two different types of algorithms based on this theorem. One class consists of the boundary methods like simplex method and its variants which stay on the boundary and proceed to the solution; the other class consists of the interior point methods that proceed towards the solution through the interior of the feasible region.

We now derive some results that establish the converse of these conditions.

3.3.4 Complementary Pair From the complementarity condition it is evident that, at optimality, there is a relationship between a variable of one problem and the slack on the constraint of the other. We call this pair of variables a complementary pair; and refer to each as the variable complementary to the other.

In the standard form of the dual, the complementary pairs are (x_j, t_j) for each $j = 1, \cdots, n$ and (y_i, s_i) for each $i = 1, \cdots, m$. Since the equality constraints have no slack variables associated with them, they generate no complementary pairs. This is no problem, since the dual variables corresponding to these constraints are unrestricted in sign.

Strict Complementarity

Given optimum solutions to a pair of duals, we say that strict complementarity holds between them if the converse of the complementarity condition holds, i.e.; given that one of the variables of the complementary pair is zero, the other must be positive. For the pair of standard duals, these conditions can be stated as follows: let (x, s) and (y, t) be solutions to a pair of standard duals. Strict complementarity is said to hold between them, if and only if the converse conditions of the Complementary Slackness Theorem, 3.3.3, hold. That is,

1. $x_j = 0$ implies $t_j > 0$ for each $j = 1, \cdots, n$.

2. $t_j = 0$ implies $x_j > 0$ for each $j = 1, \cdots, n$.

3. $y_i = 0$ implies $s_i > 0$ for each $i = 1, \cdots, m$.

4. $s_i = 0$ implies $y_i > 0$ for each $i = 1, \cdots, m$.

An important property of a pair of dual linear programs is that they have optimum solutions that satisfy the strict complementarity condition.

3.3 OPTIMALITY CONDITIONS

3.3.5 Strict Complementarity Theorem To each pair of dual linear programs that have optimum solutions, there exist optimum solutions that satisfy the strict complementarity condition.

Proof: We will show this theorem for the standard pair of duals. The result for other dual pairs can be established by a similar argument. Consider the complementary pair (x_j, t_j). We now show that $x_j = 0$ for every optimum solution to the primal if and only if $t_j > 0$ for some dual optimum solution. The part that if $t_j > 0$ for some optimum solution then $x_j = 0$ for every optimum solution of the primal follows from the Complementary Slackness Theorem, 3.3.3. To see the converse, note that if the common value of the optimum solution to the primal-dual pair is γ, then $t_j = 0$ for every optimum solution of the dual if and only if the following linear program

$$\begin{aligned} \text{maximize} \quad & u_j^T t \\ A^T y + t &= c \\ -b^T y &\leq -\gamma \\ y \geq 0 \quad t &\geq 0 \end{aligned}$$

has an optimum solution with objective value $u_j^T t = 0$, where u_j is the jth unit vector. But, from the Strong Duality Theorem 3.2.3, this is true if and only if its dual

$$\begin{aligned} \text{minimize} \quad c^T x - \lambda \gamma & \\ Ax - \lambda b &\geq 0 \\ x &\geq u_j \\ \lambda &\geq 0 \end{aligned}$$

has a solution with the objective value $c^T x - \lambda \gamma = 0$. Now, for $\lambda = 0$

$$c^T x = 0, Ax \geq 0, x \geq u_j$$

has a solution. Thus the optimum solution set of the primal is unbounded we are done (since, for every $k > 0$ and any optimum solution x^*, $x^* + kx$ is also an optimum solution and $x_j \geq 1$). In case $\lambda > 0$ we see that

$$\begin{aligned} c^T x &= \gamma \\ Ax &\geq b \\ x &\geq \lambda u_j \end{aligned}$$

has a solution, and we have the result.

We now exhibit the stipulated solution. Let $(x(j), s(j))$ and $(y(j), t(j))$ be the pair of optimum solutions for which strict complementarity holds for the pair (x_j, t_j); and $(x(n+i), s(n+i))$ and $(y(n+i), t(n+i))$ for the pair (y_i, s_i). The required pair $(\bar{x}, \bar{s})$ and $(\bar{y}, \bar{t})$ is obtained by taking a convex combination of the above $m+n$ pairs with each pair assigned the weight $\frac{1}{m+n}$, and the theorem follows from the fact that the set of optimum solutions is convex. ∎

Strict complementary solutions lie in the relative interior of the optimum face of the polyhedron. We now prove this simple fact.

3.3.6 Theorem Optimum solutions of a pair of duals satisfying the strict complementarity condition lie in the relative interior of the respective optimum faces.

Proof: We will show this for the primal of the standard pair of duals. The argument for the dual, and other pairs of duals is similar. Let (x^*, s^*) and (y^*, t^*) be the optimum dual pair satisfying the strict complementarity condition. Define

$$B_1 = \{i : y_i^* = 0\}$$
$$B_2 = \{j : t_j^* = 0\}.$$

Also, (x, s) is an optimum solution for the primal if and only if $x_j = 0$ for all $j \notin B_2$ and $s_i = 0$ for all $i \notin B_1$. Equivalently, if and only if it solves the following system of inequalities:

$$A_{B_2} x_{B_2} - I_{B_1} s_{B_1} = b$$
$$c_{B_2} x_{B_2} = \gamma$$
$$x_{B_2} \geq 0 \quad s_{B_2} \geq 0$$

where γ is the optimum value of the objective function. The strict complementary solution generates an interior solution of this system, and we are done. ∎

3.4 Complementary Pair of Variables

There is an intimate relationship between a pair of complementary variables. For example, if the value of one is bounded in the feasible region of its respective problem, the value of the other is unbounded in its respective feasible region. We will prove this relationship for the standard pair of duals. Such relationships for the other pairs are proved similarly.

3.4 COMPLEMENTARY VARIABLES

3.4.1 Theorem Let a pair of duals both have feasible solutions. Then

1. A variable is bounded in the feasible region of one problem if and only if its complementary variable is unbounded in the feasible region of the other problem.

2. A variable is positive for some feasible solution of one problem if and only if its complementary variable is bounded in the optimal solution set of the other problem.

3. A variable is zero for all feasible solutions for one problem if and only if its complementary variable is unbounded in the optimal solution set of the other problem.

Proof: We will prove each part separately. To see part (1), consider the pair (x_j, t_j), and note that x_j is unbounded on the feasible region if and only if the dual of the following linear program has no feasible solution:

$$\begin{aligned} \text{minimize} \quad & -u_j^T x \\ Ax & \geq b \\ x & \geq 0 \end{aligned}$$

i.e., $A^T y \leq u_j$, $y \geq 0$ has no feasible solution. But, from Theorem 2.5.15, this has no feasible solution if and only if $Ax \geq 0$, $u_j^T x > 0$ and $x \geq 0$ has a solution. Thus, the following linear program has a feasible solution:

$$\begin{aligned} \text{minimize} \quad & c^T x \\ Ax & \geq 0 \\ x & \geq u_j \end{aligned}$$

and, thus the optimum objective value of the its dual is bounded, or

$$\begin{aligned} \text{maximize} \quad & u_j^T t \\ A^T y \; + \; & t \; = \; c \\ y \geq 0 \quad & t \geq 0 \end{aligned}$$

and is bounded. We are done with part (1).

To see part (2) for the pair (x_j, t_j), note that x_j is bounded on the optimal solution set if and only if the following linear program has a bounded solution:

$$\begin{aligned} \text{minimize} \quad & -u_j^T x \\ Ax & \geq b \\ c^T x & = \gamma \\ x & \geq 0 \end{aligned}$$

where γ is the optimum value of the objective function. This is true, from the Strong Duality Theorem, if and only if its dual has a feasible solution, or the system
$$A^T y + \theta c \leq -u_j, \; y \geq 0$$
has a solution. From a version of Farkas' lemma, this system has a feasible solution if and only if the following system has no feasible solution:
$$Ax \geq 0, \; c^T x = 0, \; x \geq 0, \; u_j^T x > 0.$$
Thus the linear program
$$\begin{aligned}\text{minimize} \quad & c^T x \\ Ax & \geq 0 \\ x & \geq u_j\end{aligned}$$
has no feasible solution, and so its dual is unbounded, resulting in the unboundedness of t_j on the feasible region of the dual, and we are done. Part (3), can be proved by an argument identical to that of the proof of part (2), and is completed by noting that the linear program above has a solution with $c^T x \leq 0$. ∎

The following is an important corollary to the above theorem and it establishes conditions for existence of interior points.

3.4.2 Existence of Interior Points The primal has an interior point if and only if the optimal solution set of the dual is bounded, and the dual has an interior point if and only if the optimal solution set of the primal is bounded.

Proof: Follows readily from Theorem 3.4.1, part 3. ∎

3.5 Degeneracy and Uniqueness

In this section, we use the duality theorem to show that the uniqueness of the primal and/or the dual is determined by the non-degeneracy of the optimum solution of the other. We will state our results for an arbitrary pair of duals, but prove them for only the standard pair. The proofs of the results for the other pairs are similar.

We now present some preliminaries before we prove our results. Let (x^*, s^*) and (y^*, t^*) be optimum solutions for the dual pair. Define the sets:
$$B_1 = \{j : x_j^* > 0\} \qquad B_1' = \{j : x_j^* = 0\}$$

3.5 DEGENERACY AND UNIQUENESS

$$B_2 = \{i : s_i^* > 0\} \qquad B_2' = \{i : s_i^* = 0\} \qquad (3.1)$$
$$N_1 = \{j : t_j^* > 0\} \qquad N_1' = \{j : t_j^* = 0\}$$
$$N_2 = \{i : y_i^* > 0\} \qquad N_2' = \{i : y_i^* = 0\}$$

Then, defining A_1 to be the submatrix of A with columns indexed in B_1, and I_2 as the submatrix of $-I$ the negative identity matrix with columns indexed in B_2, we see that

$$A_1 x_{B_1} - I_2 s_{B_2} = b. \qquad (3.2)$$

Define the system

$$\begin{aligned} A_1 x_{B_1} - I_2 s_{B_2} &= b \\ c_{B_1}^T x_{B_1} &= c^T x^* \\ x_{B_1} \geq 0 \;,\; s_{B_2} &\geq 0. \end{aligned} \qquad (3.3)$$

We are ready to establish our results. We now present a simple lemma.

3.5.1 Lemma Let (x^*, s^*) be an optimum solution of the primal, and let B_1 be defined as above. Then c_{B_1} lies in the row space of A_1.

Proof: From the Complementary Slackness Theorem 3.3.3,

$$0 = t_{B_1}^* = c_{B_1} - A_1^T y^*$$

and we are done. ∎

3.5.2 Theorem Every linear programming problem in standard form has an optimum solution if and only if it has a basic feasible optimum solution.

Proof: Let (x^*, s^*) be an optimum solution of the primal in standard form. From the Strong Duality Theorem, 3.2.3, its dual also has an optimum solution (y^*, t^*) with $c^T x^* = b^T y^*$. Define the sets B_1 and B_2 as in (3.1) and consider the system (3.3). It has a feasible solution $(x_{B_1}^*, s_{B_2}^*)$. Thus, it also has a basic feasible solution which is optimum for the linear program. From Lemma 3.5.1, this is a basic feasible solution for the primal since the equation $c_{B_1}^T x_{B_1} = c^T x^*$ does not increase the rank of the system (3.2), and we are done. ∎

3.5.3 Uniqueness Theorem
Let a pair of dual linear programs have optimum solutions, and let every optimum solution be non-degenerate for each problem. Then they have unique basic feasible optimum solutions.

Proof: Let (x^*, s^*) and (y^*, t^*) be optimum solutions for the dual pair. Define the sets of (3.1). Since the optimum solutions are non-degenerate for their respective problems:

$$|B_1| + |B_2| \geq m \qquad |N_1| + |N_2| \geq n$$

(where $|S|$ is the number of entries in the set S.) From the Complementary Slackness Theorem, 3.3.3 $B_1 \subseteq N_1'$, $B_2 \subseteq N_2'$, $N_1 \subseteq B_1'$ and $N_2 \subseteq B_1'$. Hence, $|N_1'| + |N_2'| \geq |B_1| + |B_2| \geq m$ and $|B_1'| + |B_2'| \geq |N_1| + |N_2| \geq n$. But

$$n + m = |B_1| + |B_2| + |B_1'| + |B_2'| \geq n + m$$

$$n + m = |N_1| + |N_2| + |N_1'| + |N_2'| \geq n + m$$

forcing all inequalities to equalities. Thus $|B_1| + |B_2| = m$, $|N_1| + |N_2| = n$, $B_1' = N_1$, $B_2' = N_2$, $N_1' = B_1$ and $N_2' = B_2$.

We now show that (x^*, s^*) is a unique basic feasible solution for the primal. The argument for the dual is similar. Define $B = (A_1, I_2)$. Note that B is $m \times m$. Assume that the rank of B is less than or equal to $m - 1$. System (3.3) has a feasible solution, and thus also a basic feasible solution. But $(x_{B_1}^*, s_{B_1}^*)$ is not the basic feasible solution of this system, since from Lemma 3.5.1, c_{B_1} lies in the row space of A_1. Thus, the basic feasible solution to this system will have $m - 1$ or less positive variables. Since this solution is an optimum solution of the primal linear program, we contradict the non-degeneracy assumption. Thus (x^*, s^*) is a basic feasible optimum solution. To see the uniqueness, note that for $j \in N_1$, $t_j^* > 0$ implies $x_j^* = 0$ for all optimum solutions of the primal; and, for $i \in N_2$ $y_i^* > 0$ implies $s_i^* = 0$ for all optimum solutions of the primal. The values of the remaining variables of the primal are uniquely determined by equation (3.1), (as B has rank m), and we are done. ∎

We now prove that if one problem has a degenerate optimum solution, then the other problem has multiple solutions.

3.5.4 Non-uniqueness Theorem

1. When every optimum solution to the primal is degenerate, the dual has multiple optimum solutions.

2. When every optimum solution to the dual is degenerate, the primal has multiple optimum solutions.

Proof: We now show part (2). The proof of part (1) is identical.

From the Theorem 3.3.5, the pair of duals have solutions (x^*, s^*) and (y^*, t^*) which satisfy the strict complementarity condition. Define the sets of (3.1). Since the dual solution is degenerate,

$$|N_1| + |N_2| \leq n - 1$$

and, as the pair of solutions satisfy the strict complementarity condition, $N_1 = B'_1$, $N_2 = B'_2$, $B_1 = N'_1$ and $B_2 = N'_2$. Thus $|B'_1| + |B'_2| \leq n - 1$. Thus

$$|B_1| + |B_2| = n + m - (|B'_1| + |B'_2|) \geq m + 1.$$

Consider the system (3.3). It has the feasible solution $(x^*_{B_1}, s^*_{B_2})$ which is not basic feasible (this is readily seen from Lemma 3.5.1). This system has at least one basic feasible solution which is an optimum solution of the primal. Thus the primal has multiple solutions, and we are done. ∎

3.6 Notes

The statement of the dual is credited to John Von Neumann. The proof of the duality theorem given here follows the approach of Gale [74]. Various statements of the duality theorems appear in Dantzig and Arden [37], including the complementary slackness theorem. For more on duality theorems, see Gale [76].

Result 3.3.5 plays an important role in interior point methods, and its proof has been included here. Section 3.4 is based on the work of Williams [246]. Section 3.5 relates uniqueness to degeneracy, and is important for the study of interior point methods.

Chapter 4

Boundary Methods

4.1 Introduction

Within the last decade, there have been major advances in the ability to solve large scale linear programs. These have been primarily fueled by the fundamental work, in 1984, of Narinder K. Karmarkar, which has had considerable impact on the algorithms for solving the linear programming problems. A new class of algorithms, known as the interior point methods, have been generated. These algorithms proceed through the interior of the feasible region towards the boundary solution, and generate a converging sequence of points in the interior. The prior algorithms searched on the boundary of the feasible region, and found the solution in a finite number of steps. These boundary algorithms are variants of the simplex method discovered by George B. Dantzig in 1947.

All algorithms work with the complementary slackness optimality conditions. Each generates a sequence of solutions to the primal and the dual which violate some of them. These conditions are shown to be satisfied at termination, exactly in the case of the boundary methods, and approximately in the interior methods. The complementary slackness theorem states that a pair of primal and dual solutions are optimal to their respective problems if and only if they satisfy the following three conditions: primal feasibility, dual feasibility and complementary slackness. The table summarizes the strategy of each algorithm we will study in this book.

Algorithm	Primal Feasibility	Dual Feasibility	Complementary Slackness
Boundary Methods			
Primal Simplex	Throughout	At Optimality	Throughout
Dual Simplex	At Optimality	Throughout	Throughout
Primal-Dual Simplex	At Optimality	Throughout	Throughout
Interior Point Methods			
Primal Affine Scaling	Throughout	At Optimality	At Optimality
Dual Affine Scaling	At Optimality	Throughout	At Optimality
Primal-Dual Affine Scaling	Throughout	Throughout	At Optimality
Primal-Dual Homotopy Method	Throughout	Throughout	At Optimality

At each iteration, the algorithms in the above table satisfy all equality constraints. Feasibility is attained when the inequality constraints are also satisfied. As is evident from the table, the variants of simplex method satisfy the complementary slackness conditions at each iteration. It is this that forces them to stay on the boundary. On the other hand the interior point methods satisfy this condition only at optimality. Since this condition is not linear (it is actually quadratic), these methods only satisfy it approximately. In case the the data A, b and c is integral, it can be shown that there is no disadvantage in this, and that these algorithms are also finite.

In 1971, Victor Klee and George J. Minty produced an example demonstrating that variants of the simplex method can take an effort growing exponentially in the size of the data of the problem. On the other hand, some variants of the interior point strategy applied to the problems where the data A, b and c is integral, can be solved in time growing polynomially in the size of the data m. n, A, b, and c. For large problems, this can be an advantage. There is now a growing body of computational experience which supports this conclusion.

In this chapter we will discuss the three popular boundary methods, the primal and dual simplex methods and the primal-dual method. We will discuss the interior point methods in the next chapter.

4.2 Primal Simplex Method

In this section we will study the variants of the simplex method. These are boundary methods, which move from a vertex to an adjacent vertex of the polyhedron. We now derive the primal simplex method.

The primal simplex method, or the simplex method as it is generally referred to, works on the primal problem in canonical form

$$\begin{aligned} \text{minimize} \quad & c^T x \\ Ax &= b \\ x &\geq 0 \end{aligned}$$

where the matrix A is $m \times n$ and is assumed to have full row rank m. We emphasize that the canonical form must satisfy the following three conditions:

(a) The objective function is to minimize,

(b) All constraints are equalities,

(c) All variables are non-negative.

If the problem is not in the canonical form, it can be made canonical by adding slack variables to inequalities, replacing unrestricted variables by a difference of two restricted variables and converting 'maximize' to 'minimize' by changing the sign of the objective function. The dual of this linear program is

$$\begin{aligned} \text{maximize} \quad & b^T y \\ A^T y &\leq c. \end{aligned}$$

Simplex method requires the following assumption on the feasible solution set of the primal. This assumption will be subsequently removed.

4.2.1 Non-degeneracy Assumption We assume that every solution of the primal problem is non-degenerate.

Strategy of the Simplex Method

Given a basic feasible solution, $\hat{x}$, of the primal, define

$$\begin{aligned} B &= \{j : \hat{x}_j > 0\} \\ N &= \{j : \hat{x}_j = 0\}. \end{aligned}$$

The variables x_j with indices $j \in B$ are called the *basic* variables, and, with indices in N the *non-basic* variables. We define B as the submatrix of A with columns indexed in B, and call it the *basis* matrix. Also, we define N as the submatrix of A with columns indexed in N, and call it the *non-basic* matrix. Please note that we are using the same symbol as a set and as a matrix. If there is any chance of confusion, we will make this distinction explicit.

Because of the non-degeneracy assumption, the rank assumption on A and the fact that $\hat{x}$ is a basic feasible solution, B is an $m \times m$ nonsingular matrix. It can be readily verified that the linear system of the primal is equivalent to

$$Bx_B + Nx_N = b$$

and the corresponding dual equality constraint is equivalent to

$$B^T y + t_B = c_B$$
$$N^T y + t_N = c_N$$

where t is the vector of dual slacks.

Obtaining the Primal Solution

Since the non-basic variables x_N are zero,

$$Bx_B = b \tag{4.1}$$

has the unique solution $\hat{x}_B = B^{-1}b > 0$. The equation (4.1) is called the primal system of equations.

Obtaining the Dual Solution

Using the complementary slackness conditions, if $\hat{x}$ is an optimum solution, the fact that the basic variables with indices in B are positive implies that the corresponding slacks on the dual inequalities must be zero. This results in the dual system of equations

$$B^T y = c_B \tag{4.2}$$

and since B is nonsingular, this system has the unique solution $\hat{y} = B^{-T}c_B$ where $B^{-T} = (B^{-1})^T$. We call this the tentative solution of the dual. It satisfies m of the n dual inequalities with equality.

Pricing the Non-Basic Columns

Let
$$\bar{c}_j = c_j - \hat{y}^T A_{.j} \text{ for each } j \in N$$
where $\bar{c}_j$ are the computed values of the dual slacks of the remaining $(n-m)$ inequalities of the dual.

Optimality Condition

In case
$$\bar{c}_j \geq 0 \text{ for each } j \in N \tag{4.3}$$
we can declare the values of the slacks t_j to be
$$\hat{t}_j = \begin{cases} \bar{c}_j & j \in N \\ 0 & j \in B. \end{cases}$$

Since the primal feasible solution $\hat{x}$ and the dual feasible solution $(\hat{y}, \hat{t})$ satisfy the conditions of the Complementary Slackness Theorem 3.3.3, they are optimal solutions for their respective problems, and thus both problems are solved.

Min $\bar{c}_j$ rule

In case
$$\bar{c}_s = \min\{\bar{c}_j : j \in N\} < 0 \tag{4.4}$$
the strategy of the simplex method is to satisfy the most violated dual constraint during the next iteration.

Entering Variable

This is achieved by the use of complementary slackness condition. By assuring that during the next iteration variable x_s (which is complementary to t_s) is positive, t_s is forced to be zero, and thus most violated constraint is satisfied. The variable x_s is called the *entering* variable.

Min Ratio Test

As the entering variable x_s is increased from its present value zero, the values of the other basic variables must be adjusted to satisfy the equality constraint. This is done in the following manner: By keeping all the other non-basic variables other than x_s at value zero, the following equivalent system of m equations in $m+1$ variables is generated

$$Bx_B + A_{.s}x_s = b. \tag{4.5}$$

This system is readily seen equivalent to

$$x_B + B^{-1}A_{.s}x_s = B^{-1}b.$$

By defining $\bar{A}_s = B^{-1}A_{.s}$ and $\hat{x}_B = B^{-1}b$ and letting $x_s = \theta$ we see that the above system is equivalent to

$$x_B(\theta) = \hat{x}_B - \bar{A}_s\theta$$

and all values of θ are acceptable as long as $x_B(\theta) \geq 0$. Define

$$x_j(\theta) = \begin{cases} 0 & j \in N\setminus\{s\} \\ \theta & j = s \end{cases}$$

and note that $Ax(\theta) = b$. Also

$$\begin{aligned}
c^T x(\theta) &= c_B^T x_B(\theta) + c_s \theta \\
&= c_B^T \hat{x}_B + (c_s - c_B^T \bar{A}_s)\theta \\
&= c_B^T \hat{x}_B + (c_s - c_B^T B^{-1} A_{.s})\theta \\
&= c_B^T \hat{x}_B + (c_s - \hat{y}^T A_{.s})\theta \\
&= c_B^T \hat{x}_B + \bar{c}_s \theta
\end{aligned}$$

and since $\bar{c}_s < 0$, the objective function strictly decreases as θ increases. The strategy in the simplex method is to let θ take the largest value that keeps $x(\theta) \geq 0$.

If $\bar{A}_s = (\bar{a}_{1s}, \cdots, \bar{a}_{1s})^T$ then it is easily seen that the largest value of θ that preserves primal feasibility is

$$\bar{\theta} = \begin{cases} \infty & \text{if } \bar{A}_s \leq 0 \\ \min\{\hat{x}_{B_i}/\bar{a}_{is} : \bar{a}_{is} > 0\} & \text{otherwise} \end{cases}$$

4.2 PRIMAL SIMPLEX METHOD

Leaving Variable

Assume that $\bar{\theta} < \infty$. As $\hat{x}_B > 0$ so is $\bar{\theta}$, as it is a ratio of two positive numbers. Assume that $\bar{\theta} = \hat{x}_{B_l}/\bar{a}_{ls}$ for some index l. Then $x(\bar{\theta})_l = 0$. Thus the basic variable x_l becomes zero when $\theta = \bar{\theta}$ and should be removed from the basic set of variables and x_s introduced in its place. Thus the next basic set is $\bar{B} = B \cup \{s\} \setminus \{B_l\}$ and the non-basic set is $\bar{N} = N \cup \{B_l\} \setminus \{s\}$. We call x_l the leaving variable.

4.2.2 Theorem Let $\bar{A}_s \not\leq 0$. Then a unique index solves the min ratio test $\min\{\hat{x}_{B_i}/\bar{a}_{is} : \bar{a}_{is} > 0\}$ if and only if $x(\bar{\theta})$ is non-degenerate.

Proof: Follows from the fact that several indices achieve the minimum if and only if the next feasible solution $x(\bar{\theta})$ is degenerate. ∎

Unbounded Solution

If $\bar{\theta} = \infty$, or equivalently, $\bar{A}_s \leq 0$, the primal solution is unbounded, and thus the dual has no feasible solution.

We now summarize the main theorem of the simplex method:

4.2.3 Theorem Let $\bar{x}$ be the next basic feasible solution in the iterate of the simplex method. If $\bar{\theta} < \infty$ and the basic feasible solution $\hat{x}$ is non-degenerate then $c^T \bar{x} < c^T \hat{x}$.

4.2.4 Revised Simplex Method What we have described above is referred to as the *revised* simplex method. We summarize it below:

Step 0 Let $\hat{x}$ be a basic feasible solution, and B the corresponding basis with $\hat{x}_B > 0$.

Step 1 Calculate the primal solution

$$\hat{x}_B = B^{-1} b.$$

Step 2 Calculate the tentative solution to the dual

$$\hat{y} = B^{-T} c_B.$$

Step 3 Calculate the values of the dual slacks for the non-basic variables

$$\bar{c}_j = c_j - \hat{y}^T A_j \text{ for } j \in N.$$

Step 4 Optimality Condition: If $\bar{c}_j \geq 0$ for each $j \in N$ then $\hat{x}$ is optimal for the primal and $\hat{y}$ is optimal for the dual. STOP

Step 5 Define
$$\bar{c}_s = \min\{\bar{c}_j : j \in N\} < 0$$
and declare x_s as the entering variable. Update the entering column,
$$\bar{A}_s = B^{-1} A_{.s}.$$

Step 6 If $\bar{A}_s \leq 0$, STOP. The dual is infeasible.

Step 7 Min Ratio Test: Define
$$\begin{aligned}\bar{\theta} &= \min\{\hat{x}_{B_i}/\bar{a}_{is} : \bar{a}_{is} > 0\} \\ &= \hat{x}_{B_l}/\bar{a}_{ls}.\end{aligned}$$
and declare x_l as the leaving variable.

Step 8 $x(\bar{\theta})$ is a basic feasible solution with the basic variables with indices in $\bar{B} = B \cup \{s\} \setminus \{B_l\}$ and the non-basic variables with indices in $\bar{N} = N \cup \{B_l\} \setminus \{s\}$. Go to Step 1.

4.2.5 Basis Inverse of Next Iteration There is one more point that deserves special attention. As is evident from the above description of the revised simplex method, the subsequent basic feasible solutions differ by only one index. Thus the basis matrices also differ by one column. Given the inverse of a basis matrix at an iteration of the method, the inverse during the next iteration should be readily obtainable. We now show that this is the case.

Let x_s be the entering and x_l the leaving variable during a step of the simplex method, and assume B^{-1} is known where $B = (B_1, \cdots, B_m)$ is the existing basis matrix. Also, let
$$\bar{B} = (B_1, \cdots, B_{l-1}, A_{.s}, B_{l+1}, \cdots, B_m)$$
be the next basis matrix. How can we obtain the inverse of $\bar{B}$ from B^{-1} ? Note that we have computed, at step 5 of the revised simplex method,
$$B\bar{A}_s = A_{.s}.$$
Thus
$$\bar{B} = BM$$

4.2 PRIMAL SIMPLEX METHOD

for the elementary matrix

$$M = I + (\bar{A}_s - u_l)u_l^T$$

where u_l is the lth unit vector. Hence

$$\bar{B}^{-1} = M^{-1}B^{-1}$$

where, from section 2.3.9

$$M^{-1} = I - \frac{1}{\bar{a}_{ls}}(\bar{A}_s - u_l)u_l^T$$

and thus

$$\bar{B}^{-1} = B^{-1} - \frac{1}{\bar{a}_{ls}}(\bar{A}_s - u_l)u_l^T B^{-1}$$

It can be readily verified that the rows of the matrices are related by the following formula:

$$(\bar{B}^{-1})_{i.} = \begin{cases} \frac{1}{\bar{a}_{ls}}(B^{-1})_{l.} & i = l \\ (B^{-1})_{i.} - \frac{\bar{a}_{is}}{\bar{a}_{ls}}(B^{-1})_{l.} & i \neq l \end{cases} \quad (4.6)$$

and thus getting the next inverse is considerably simpler than computing the inverse from scratch, and involves adding multiples of the lth row to the other rows.

4.2.6 Finite Convergence Theorem Under the non-degeneracy assumption, after a finite number of iterations the simplex method will terminate with either a solution or an indication that the dual has no feasible solution.

Proof: As is evident, the simplex method goes from one basic feasible solution to another. Under non-degeneracy of all solutions, from Theorem 4.2.3, no basic feasible solution can repeat. But there are at most $\binom{n}{m}$ basic feasible solutions, and we are done. ∎

4.2.7 Degeneracy Resolution When the basic feasible solution is degenerate, the objective function does not strictly decrease, and a basic feasible solution can repeat. There are, now classical, examples of degenerate linear programs that cycle. This is exactly what happens in these examples. We now present a scheme for resolving this problem, and thus remove the non-degeneracy assumption.

4.2.8 Lexico-graphic Ordering
Given two vectors x and y in R^m, we say x is lexico-greater than y, and represent it by $x \succ y$, if and only if there is an index k such that

$$x_i = y_i \quad i \leq k$$
$$x_k > y_k$$

and we say x is lexico-positive if and only if $x \succ 0$, i.e., the first nonzero component of x (from the top) is positive.

The following theorem introduces the lexico-min operation.

4.2.9 Theorem
Let $\{x_i\}_{i=1}^k$ be a set of vectors such that $x_i \neq x_j$ for each $i \neq j$. Then there is a unique index s such that

$$x_s = \text{lexico-min}\{x_i : i = 1, \cdots, k\}$$

i.e., $x_i \succ x_s$ for each $i \neq s$. ∎

We now modify the revised simplex method such that we can work with lexico-positive vectors. Consider the equation (4.1) where the right hand side has been replaced by an $m \times (m + 1)$ matrix $\vec{b} = (b, D)$ for some non-singular matrix D. Thus its solution will also be an $m \times (m + 1)$ matrix, given by

$$B\vec{x} = \vec{b}.$$

4.2.10 Definition
We say that a basic feasible solution of the primal is lexico basic feasible if and only if the rows of the matrix $(x_B, B^{-1}D)$ are lexico-positive.

The idea now is to start with a lexico basic feasible solution and to preserve this property throughout the iterations of the simplex method. Since only the right hand side of the system has been modified, the steps 2 through 6 of the revised simplex method are not affected. Step 7 must be modified, since the solution is a matrix. The modified step 7 is the following:

Step 7' Lexico-Graphic Min Ratio Test: Define

$$\vec{\tilde{x}}_{B_l}/\bar{a}_{ls} = \text{lexico-min}\{\vec{\tilde{x}}_{B_i}/\bar{a}_{is} : \bar{a}_{is} > 0\}$$

and declare x_l as the leaving variable.

It is straight forward to check that the next solution is also lexico basic feasible, and that the value of the objective function lexico-graphically decreases. Thus no basic feasible solution can repeat, and thus the finiteness follows.

4.2 PRIMAL SIMPLEX METHOD

To obtain a matrix D such that the starting basic feasible solution becomes lexico basic feasible, let B_0 be the initial basis matrix. Then $D = B_0$ will do the trick, since the initial basic feasible solution matrix is $B_0^{-1}(b, B_0) = (B_0^{-1}b, I)$ which is readily seen to be lexico basic feasible.

Obtaining an Initial Basic Feasible Solution

We see that the simplex method requires an initial basic feasible solution to start. Obtaining such a solution can be hard, since a priori we do not know that the primal has a feasible solution. In many situations, an initial basic feasible solution can be readily identified. But in general this is not the case. We now discuss two popular methods to get such a solution.

4.2.11 Big M Method In this technique, we add an *artificial* variable to each constraint that does not have obvious choice of a variable to which assigning a value satisfies this constraint. This artificial variable is assigned a very large cost, M, so that its use will result in a huge increase in the value of the objective function. This will then discourage the use of these variables, and if the problem has a feasible solution, the optimum solution will not involve any of these artificial variables. On the other hand if the optimal solution involves these variables at positive values, we will conclude that the problem has no feasible solution.

Assuming that the canonical form of the primal has no obvious choice of variables for each constraint, the Big M Method sets up the problem as:

$$\begin{aligned} \text{minimize} \quad & c^T x + M e^T u \\ & Ax + Du = b \\ & x \geq 0 \quad\quad u \geq 0 \end{aligned}$$

Here $e = (1, \cdots, 1)^T$ a m vector of all 1's, and D a diagonal matrix with diagonal entries D_{ii} selected such that for each i

$$D_{ii} = \begin{cases} +1 & \text{if } b_i \geq 0 \\ -1 & \text{if } b_i < 0. \end{cases}$$

Then the solution $x = 0$ and $u = D^{-1}b \geq 0$ is basic feasible, and the simplex method can be initiated with this.

M here is chosen as a very large positive number. The reasoning here is that if the primal has a feasible solution then the modified problem has a solution with $u = 0$; and thus the simplex method will remove these variables from the basis. It will thus find an optimum basic feasible solution with $u = 0$, and thus an optimum solution of the primal problem.

4.2.12 Two Phase Method

A serious disadvantage of the Big M method is that there is no indication of how large M should be. This is resolved by breaking the solution of the problem into two phases, with the goal of the first phase, called Phase 1, is finding a basic feasible solution if one exists; and the goal of the second phase, called the Phase 2, is finding the optimum solution starting with the basic feasible solution found in the first phase.

4.2.13 Phase 1

The following problem with added artificial variables, is called the Phase 1 linear program:

$$
\begin{aligned}
\text{minimize} \quad & e^T u \\
Ax \;+\; & Du \;=\; b \\
x \geq 0 \quad & u \geq 0
\end{aligned}
$$

where D is exactly as in the Big M method. The dual of this Phase 1 linear program is

$$
\begin{aligned}
\text{maximize} \quad & b^T y \\
A^T y \;+\; t \quad & \;=\; 0 \\
Dy \quad\quad\quad\;\; +\; v & \;=\; e \\
t \geq 0 \quad\quad v & \geq 0
\end{aligned}
$$

The following result summarizes the relationship between these two problems. Let $P = \{x : Ax = b, x \geq 0\}$.

4.2.14 Theorem

Both problems of the Phase 1 dual pair have optimum solutions, with the values of the respective objective functions the same. Also, if $(\hat{x}, \hat{u})$ and $(\hat{y}, \hat{t}, \hat{v})$ are the respective optimum solutions, then

1. $e^T \hat{u} > 0$ if and only if $P = \emptyset$.

2. $e^T \hat{u} = 0$ if and only if $P \neq \emptyset$.

3. If $P \neq \emptyset$ and $\hat{t}_j > 0$ then $x \in P$ implies $x_j = 0$.

Proof: It is readily verified that $(0, D^{-1}b)$ and $(0, 0, e)$ are feasible solutions for the dual pair. Thus both have optimum solutions with the objective function values the same. This follows from the Strong Duality Theorem, 3.2.3.

4.2 PRIMAL SIMPLEX METHOD

Parts (1) and (2) follow readily. To see part (3), consider any feasible solution (x, u) of the primal, and the optimum solution $(\hat{y}, \hat{t}, \hat{v})$. Then

$$\begin{aligned} e^T u &= \hat{y}^T D u + \hat{v}^T u \\ &\geq \hat{y}^T (b - Ax) \\ &= \hat{y}^T b - \hat{y}^T A x \\ &= \hat{t}^T x \\ &\geq \hat{t}_j x_j \end{aligned}$$

Thus, as $x \in P$ implies $e^T u = 0$, we see that $\hat{t}_j > 0$ implies $x_j = 0$ and we are done. ∎

4.2.15 Transition to Phase 2 Starting with the basic feasible solution $(0, D^{-1})$, simplex method will solve the Phase 1 linear program. If on termination $e^T \hat{u} > 0$ we declare that $P = \emptyset$ and terminate. If $e^T \hat{u} = 0$ then a basic feasible solution for the Phase 2 has been found. During Phase 2, the linear program with all artificial variables equal to zero is solved. If all artificial variables u are non-basic and thus zero, the transition to Phase 2 is simple. Drop all artificial variables and their columns from the Phase 1 linear program; and drop all variables x_j and their columns for which $\hat{t}_j > 0$. Introduce the Phase 2 objective function, minimize $c^T x$, and proceed to solve this problem.

On the other hand, if some artificial variables are basic they must be at value zero. As before, drop all non-basic artificial variables, and all variables with $\hat{t}_j > 0$. If possible, remove the artificial variables from the basis by introducing the non-basic variables x_j. Since the value of the basic artificial variables is zero, any non-basic variable x_j which preserves the non-singularity of the basis matrix, when its column replaces the column of the basic artificial variable, can be used for this exchange. If such an exchange cannot be made, the system of equations is rank deficient. In this case the artificial variable can be kept in the basis, and will remain zero through out the Phase 2 iterations.

4.2.16 Geometry of the Simplex Method As we have seen, the simplex method generates a finite sequence of basic feasible solutions. We now show that

4.2.17 Theorem The simplex method moves from a vertex, along an edge, to an adjacent vertex.

Proof: Simplex method generates a sequence of basic feasible solutions. As is seen in Theorem 2.6.18, these correspond to vertices of the convex polyhedron. This transition is controlled by the system of equations (4.5) which is an $m \times (m+1)$ system of rank 1. From Theorem 2.6.19, this system defines an edge of the polyhedron. Also, the next basic feasible solution (if one exists) lies on this edge, and is thus adjacent to the previous vertex, and we are done. ∎

4.3 Bounded Variable Simplex Method

The simplex method specializes for many important and specially structured linear programs. In this section we show how it specializes and handles constraints that specify bounds on variables. We consider here the linear program

$$\begin{aligned} \text{minimize} \quad & c^T x \\ & Ax = b \\ & l \leq x \leq u \end{aligned}$$

where A is an $m \times n$ matrix, b is an m vector, c is an n vector and $l < u$ are n vectors called the bounds on the variables, with l the lower bounds and u the upper bounds. Simplex method specializes nicely for this linear program. We now derive this variant.

The dual of this linear program is:

$$\begin{aligned} \text{maximize} \quad & b^T y + l^T \Phi - u^T \Psi \\ & A^T y + \Phi - \Psi = c \\ & \Phi \geq 0 \quad \Psi \geq 0 \end{aligned}$$

It is normal practice to represent the dual variables associated with the bounds by upper case greek letters. This is done to express that these variables are implicitly handled by the simplex method. In writing the dual, we assumed that x is unrestricted in sign. This is true as there are no sign restrictions on l and u.

4.3.1 Theorem Let $\hat{x}$ solve the primal and $(\hat{y}, \hat{\Phi}, \hat{\Psi})$ solve the dual. Then

1. $\hat{x}_j = l_j$ implies $(A^T \hat{y})_j \leq c_j$.

2. $\hat{x}_j = u_j$ implies $(A^T \hat{y})_j \geq c_j$.

3. $l_j < \hat{x}_j < u_j$ implies $(A^T \hat{y})_j = c_j$.

4.3 BOUNDED VARIABLE METHOD

Proof: Follows readily from the Complementary Slackness Theorem 3.3.3, and we only show this for part (1). The other parts are proved in the same way.

$$\hat{x}_j = l_j \text{ implies } \hat{x}_j < u_j \text{ implies } \Psi_j = 0 \text{ implies } (A^T\hat{y})_j + \Phi_j = c_j$$

and we are done. ∎

We now derive a variant of the simplex method that will solve this problem without explicitly using the upper and lower bound constraints.

4.3.2 Basic Feasible Solution

We say a feasible solution $\hat{x}$ is a basic feasible solution if $B = \{j : l_j < x_j < u_j\}$ is a basis, i.e., $|B| = m$ and the matrix, B, of columns of A indexed in B is non-singular. The variables x_j with $j \in B$ are called basic variables, and the others non-basic.

Under the usual non-degeneracy assumption, the bounded variable simplex method starts with a basic feasible solution, and proceeds exactly as in the simplex method. The variation occurs at the optimality step, the entering variable step, and the min-ratio test. We now discuss these variations in some detail. Compute the tentative solution to the dual $\hat{y}$, by solving

$$B^T \hat{y} = c_B$$

and the dual slacks by

$$\bar{c}_j = c_j - \hat{y}^T A_{.j} \text{ for each } j \in N$$

4.3.3 Optimality Test

Let $N_u = \{j : \hat{x}_j = u_j\}$ and $N_l = \{j : \hat{x}_j = l_j\}$. From Theorem 4.3.1 we say $\hat{x}$ is an optimal solution if and only if

$$\bar{c}_j \geq 0 \quad \text{for all} \quad j \in N_l$$
$$\bar{c}_j \leq 0 \quad \text{for all} \quad j \in N_u.$$

In case the optimality conditions are violated, define

$$\bar{c}_s = \min\{\bar{c}_{s_1}, -\bar{c}_{s_2}\} < 0$$

where $\bar{c}_{s_1} = \min\{\bar{c}_j : j \in N_l\}$, $\bar{c}_{s_2} = \max\{\bar{c}_j : j \in N_u\}$ and s is the index of the most violated dual constraint. If $s = s_1$, then the entering variable x_s enters at lower bound and thus its value must be increased. On the other hand if $s = s_2$, the entering variable is at upper bound and its value must be decreased. This is achieved by modifying the min ratio test. Another variation is that the value of the entering variable must remain within its upper and lower bounds. We now implement these variations in the min ratio test.

4.3.4 Min Ratio Test

Note that the value of the basic variables are computed by the equation

$$B\hat{x}_B = b - \sum_{j \in N_l} A_{.j}l_j - \sum_{j \in N_u} A_{.j}u_j = \bar{b}$$

As is evident, there are two cases to consider:

1. Entering variable enters at lower bound.
2. Entering variable enters at upper bound.

We now consider each of these cases:

Entering Variable at Lower Bound Let the value of the entering variable be

$$x_s(\theta) = l_s + \theta, \ 0 \leq \theta \leq u_s - l_s$$

Then it is readily seen that the linear system is equivalent to

$$Bx_B(\theta) + A_{.s}\theta = \bar{b}$$

or,

$$x_B(\theta) = \hat{x}_B - \bar{A}_s\theta$$

with $\hat{b} = \hat{x}_B$, we can define

$$\theta_1 = \begin{cases} u_s - l_s & \bar{A}_s \leq 0 \\ \frac{\hat{b}_r - l_r}{\bar{a}_{rs}} = \min\{\frac{\hat{b}_i - l_i}{\bar{a}_{is}} : \bar{a}_{is} > 0\} & \text{otherwise} \end{cases}$$

$$\theta_2 = \begin{cases} u_s - l_s & \bar{A}_s \geq 0 \\ \frac{u_t - \hat{b}_t}{-\bar{a}_{ts}} = \min\{\frac{u_i - \hat{b}_i}{-\bar{a}_{is}} : \bar{a}_{is} < 0\} & \text{otherwise} \end{cases}$$

Define

$$\hat{\theta} = \min\{\theta_1, \theta_2\}$$

If $\hat{\theta} \geq u_s - l_s$ then the entering variable hits its upper bound before any basic variable hits its bounds, and the set of basic variables does not change. The entering variable, x_s, becomes non-basic at upper bound, and

$$N_l = N_l \setminus \{s\}$$
$$N_u = N_u \cup \{s\}.$$

4.3 BOUNDED VARIABLE METHOD

In the contrary case, if $\hat{\theta} = \theta_1$, the variable x_r leaves the basis, at its lower bound, and

$$N_l = N_l \cup \{r\}$$
$$N_l = N_l \setminus \{s\}$$
$$B = B \cup \{s\}.$$

and, if $\theta = \theta_2$, the variable x_t leaves the basis at its upper bound, and

$$N_u = N_u \cup \{t\}$$
$$N_l = N_l \setminus \{s\}$$
$$B = B \cup \{s\},$$

Entering Variable at Upper Bound Let the value of the entering variable be

$$x_s(\theta) = u_s - \theta, \ 0 \le \theta \le u_s - l_s,$$

that is, it is decreased from its upper bound.

Then it is readily seen that the linear system is equivalent to

$$Bx_B(\theta) - A_{.s}\theta = \bar{b}$$

or,

$$x_B(\theta) = \hat{x}_B + \bar{A}_{.s}\theta$$

with $\hat{b} = \hat{x}_B$, we can define

$$\theta_1 = \begin{cases} u_s - l_s & \bar{A}_s \ge 0 \\ \frac{\hat{b}_r - l_r}{-\bar{a}_{rs}} = \min\{\frac{\hat{b}_i - l_i}{-\bar{a}_{is}} : \bar{a}_{is} < 0\} & \text{otherwise} \end{cases}$$

$$\theta_2 = \begin{cases} u_s - l_s & \bar{A}_s \le 0 \\ \frac{u_t - \hat{b}_t}{\bar{a}_{ts}} = \min\{\frac{u_i - \hat{b}_i}{\bar{a}_{is}} : \bar{a}_{is} > 0\} & \text{otherwise} \end{cases}$$

Define

$$\hat{\theta} = \min\{\theta_1, \theta_2\}$$

If $\hat{\theta} \ge u_s - l_s$ then the entering variable hits its lower bound before any basic variable hits its bound, and the set of basic variables does not change. The entering variable becomes non-basic at lower bound, and

$$N_u = N_u \setminus \{s\}$$
$$N_l = N_l \cup \{s\}.$$

In the contrary case, if $\hat{\theta} = \theta_1$, the variable x_r leaves the basis, at its lower bound, and

$$N_l = N_l \cup \{r\}$$
$$N_u = N_u \backslash \{s\}$$
$$B = B \cup \{s\}.$$

and, if $\theta = \theta_2$, the variable x_t leaves the basis at its upper bound, and

$$N_u = N_u \cup \{t\}$$
$$N_u = N_u \backslash \{s\}$$
$$B = B \cup \{s\},$$

The inverse of the basis matrix is updated in exactly the same manner as in the simplex method.

4.3.5 Bounded Variable Simplex Method

Step 0 Let $\hat{x}$ be a basic feasible solution, B the corresponding basis with $\hat{x}_B > 0$. Also let N_l be the indices of non-basic variables at lower bound and N_u the indices of variables at upper bound.

Step 1 Calculate the primal solution

$$\hat{x}_B = B^{-1}b = \hat{b}.$$

Step 2 Calculate the tentative solution to the dual

$$\hat{y} = B^{-T}c_B.$$

Step 3 Calculate the values of the dual slacks for the non-basic variables

$$\bar{c}_j = c_j - \hat{y}A_j \text{ for } j \in N.$$

Step 4 Optimality Condition: Compute

$$\bar{c}_{s_1} = \min\{\bar{c}_j : j \in N_l\}$$

$$\bar{c}_{s_2} = \max\{\bar{c}_j : j \in N_u\}$$

and $\bar{c}_s = \min\{\bar{c}_{s_1}, -\bar{c}_{s_2}\}$. If $\bar{c}_s \geq 0$, then $\hat{x}$ is an optimal solution to the primal and $\hat{y}$ is an optimal solution of the dual. STOP.

4.4. DUAL SIMPLEX METHOD

Step 5 x_s is the entering variable. Let

$$\bar{A}_s = B^{-1} A_{\cdot s}.$$

If $\bar{c}_s = \bar{c}_{s_1}$, then x_s is non-basic at lower bound, go to Step 6. Otherwise x_s is non-basic at upper bound and go to Step 7.

Step 6 Entering Variable is at Lower Bound. Define

$$\theta_1 = \begin{cases} u_s - l_s & \bar{A}_s \leq 0 \\ \frac{\hat{b}_r - l_r}{\bar{a}_{rs}} = \min\{\frac{\hat{b}_i - l_i}{\bar{a}_{is}} : \bar{a}_{is} > 0\} & \text{otherwise} \end{cases}$$

$$\theta_2 = \begin{cases} u_s - l_s & \bar{A}_s \geq 0 \\ \frac{u_t - \hat{b}_t}{-\bar{a}_{ts}} = \min\{\frac{u_i - \hat{b}_i}{-\bar{a}_{is}} : \bar{a}_{is} < 0\} & \text{otherwise} \end{cases}$$

and

$$N_l = N_l \backslash \{s\}.$$

If $\hat{\theta} \geq u_s - l_s$ then $N_u = N_u \cup \{s\}$, and go to Step 1. Otherwise, go to Step 8.

Step 7 Entering Variable at Upper Bound. Define

$$\theta_1 = \begin{cases} u_s - l_s & \bar{A}_s \geq 0 \\ \frac{\hat{b}_r - l_r}{-\bar{a}_{rs}} = \min\{\frac{\hat{b}_i - l_i}{-\bar{a}_{is}} : \bar{a}_{is} < 0\} & \text{otherwise} \end{cases}$$

$$\theta_2 = \begin{cases} u_s - l_s & \bar{A}_s \leq 0 \\ \frac{u_t - \hat{b}_t}{\bar{a}_{ts}} = \min\{\frac{u_i - \hat{b}_i}{\bar{a}_{is}} : \bar{a}_{is} > 0\} & \text{otherwise} \end{cases}$$

and

$$N_u = N_u \backslash \{s\}.$$

If $\hat{\theta} \geq u_s - l_s$ then $N_l = N_l \cup \{s\}$, and go to Step 1. Otherwise, continue.

Step 8 $B = B \cup \{s\}$. If $\hat{\theta} = \theta_1$, then $N_l = N_l \cup \{r\}$ otherwise $N_u = N_u \cup \{t\}$ and go to Step 1.

4.4 Dual Simplex Method

In several important formulations of linear programs, it is hard to obtain a primal basic feasible solution, but there is a readily available dual basic feasible solution. This case can also arise when a constraint is added to an already solved (by the primal simplex method) linear program. In this case a variant called the dual simplex method can be employed. This variant works with the equality constraints and a basis for the primal instead of the constraints of the dual which has some variables unrestricted in sign. We now derive this variant.

4.4.1 Dual Feasible Basis
A nonsingular submatrix B of A is called dual feasible if and only if for $\hat{y} = B^{-T}c_B$

$$\bar{c}_j = c_j - \hat{y}^T A_{.j} \geq 0 \text{ for all } j = 1, \cdots, n$$

and primal feasible if $B^{-1}b \geq 0$.

Strategy of the Dual Simplex Method

On can view the primal simplex method as a procedure that starting with a primal feasible basis (rather that a primal feasible solution), generates a sequence of primal feasible bases, and terminates with a primal and a dual feasible basis if one exists; or indicates that no such basis exists. The strategy of the dual simplex method, on the other hand, is to start with a dual feasible basis, generate a sequence of dual feasible bases and terminate with a primal and a dual feasible basis if one exists; or indicate that no such basis exists.

The dual simplex method starts with a basic primal solution $\hat{x}$ (which is not necessarily feasible), with $B = \{j : \hat{x}_j \neq 0\}$ and $N = \{j : \hat{x}_j = 0\}$; and, with the basis B dual feasible.

4.4.2 Dual Non-Degeneracy Assumption
For each dual feasible basis B, $\bar{c}_j > 0$ for all $j \in N$, and the basis B is non-singular.

We now describe the dual simplex method.

Obtaining the Dual Solution

Using the non-degeneracy assumption, define

$$\hat{y} = B^{-T}c_B$$

and, for each $j \in N$,

$$\bar{c}_j = c_j - \hat{y}^T A_{.j} > 0.$$

Obtaining the Tentative Solution of the Primal

Define

$$x_B = B^{-1}b, \quad x_N = 0$$

as the tentative solution to the primal.

4.4 DUAL SIMPLEX METHOD

Optimality Condition

Define
$$x_r = x_{B_l} = \min\{x_{B_i} : i = 1, \cdots, m\}.$$

If $x_r \geq 0$ then B is primal feasible as well, and we are done. Otherwise, remove x_r from the basic set by bringing in some non-basic variable x_s. This is different from the primal simplex method, where the entering variable is determined first, and then the leaving variable.

Entering Variable

As in the primal simplex method, we will guarantee that the most violated primal inequality constraint (on the variable x_r) is satisfied during the next iteration. This is done by increasing the value of its complementary variable t_r. The value of t_r in the existing dual feasible solution is zero. We now show how to increase its value. Consider the dual constraint corresponding to x_r

$$y^T A_{.r} + t_r = c_r.$$

It is clear from this equation that increasing the value of t_r by a constant θ is equivalent to decreasing c_r by the same constant θ. If t_r is the only variable in the set B which is increased by θ, its effect on the new dual solution can be calculated by setting $c_B(\theta) = c_B - \theta u_l$, where u_l is the lth unit vector. Note that x_r occupies the lth position in the basis B. The value of the dual variable y corresponding to this change is

$$\begin{aligned} \hat{y}(\theta) &= B^{-T} c_B(\theta) \\ &= B^{-T} c_B - \theta B^{-T} u_l \end{aligned}$$

and, the values of dual slacks associated with x_j, $j \in N$ are

$$\begin{aligned} \bar{c}_j(\theta) &= c_j - \hat{y}(\theta) A_{.j} \\ &= \bar{c}_j + \theta u_l^T B^{-1} A_{.j} \\ &= \bar{c}_j + \theta u_l^T \bar{A}_{.j} \\ &= \bar{c}_j + \theta \bar{a}_{lj} \end{aligned}$$

Thus the largest θ for which dual feasibility is maintained is obtained by the following min-ratio test:

$$\hat{\theta} = \begin{cases} \infty & \text{if } \bar{a}_{lj} \geq 0 \text{ for all } j \in N \\ \bar{c}_s/(-\bar{a}_{ls}) = \min\{\bar{c}_j/(-\bar{a}_{lj}) : \bar{a}_{lj} < 0\} & \text{otherwise} \end{cases}$$

When $\hat{\theta} < \infty$, x_s is the entering variable, and, as in the primal simplex method, we define a new dual feasible basis

$$\bar{B} = B\backslash\{r\} \cup \{s\}$$

and proceed as before. In case $\hat{\theta} = \infty$, the dual is unbounded, and we can declare that the primal has no feasible solution.

4.4.3 Theorem Under the non-degeneracy assumption, the value of the dual objective function is strictly increasing on the sequence of dual feasible bases generated by the dual simplex method.

Proof: For some basis, let $\hat{y}$ be the dual solution, and let $\hat{\theta} < \infty$ that determines the next dual feasible basis. Then,

$$\begin{aligned} b^T \hat{y}(\hat{\theta}) &= b^T \hat{y} - \hat{\theta} b^T B^{-T} u_l \\ &= b^T \hat{y} - \hat{\theta} x_r. \end{aligned}$$

From the non-degeneracy assumption, $\hat{\theta} > 0$. Since $x_r < 0$ we are done. ∎

4.4.4 Theorem Under the non-degeneracy assumption, the dual simplex method is finite.

Proof: Follows readily since there are only a finite number of bases, and from Theorem 4.4.3, none can repeat. ∎

A perturbation scheme using lexico-graphic ordering can be readily developed for this variant. We leave development of this to the reader.

4.4.5 Revised Dual Simplex Method We now summarize steps of revised dual simplex method.

Step 0 Let $\hat{x}$ be a basic solution to the primal, not necessarily feasible, $B = \{j : \hat{x}_j \neq 0\}$ be the corresponding basis which is dual feasible, and $N = \{j : \hat{x}_j = 0\}$.

Step 1 Calculate the dual feasible solution

$$\hat{y} = B^{-T} c_B;$$

and, for each $j \in N$,

$$\bar{c}_j = c_j - \hat{y}^T A_{\cdot j} \geq 0$$

Step 2 Calculate the tentative solution to the primal

$$\hat{x}_B = B^{-1} b.$$

4.5. PRIMAL - DUAL METHOD

Step 3 Optimality Condition: If $\hat{x}_{B_i} \geq 0$ for each $B_i \in B$ then $(\hat{x}_B, 0)$ is optimum solution of the primal and $\hat{y}$ is optimum solution of the dual. STOP

Step 4 Define
$$\hat{x}_r = \hat{x}_{B_l} = \min\{\hat{x}_{B_i} : B_i \in B\} < 0$$
and declare x_r as the leaving variable. Update the row associated with the leaving variable:
$$\bar{N}_{l.} = u_l^T B^{-1} N$$

Step 5 If $\bar{N}_{r.} \geq 0$, STOP. The primal is infeasible.

Step 6 Min Ratio Test: Define
$$\hat{\theta} = \bar{c}_s/(-\bar{a}_{ls}) = \min\{\bar{c}_j/(-\bar{a}_{lj}) : \bar{a}_{lj} < 0\}$$
and declare x_s as the entering variable.

Step 7 The next basis $\bar{B} = B \cup \{s\} \setminus \{r\}$ is dual feasible. Go to Step 1.

4.5 Primal - Dual Method

In certain special situations it is hard to obtain primal or dual basic feasible solutions, but a dual feasible solution is relatively easy to obtain. An example of this is the case where a number of constraints are added to an already solved linear program. In this case a ready dual feasible solution can be obtained by expanding the dual solution and assigning the value zero to dual variables associated with the added constraints. In such cases the primal-dual method can be employed.

The strategy of this method is to find a feasible solution to a *restricted* primal problem, consisting of only those primal variables which can be assigned positive values to satisfy the Complementary Slackness Condition 3.3.3 with the given dual feasible solution. The feasibility of the restricted problem can be checked by employing the Phase 1 procedure. In case a feasible solution to the restricted primal problem is found, the dual pair of linear programs has been solved. Otherwise, the dual solution is modified such that at least one more primal variable is added to the restricted problem, and the value of the objective function of the dual strictly decreases. We now discuss the details of this method.

Initial Dual Feasible Solution

Let $\hat{y}$ be a dual feasible solution, and define $\bar{c}_j = c_j - \hat{y}^T A_{.j}$ for every j, $B = \{j : c_j - \hat{y}^T A_{.j} = 0\}$ and $N = \{j : c_j - \hat{y}^T A_{.j} > 0\}$. This, using the Complementary Slackness Theorem, defines the following constraints of the restricted primal problem:

$$A_B x_B = b$$
$$x_B \geq 0$$

Optimality Condition

If the restricted primal has a feasible solution $\hat{x}_B$, we are done since we can declare $\hat{y}$ as the optimum solution of the dual and $(\hat{x}_B, 0)$ as the optimum solution of the primal. The task of checking if the restricted primal has a feasible solution is performed by solving (by the simplex method) the following Phase 1 linear program:

$$\text{minimize} \quad e^t u$$
$$A_B x_B + Du = b$$
$$x_B \geq 0 \quad u \geq 0;$$

with its dual

$$\text{maximize} \quad b^T y$$
$$A_B^T y \leq 0$$
$$Dy \leq e,$$

where D is a diagonal matrix with $D_{ii} = \text{sign}(b_i)$ for each i and e is the m vector of all 1's.

Let (x_B^*, u^*) and y^* be the optimum solution of the pair of duals of the Phase 1. Then, if $e^T u^* = 0$, x_B^* is a feasible solution of the restricted primal, and we are done. Thus assume $e^T u^* > 0$.

New Dual Feasible Solution

For each $j \in N$, consider
$$c_j^* = A_{.j}^T y^*,$$
and, define
$$y(\theta) = \hat{y} + \theta y^*.$$

4.5 PRIMAL DUAL METHOD

4.5.1 Theorem For all $\theta > 0$, $b^T y(\theta) > b^T \hat{y}$. Also, $y(\theta) \to \infty$ as $\theta \to \infty$.

Proof: This readily follows, as from the Strong Duality Theorem 3.2.3 $b^T y^* = e^T u^* > 0$. ∎

If $c_j^* \leq 0$ for each $j \in N$, $y(\theta)$ remains feasible for all $\theta > 0$. Thus from Theorem 4.5.1, the dual is unbounded, the primal has no feasible solution, and we are done. Otherwise, we show that there exists a $\hat{\theta} > 0$ such that $y(\hat{\theta})$ is dual feasible. Consider:

$$\hat{\theta} = \min \{\bar{c}_j / c_j^* : c_j^* > 0, j \in N\}.$$

The primal-dual method defines a new restricted primal using the new dual solution $y(\hat{\theta})$. It is readily confirmed that basic variables of the previous restricted primal remain basic in the next restricted problem.

4.5.2 Theorem If the Phase 1 problem can be solved in finite number of iterations, the primal-dual method is finite.

Proof: The result follows if the number of changes of the dual solution is finite. At each change of the dual solution, the dual objective function strictly decreases; and, each occurs at some basic feasible solution of the following system:
$$\begin{array}{rcl} Ax + Du & = & b \\ x \geq 0 \quad u \geq 0 & & \end{array}$$
which has only a finite number of basic feasible solutions. Thus we are done. ∎

4.5.3 Primal - Dual Method Summary We now summarize the primal-dual method developed earlier.

Step 0 Let $\hat{y}$ be an arbitrary dual feasible solution.

Step 1 For each $j = 1, \cdots, n$ define

$$\bar{c}_j = c_j - \hat{y}^T A_{\cdot j} \geq 0,$$

$B = \{j : \bar{c}_j = 0\}$ and $N = \{j : \bar{c}_j > 0\}$; and, solve the Phase 1 of the restricted primal:

$$\begin{array}{ll} \text{minimize} & e^t u \\ & A_B x_B + Du = b \\ & x_B \geq 0 \quad u \geq 0. \end{array}$$

Let its solution be (x_B^*, u^*) and the solution to its dual be y^*. If $e^T u^* = 0$ then $(x_B^*, 0)$ is an optimum solution of the primal and $\hat{y}$ of the dual. STOP.

Step 2 For each $j = 1, \cdots, n$, define

$$c_j^* = A_{.j}^T y^*$$

Step 3 If $c_j^* \leq 0$ for all $j = 1, \cdots, n$, STOP. The primal has no feasible solution.

Step 4 Define

$$\hat{\theta} = \min \{\bar{c}_j / c_j^* : c_j^* > 0, j \in N\}.$$

and

$$y(\hat{\theta}) = \hat{y} + \hat{\theta} y^*.$$

Go to Step 1 with the dual feasible solution $y(\hat{\theta})$.

4.6 Notes

Simplex method traces its roots to the work of George B. Dantzig in 1947. The version we give here is derived from complementary slackness theorem and is called the revised simplex method. Cycling in the simplex method under the presence of degeneracy was demonstrated by Hoffman [101] and Beale [18]. Its resolution by perturbation and lexicographic method is due to Charnes [27] and Dantzig [36]. The smallest example of a linear program on which the simplex method cycles is due to Marshall and Suurballe [139]. An example on which the simplex method takes an exponential number of steps has been produced by Klee and Minty [117].

The dual simplex method is due to Lemke [129], and the primal-dual method is due to Dantzig, Ford, and Fulkerson [38].

Chapter 5

Interior Point Methods

In this chapter we will describe the methods that start with a point in the interior of the feasible region and continue through the interior towards the boundary solution. The study of these methods was started by the work of Karmarkar, and has been an area of intense international activity during the past decade.

There are basically two strategies for tracing the sequence of points through the interior. In one strategy, which includes Karmarkar's method, an approximating problem defines the direction in which the next point will be generated. The algorithms based on this strategy include the affine scaling and the projective transformation methods. The second class of methods deal with the optimality conditions (the complementary slackness conditions) either directly, or derived from an auxiliary problem using a barrier or penalty strategy. These conditions are solved using a variant of Newton's method. These methods generally go under the name of path-following or homotopy methods; and their implementation usually involves a predictor-corrector strategy.

Since the interior of any set in $\boldsymbol{R}^n$ has, locally, all the properties of $\boldsymbol{R}^n$, these methods are able to avoid the combinatorial structure of the boundary, and are thus able to adapt methods of unconstrained optimization. As the solution is on the boundary, some boundary properties like degeneracy, play a role in both the convergence rates and the proofs of convergence of these methods. It is also important that the interior points generated stay 'away' from the boundary.

In this chapter we discuss the primal, the dual and the primal-dual affine scaling methods, as well as the primal-dual homotopy or path following methods.

5.1 Primal Affine Scaling Method

In this section we discuss the first interior point method, the primal affine scaling method. As in the case of the boundary methods, we will assume that the linear program is given in the following canonical form:

$$\begin{aligned} \text{minimize} \quad & c^T x \\ Ax &= b \\ x &\geq 0. \end{aligned} \tag{5.1}$$

Like the boundary primal methods, the interior primal methods generate a sequence of primal feasible interior solutions, along with tentative dual solutions. Unlike the boundary methods, these solutions to the dual pair do not necessarily satisfy the complementary slackness conditions. Both dual feasibility and complementary slackness condition are investigated at termination.

5.1.1 Interior Feasible Solution A feasible solution x is called an interior solution if $x > 0$; i.e., all the inequalities of the problem are strictly satisfied.

5.1.2 Ellipsoidal Approximating Problem Given an interior point, strategy of affine scaling methods is to replace the inequalities, here $x \geq 0$, by an ellipsoidal constraint generated by the given interior point. The resulting approximating problem is readily solvable, and has an interior point for a solution. Starting with the solution of this problem, a new approximating problem is generated and solved, thus generating a sequence of interior solutions. Under certain conditions, this sequence can be shown to converge to the solution of the original problem. We now describe, in detail, this strategy of the primal affine scaling method.

Given an interior feasible solution $x^0 > 0$, the non-negative orthant, R^n_+, is approximated by the ellipsoid

$$E = \{x : \sum_{i=1}^{n} (\frac{x_i - x_i^0}{x_i^0})^2 \leq \rho^2\}$$

where $\rho \leq 1$. The following lemma establishes the conditions under which $E \subset R^n_+$.

5.1.3 Lemma Let $x \in E$. Then $x > 0$ if $\rho < 1$ and $x \geq 0$ if $\rho = 1$.

5.1 PRIMAL AFFINE SCALING METHOD

Proof: Let x be such that $x_i < 0$ for some i. As $x_i^0 > 0$, we see that $|x_i - x_i^0| > x_i^0$ implying that $x \notin E$ when $\rho \leq 1$. Also, let x be such that

$$x_i = \begin{cases} 0 & i = 1 \\ x_i^0 & i \neq 1 \end{cases}$$

Note that $x \in E$ if and only if $\rho = 1$ and we are done. ∎

The approximating problem is obtained by replacing the non-negativity constraints by an ellipsoidal constraint. Given an interior point x^0, and $\rho \leq 1$, this problem is:

$$\begin{aligned} \text{minimize} \quad & c^T x \\ Ax &= b \\ \sum_{i=1}^{n} \left(\frac{x_i - x_i^0}{x_i^0}\right)^2 &\leq \rho^2 \end{aligned} \tag{5.2}$$

The solution to this problem generates an iterate of the primal affine scaling method. We will now derive this solution by solving a sequence of simpler problems, thus highlighting the details of the affine scaling strategy.

Minimizing a Linear Function on a Hypersphere

The first problem we consider is the following:

$$\begin{aligned} \text{minimize} \quad & c^T x \\ \sum_{i=1}^{n} (x_i - x_i^0)^2 &\leq \rho^2 \end{aligned} \tag{5.3}$$

which minimizes the linear function, $c^T x$, on the hypersphere $\sum_{i=1}^{n} (x_i - x_i^0)^2 \leq \rho^2$, with its center at x^0. The solution to this problem is

$$x^* = x^0 - \rho \frac{c}{\|c\|}. \tag{5.4}$$

This can be intuitively verified by noting that since the constraints represent a hypersphere with center x^0, the function $c^T x$ is minimized when the vector $x^0 - \theta c$, for some $\theta \geq 0$, meets the boundary of the hypersphere. This vector is precisely x^*. Though the solution of this problem lies on the boundary of the feasible region of this problem, we are able to write its solution as an explicit formula. This is not the case for the original problem, where the combinatorial structure of the boundary complicates the search for the solution in boundary methods. The simple structure of the boundary of a hypersphere appears to be responsible for this.

This simple fact will enable us to solve the approximating problem efficiently. Before we do this, we prove that x^* indeed solves the problem (5.3).

5.1.4 Theorem The vector x^* given by equation (5.4) solves the problem (5.3).

Proof: Let x be any other solution. Thus, $\|x - x^0\| \leq \rho$. Then

$$\begin{aligned} c^T x - c^T x^* &= c^T x - c^T (x^0 - \rho \frac{c}{\|c\|}) \\ &= c^T(x - x^0) + \rho\|c\| \\ &\geq -\|x - x^0\|\|c\| + \rho\|c\| \\ &\geq 0 \end{aligned}$$

and we are done. ∎

Minimizing a Linear Function on an Ellipsoid

Now consider the problem

$$\text{minimize} \quad c^T x$$
$$\sum_{i=1}^n \left(\frac{x_i - x_i^0}{x_i^0}\right)^2 \leq \rho^2 \quad (5.5)$$

which has a slightly more complicated geometry. But by a simple change of variables, we can convert the ellipsoid into a hypersphere. This is the affine scaling (actually a linear transformation is used) strategy.

Define

$$x_i' = x_i / x_i^0.$$

This transformation (change of variables) simply scales the variables. The above problem then is equivalent to

$$\text{minimize} \quad c^T D x'$$
$$\sum_{i=1}^n (x_i' - 1)^2 \leq \rho^2$$

where D is the diagonal matrix whose ith diagonal entry $D_{ii} = x_i^0$, and the transformation is $x = Dx'$. From Theorem 5.1.4 the solution to the transformed problem is

$$x'^* = e - \rho \frac{Dc}{\|Dc\|}$$

5.1 PRIMAL AFFINE SCALING METHOD

where $e = (1, \cdots, 1)^T$ the n vector of all 1's. Thus the solution is obtained by transforming the above solution into the variables x:

$$x^* = x^0 - \rho \frac{D^2 c}{\|Dc\|}.$$

Proving that x^* solves the problem (5.5) is identical to the proof of Theorem 5.1.4.

Solution of the Approximating Problem

We have all the tools we need to solve the approximating problem. By the change of variables $x' = Dx$, where D is the diagonal matrix as before, transform the problem into the equivalent problem:

$$\begin{aligned} \text{minimize} \quad & c^T Dx' \\ ADx' &= b \\ \sum_{i=1}^n (x'_i - 1)^2 &\leq \rho^2 \end{aligned}$$

and note that $De = x^0$. Since x^0 is an interior point $Ax^0 = b$ or $ADe = b$. Thus e lies in the affine space $H = \{x' : ADx' = b\}$. Also, e is the center of the hypersphere. Since the intersection of a hypersphere of radius ρ with an affine space passing through its center is a lower dimension hypersphere of radius ρ, we can solve this problem by the following steps:

(a) Project Dc into the null space $\mathcal{N}(AD)$ of AD, thus ensuring that for every θ the vector $e - \theta P Dc$ lies in the affine space H. Here P is the matrix that projects $\mathbf{R}^n$ into $\mathcal{N}(AD)$.

(b) Find $\theta' > 0$ such that $e - \theta' P Dc$ lies in the boundary of the hypersphere of radius ρ.

It is readily verified that

$$PDc = (I - DA^T(AD^2A^T)^{-1}AD)Dc \tag{5.6}$$

$$x'^* = e - \rho \frac{PDc}{\|PDc\|}.$$

Thus, transforming back to the original variables, the solution to the approximating problem is:

$$x^* = x^0 - \rho \frac{DPDc}{\|PDc\|}. \tag{5.7}$$

5.1.5 Theorem x^* solves the approximating problem (5.2).

Proof: Let x be any feasible solution of the approximating problem. Then $D^{-1}(x - x^0)$ lies in the null space of AD. Hence

$$\begin{aligned} c^T(x - x^0) &= c^T D . D^{-1}(x - x^0) \\ &= c^T DPD^{-1}(x - x^0), \end{aligned}$$

as P projects into the null space of AD. Also, since $P^2 = P$, $P^T = P$, and $\|D^{-1}(x - x^0)\| \le \rho$,

$$\begin{aligned} c^T x - c^T x^* &= c^T x - c^T x^0 + \rho \frac{c^T DPDc}{\|PDc\|} \\ &= c^T DPD^{-1}(x - x^0) + \rho \frac{c^T DP^T PDc}{\|PDc\|} \\ &\ge -\|PDc\| \|D^{-1}(x - x^0)\| + \rho \|PDc\| \\ &\ge 0 \end{aligned}$$

and we are done. ∎

The obvious question now is the relationship between the solution of the approximating problem and the linear program. There is one case where this relationship is easy to establish. This happens when the solution to the approximating problem is on the boundary of the feasible region of the linear program.

5.1.6 Theorem If $x_j^* = 0$ for some j, then x^* solves the linear program (5.1).

Proof: From Lemma 5.1.3 this can only happen if $\rho = 1$. Rewriting equation (5.6), we get

$$\begin{aligned} PDc &= D(I - A^T(AD^2 A^T)^{-1} AD^2)c \\ &= D(c - A^T y) \\ &= Ds \end{aligned} \qquad (5.8)$$

for $y = (AD^2 A^T)^{-1} AD^2 c$ and $s = c - A^T y$; we note that equation (5.7) then becomes

$$x^* = x^0 - \frac{D^2 s}{\|Ds\|}. \qquad (5.9)$$

5.1 PRIMAL AFFINE SCALING METHOD

Thus,
$$x_j^* = x_j^0 - \frac{(x_j^0)^2 s_j}{\|Ds\|} = 0$$
implies that $x_j^0 s_j = \|Ds\|$. Thus $x_i^0 s_i = 0$ for all $i \neq j$. Since $x_i^0 > 0$ we see that $s_i = 0$ for all $i \neq j$. Thus, using y as a solution to the dual with s the dual slacks, we note that the pair of solutions x^* and (y, s) satisfy the conditions of the Complementary Slackness Theorem 3.3.3, and are thus optimal for the respective linear programs in the dual pair. ∎

Primal Affine Scaling Method

Starting with an interior solution $x^0 > 0$ of the linear program, the solution of the approximating problem may also be an interior solution. The primal affine scaling method then continues the process by solving the approximating problem defined by the next interior solution. As we have seen in Theorem 5.1.6, if the solution to the approximating problem is not an interior solution, the process can stop, and an optimum solution to the linear program has been found.

5.1.7 Definition Given a positive vector x, by X (i.e., upper case of the letter representing the vector) we represent the diagonal matrix whose jth diagonal entry is x_j for each $j = 1, \cdots, n$. Thus
$$X_{ij} = \begin{cases} 0 & \text{if } i \neq j \\ x_j & \text{if } i = j. \end{cases}$$

5.1.8 The Method A hint on how to relate the steps of the procedure to duality theory is given in the proof of Theorem 5.1.6. Thus it is customary to break up the projection5.6 operation into the following sequence of steps:

Step 0 Let x^0 be an interior point solution, X_0 the diagonal matrix generated by x^0, and let $k = 0$.

Step 1 Tentative Solution to the Dual:
$$y^k = (AX_k^2 A^T)^{-1} AX_k^2 c$$

Step 2 Tentative Dual Slack:
$$s^k = c - A^T y^k$$

Step 3 Next Interior Point:
$$x^{k+1} = x^k - \frac{X_k^2 s^k}{\|X_k s^k\|}$$

Step 4 Iterative Step: If $x_j^{k+1} = 0$ for some j, then STOP. x^{k+1} is an optimal solution. Otherwise set $k = k + 1$ and go to step 1.

More on the Method

In section 5.1.8, the solution of the approximating problem is used as the next iterate in method. Thus the next iterate is chosen on the surface of the ellipsoid. In practice, it has been observed that the convergence can be considerably enhanced if the next iterate is chosen outside the ellipsoid, but still in the interior of the non-negative orthant. Such a modification of the theoretical method is called a large step method. We now present the large step primal affine scaling method.

5.1.9 Large Step Primal Affine Scaling Method

Step 0 Let x^0 be an interior point, X_0 the respective diagonal matrix, $0 < \alpha < 1$ and let $k = 0$. Typically $\alpha = 0.95$.

Step 1 Tentative Solution to the Dual:
$$y^k = (AX_k^2 A^T)^{-1} AX_k^2 c$$

Step 2 Tentative Dual Slack:
$$s^k = c - A^T y^k$$

Step 3 Unbounded Solution: If
$$s^k \leq 0$$
STOP. The solution is unbounded.

Step 4 Min Ratio Test:
$$\theta_k = \min \{\frac{\|X_k s^k\|}{x_j^k s_j^k} : s_j^k > 0\}$$

If $\theta_k = 1$ then set $\alpha = 1$.

5.1 PRIMAL AFFINE SCALING METHOD

Step 5 Next Interior Point:

$$x^{k+1} = x^k - \alpha\theta_k \frac{X_k^2 s^k}{\|X_k s^k\|}$$

Step 6 Iterative Step: If $x_j^{k+1} = 0$ for some j, then STOP. x^{k+1} is an optimal solution. Otherwise set $k = k + 1$ and go to step 1.

From Theorem 5.1.6, we can conclude that if the procedure stops at Step 6, then we can declare the last solutions found as the optimum solutions to the dual pair. We now show that if the procedure stops at Step 3, then the linear program (5.1) is unbounded below, and thus its dual has no feasible solution.

5.1.10 Theorem Assume that large step primal affine scaling method terminates at the Step 3. Then the linear program (5.1) has no minimum, i.e., its dual has no feasible solution.

Proof: Assume the procedure stops at Step 3 and consider

$$x(\theta) = x^k - \theta X_k^2 s^k.$$

Then $c^T x(\theta) = c^T x^k - \theta c^T X_k^2 s^k = c^T x^k - \theta c^T X_k P_k X_k c$ for $P_k = I - X_k A^T (A X_k^2 A^T)^{-1} A X_k$. But $c^T X_k P_k X_k c = \|P_k X_k c\|^2 > 0$. Thus as $\theta \to \infty$, $c^T x(\theta) \to -\infty$, and we are done since $x(\theta) > 0$ and $Ax(\theta) = b$ for all $\theta \geq 0$. ∎

Note that the min ratio test at Step 4 generates the largest allowable step outside the ellipsoid preserving the non-negativity of the next iterate. It is clear that $\theta_k \geq 1$, and is equal to 1 if and only if the next iterate x^{k+1} has some component equal to zero.

Connection to a Least Squares Problem

To get a clearer understanding of the method, given an interior point $x > 0$, consider the problem of finding a dual solution which may satisfy the optimality conditions. In the simplex method, one finds a dual solution for a given basic feasible solution of the primal which satisfies the complementary slackness conditions 3.3.3, and the strategy is to ignore the non-negativity constraints on the dual slacks. This same strategy is adapted by the affine scaling method. Here, as we will see, the computed dual solution reduces the complementary slackness violation the most; i.e., the following problem,

$$\begin{align} \text{minimize} \quad & \|Xs\| \\ A^T y + s &= c \end{align} \tag{5.10}$$

whose solution is readily seen as

$$y = (AX^2 A^T)^{-1} AX^2 c \tag{5.11}$$
$$s = c - A^T y \tag{5.12}$$

is solved. Since $x > 0$, the complementary slackness violation cannot be reduced to zero unless c is in the row space of A.

If x is a non-degenerate basic feasible solution, this formula gives the same dual solution computed in the simplex method, namely $B^{-T} c_B$, where B is the corresponding basis matrix. The simplex method then uses the resulting dual slack vector s to move to an adjacent vertex by changing only $m+1$ components of the primal solution vector. In the interior point strategy, every component of the primal solution can be potentially changed.

Boundedness of y and s

For any given $x > 0$, let $y(x)$ and $s(x)$ be the solutions obtained by solving the least squares problem (5.10) for the given vector x. We now show that, for every $x > 0$, the solutions $y(x)$ and $s(x)$ are bounded above by a function of A and c, which is independent of x.

5.1.11 Theorem Let A be $m \times n$, and have full row rank m. Then, there is a constant $q(A)$ which is a function only of A, such that for every $x > 0$, $\|y(x)\| \leq q(A)\|c\|$.

Proof: For each $j = 1, \cdots, n$, consider the hyperplanes

$$A_{\cdot j}^T y - c_j = 0$$

where $A_{\cdot j}$ is the jth column of the matrix A. These hyperplanes partition $\mathbf{R}^m$ into at most 3^n polyhedral sets, each generated by choosing disjoint index sets J_1, J_2, J_3 from $\{1, \cdots, n\}$ such that $\cup J_i = \{1, \cdots, n\}$ and

$$\begin{align} A_{\cdot j}^T y - c_j &\geq 0 \quad j \in J_1 \\ A_{\cdot j}^T y - c_j &= 0 \quad j \in J_2 \\ A_{\cdot j}^T y - c_j &\leq 0 \quad j \in J_3. \end{align}$$

5.1 PRIMAL AFFINE SCALING METHOD

Let y be such that it belongs to an unbounded polyhedron in this partition, and let this polyhedron be defined by $J_1, J_2,$ and J_3. Since it is unbounded, from Theorem 2.6.16, there is a $0 \neq z \in \mathbf{R}^m$ such that

$$A_{\cdot j}^T z \geq 0 \quad j \in J_1$$
$$A_{\cdot j}^T z = 0 \quad j \in J_2$$
$$A_{\cdot j}^T z \leq 0 \quad j \in J_3.$$

and since A has full row rank, $A^T z \neq 0$. Thus for all sufficiently small $\epsilon > 0$,

$$|A_{\cdot j}^T(y - \epsilon z) - c_j| \leq |A_{\cdot j}^T y - c_j|$$

with strict inequality for each j with $A_{\cdot j} z \neq 0$. Hence y cannot solve the least squares problem (5.10). Thus, for each $x > 0$, $y(x)$ belongs to a bounded polyhedron of this partition. Our result now follows since for any bounded polyhedron the function $\|y\|$ attains its maximum at an extreme point, extreme points are linear functions of the right hand side of the system defining the polyhedron, and involve the inverse of a submatrix of A. ∎

5.1.12 Corollary Let A be $m \times n$, and have full row rank m. Then, there is a constant $q(A)$ which is a function only of A, such that for every $x > 0$, $\|s(x)\| \leq q(A)\|c\|$.

Proof: Follows readily from the equation (5.12) and the Theorem 5.1.11. ∎

We will now establish some additional properties of the least squares problem and the affine scaling method.

5.1.13 Definition Let $P = \{x : Ax = b, x \geq 0\}$ be the convex polyhedron of the linear program, and let $x \in P$. We define $F(x)$ as the *smallest face* of P containing x. Note that $x > 0$ if and only if $F(x) = P$ and x is an extreme point if and only if $F(x) = x$. We caution the reader that we are defining the polyhedron and a permutation matrix by the same symbol P. Since their use is always in different contexts, there is no chance of confusion.

5.1.14 Lemma $c^T x = \gamma$ for all $x \in H = \{x : Ax = b\}$ if and only if c belongs to the row space $\mathcal{R}(A^T)$ of A.

Proof: If c belongs to the row space of A, then $c = A^T y$ for some y. Thus $c^T x = y^T A x = y^T b = \gamma$ and is thus a constant on for all $x \in H$. Now assume that c does not belong to $\mathcal{R}(A^T)$. Then there is a unique decomposition $c = c_n + c_r$ where $0 \neq c_n \in \mathcal{N}(A)$ and $c_r \in \mathcal{R}(A^T)$. Thus $c^T c_n = c_r^T c_n + c_n^T c_n = c_n^T c_n > 0$. Let $x \in H$ and consider $x + \theta c_n$. It is readily seen that for every θ, $x + \theta c_n \in H$ and $c^T(x + \theta c_n) = c^T x + \theta c_n c_n \neq c^T x$ for all $\theta \neq 0$, and thus $c^T x$ is not constant on H. ∎

5.1.15 Theorem

(a) Let $\hat{x}$ be an optimal solution of the linear program. Then

$$\text{minimum } \|\hat{X}(c - A^T y)\| = 0.$$

(b) Let $\hat{x}$ be such that $c^T x$ is not constant on $F(\hat{x})$. Then

$$\text{minimum } \|\hat{X}(c - A^T y)\| > 0.$$

(c) Let $\hat{x}$ be such that $c^T x$ is constant on $F(\hat{x})$. Then

$$\text{minimum } \|\hat{X}(c - A^T y)\| = 0.$$

Proof:

(a) Since $\hat{x}$ is an optimal solution to (5.1), using an optimal dual solution $\hat{y}$ the result follows from the complementary slackness theorem.

(b) Without loss of generality, assume that $\hat{x}^T = (\bar{x}^T, 0)$, with $\bar{x} > 0$. Since $c^T x$ is not constant on the set $F(\hat{x}) = \{x : \bar{A}x = b, x \geq 0\} \subseteq P$, $\bar{c}$ is not in the row space $\mathcal{R}(\bar{A}^T)$ of $\bar{A}$; or equivalently, $\bar{X}\bar{c}$ is not in $\mathcal{R}(\bar{X}\bar{A}^T)$. But, for $\bar{y}$ from Step 1,

$$\begin{aligned} \bar{X}\bar{s} &= \bar{X}(\bar{c} - \bar{A}^T \bar{y}) \\ &= \bar{X}(\bar{c} - \bar{A}^T(\bar{A}\bar{X}^2\bar{A}^T)^{-1}\bar{A}\bar{X}^2\bar{c}) \\ &= (I - \bar{X}\bar{A}^T(\bar{A}\bar{X}^2\bar{A}^T)^{-1}\bar{A}\bar{X})\bar{X}\bar{c} \\ &= \bar{P}\bar{X}\bar{c} \end{aligned}$$

where $\bar{P}$ is the projection matrix into the null space $\mathcal{N}(\bar{A}\bar{X})$ of $\bar{A}\bar{X}$. But since $\bar{X}\bar{c}$ does not belong to $\mathcal{R}(\bar{X}\bar{A}^T)$, $\bar{X}\bar{s} \neq 0$. Hence our result.

(c) Let $\bar{x}, \bar{c}$ and $\bar{A}$ be as defined in (b). Since $\bar{c}^T \bar{x}$ is constant on $F(\hat{x})$, from Lemma 5.1.14, $\bar{X}\bar{c}$ belongs to $\mathcal{R}(\bar{X}\bar{A}^T)$. Thus the result follows.

Connection to Quadratic Programming

By making a change of variables $v = x - x^0$ in the problem (5.2), we obtain the equivalent approximating problem:

$$\begin{aligned} \text{minimize} \quad & c^T v \\ Av &= 0 \\ \|X_0^{-1} v\|^2 &\leq 1. \end{aligned} \quad (5.13)$$

The solution to this problem has an intimate relation to the following quadratic programming problem:

$$\begin{aligned} \text{minimize} \quad & \tfrac{1}{2} v^T X_0^{-2} v - c^T v \\ Av &= 0, \end{aligned} \quad (5.14)$$

which we will explore in the next result. We have already seen that the solution to the ellipsoidal approximating problem is

$$\begin{aligned} v^* &= \frac{X_0^2 s}{\|X_0 s\|} \\ s &= c - A^T y \\ y &= (A X_0^2 A^T)^{-1} A X_0^2 c. \end{aligned}$$

The next result shows the connection between the solutions of the two problems:

5.1.16 Quadratic Programming Connection The solution to the quadratic program (5.14) is

$$\begin{aligned} v^* &= X_0^2 s \\ s &= c - A^T y \\ y &= (A X_0^2 A^T)^{-1} A X_0^2 c. \end{aligned}$$

Proof: By setting up the Lagrangian, we can show that v solves (5.14) if and only if there is a y such that

$$\begin{aligned} -c + X_0^{-2} v + A^T y &= 0 & (5.15) \\ Av &= 0. & (5.16) \end{aligned}$$

Multiplying (5.15) by $A X_0^2$ and substituting (5.16), we obtain

$$y = (A X_0^2 A^T)^{-1} A X_0^2 c,$$

and; from (5.15)
$$v = X_0^2(c - A^T y)$$
and we are done. ∎

The Theorem 5.1.16 shows that the Lagrange multiplier or the dual solution is the same in both the problems; and the directions are scaled multiples of each other.

Linear Functions and Ellipsoids

We now investigate an important property of a linear function on an ellipsoid. This property is related to the following problem: given a $k \times l$ matrix Q of rank k, a k vector $c \neq 0$ and an $l \times l$ diagonal matrix D with positive diagonal entries, define

$$\text{maximize } c^T x$$
$$x^T Q D Q^T x \leq 1. \qquad (5.17)$$

The following establishes a result that relates the objective function value to the solution vector of this problem.

5.1.17 Theorem Let $\hat{x}$ solve the problem 5.17. There exists a constant $p(Q, c) > 0$, independent of D, such that

$$|c^T \hat{x}| \geq p(Q, c) \|\hat{x}\|.$$

Proof: Let $c, B_1, \cdots, B_{k-1}$ be a basis for $\mathbf{R}^k$ with $B = (B_1, \cdots, B_{k-1})$ an orthonormal basis for the orthogonal complement of the one dimensional subspace spanned by c. Thus, $c^T B = 0$ and $B^T B = I$. Expressing the columns of Q in this basis we obtain

$$Q = ca^T + BR$$

where a is a l vector and R is a $(k-1) \times l$ matrix. Since Q has full rank k, so does (a, R). Also, let

$$x = uc + Bv$$

where u is a scalar. Rewriting the problem 5.17 in the variables u and v we get

$$c^T x = u c^T c$$
$$x^T Q D Q^T x = (u c^T c)^2 a^T D a + 2(u c^T c) a^T D R^T v + v^T R D R^T v$$

5.1 PRIMAL AFFINE SCALING METHOD

and if $w = c^T cu$, problem 5.17 is equivalent to

$$\text{maximize } w$$
$$w^2 a^T Da + 2w a^T DR^T v + v^T RDR^T v \leq 1.$$

Let λ be the multiplier on the constraint. Since the solution of this problem is on the boundary of the constraint, the optimality conditions for it can be stated as follows:

$$1 + \lambda(2wa^T Da + 2a^T DR^T v) = 0 \qquad (5.18)$$
$$wRDa + RDR^T v = 0 \qquad (5.19)$$
$$w^2 a^T Da + 2wa^T DR^T v + v^T RDR^T v = 1 \qquad (5.20)$$
$$\lambda > 0$$

and we note that Equation 5.19 has the solution

$$\hat{v} = -w(RDR^T)^{-1} RDa$$

and let

$$\hat{w} = \frac{1}{\sqrt{a^T D^{\frac{1}{2}}(I - D^{\frac{1}{2}} R^T (RDR^T)^{-1} RD^{\frac{1}{2}}) D^{\frac{1}{2}} a}}$$

be the solution obtained by substituting $\hat{v}$ into Equation 5.20. Then, after computing $\hat{x}$ we note that

$$c^T \hat{x} = \hat{w}$$
$$\hat{x} = \hat{u}c + B\hat{v}$$
$$= \frac{\hat{w}}{c^T c} c - \hat{w} B (RDR^T)^{-1} RDa.$$

Thus

$$\frac{\hat{x}}{c^T \hat{x}} = \frac{c}{c^T c} - B(RDR^T)^{-1} RDa,$$

and, noting that $\|B\| \leq 1$ we get

$$\frac{\|\hat{x}\|}{|c^T \hat{x}|} \leq \frac{1}{\|c\|} + \|(RDR^T)^{-1} RDa\|$$

and if, from Theorem 5.1.11, $q(R, a)$ bounds the second term of the right hand side above, since both R and a are only functions of Q and c, we get our result with $p(Q, c) = 1/(\frac{1}{\|c\|} + q(R, a))$. ∎

Convergence of Primal Affine Scaling Method

We will prove the convergence of the primal affine scaling method, presented in the section 5.1.9, with the following assumptions on the underlying linear program:

1. The matrix A is $m \times n$ and has full rank m.
2. The linear program has an interior solution.
3. The objective function is not constant on the feasible region.
4. Every feasible solution to the linear program is non-degenerate.

A note on assumption (4) is in order here. In this section, we will prove the convergence of the affine scaling method with this non-degeneracy assumption. Later in this chapter we will remove this assumption, and establish convergence by controlling the stepsize α.

Let $\{x^k\}$ be the sequence generated by the primal affine scaling algorithm. We have already seen that if this sequence is finite, then the last point of this sequence solves the linear program, Theorem 5.1.6, or we can declare that the linear program has no solution, Theorem 5.1.10. Thus, we henceforth assume that the sequence is not finite. We now establish some important properties of this sequence.

5.1.18 Theorem $x^k > 0$ and $\theta_k \geq 1$ for all k.

Proof: Can be readily seen as a consequence of the min-ratio test of Step 4, and the fact that $\alpha < 1$.

5.1.19 Theorem $\{c^T x^k\}$ is a strictly decreasing sequence with

$$c^T x^{k+1} = c^T x^k - \alpha \theta_k \|X_k s^k\|.$$

Either this sequence diverges to $-\infty$, or is bounded. In the later case, it converges, and $\|X_k s^k\| \to 0$ as $k \to \infty$. Also, $\infty > \sum_{k=1}^{\infty} c^T(x^k - x^{k+1})$.

Proof: From Step 5 and Equation (5.8) with $D = X_k$, $P = P_k = I - X_k A^T (A X_k^2 A^T)^{-1} A X_k = P_k^2 = P_k^T$, we have

$$\begin{aligned}
c^T x^{k+1} &= c^T x^k - \alpha \theta_k \frac{c^T X_k P_k X_k c}{\|P_k X_k c\|} \\
&= c^T x^k - \alpha \theta_k \frac{c^T X_k P_k P_k^T X_k c}{\|P_k X_k c\|}
\end{aligned}$$

5.1 PRIMAL AFFINE SCALING METHOD

$$= c^T x^k - \alpha \theta_k \|P_k X_k c\|$$
$$= c^T x^k - \alpha \theta_k \|X_k s^k\|.$$

Thus from Theorem 5.1.15, Assumption (3) and the fact that $\alpha > 0$, $\theta_k \geq 1$, it can be seen that $\{c^T x^k\}$ is strictly monotone decreasing. If the dual has a feasible solution, $c^T x^k$ is bounded on P. Since every bounded monotone sequence converges, if the dual has a feasible solution, $c^T x^k \to c^*$, and $\infty > c^T x^1 - c^* = \sum_{k=1}^{\infty} c^T (x^k - x^{k+1})$. The theorem follows, since $c^T (x^k - x^{k+1}) \to 0$, and $\alpha \theta_k > 0$ implies $\|X_k s^k\| \to 0$. ∎

We now show that the sequences $\{x^k\}$, $\{y^k\}$ and $\{s^k\}$ converge to respective optimum solutions of the dual pair. We first show that the primal sequence converges independent of the non-degeneracy assumption. We will then use the non-degeneracy assumption to prove that this limit point is an optimum solution of the linear program; and, in that case, the dual sequences also converge.

5.1.20 Primal Convergence Theorem Let the assumptions (1) through (3) hold, and the sequence $\{x^k\}$ be generated by the primal affine scaling method. If the sequence $\{c^T x^k\}$ is bounded, then $x^k \to x^*$.

Proof: Consider the approximating problem of Section 5.1.2. From Step 5 of the large step affine scaling method, by defining

$$\alpha \theta_k v^k = x^k - x^{k+1}$$

we note that $v^k = \frac{X_k^2 s^k}{\|X_k s^k\|}$ solves the following problem:

$$\begin{aligned} \text{maximize} \quad & c^T v \\ Av &= 0 \\ \|X_k^{-1} v\| &\leq 1 \end{aligned} \quad (5.21)$$

We will now show that $\sum_{k=1}^{\infty} \alpha \theta_k \|v^k\| < \infty$ and thus the sequence $\{x^k\}$ converges. Let the set of vectors $\{Q_1, \cdots, Q_{n-m}\}$ be an orthonormal basis for the null space $\mathcal{N}(A)$; i.e., for $Q = (Q_1, \cdots, Q_{n-m})$, $Q^T Q = I$. Since, in the above problem, $v \in \mathcal{N}(A)$, we note that $c^T v = c_n^T v$ where c_n is the projection of c into the null space of A. Define $\bar{c}$ and u such that

$$c_n = Q\bar{c}$$
$$v = Qu.$$

Expressing the problem in variables u we obtain an equivalent problem:

$$\text{maximize } \bar{c}^T u$$
$$u^T Q^T X_k^{-2} Q u \leq 1.$$

Let u^k solve this problem. Theorem 5.1.17, gives a constant $p(Q, \bar{c}) > 0$ such that

$$\begin{aligned} c^T v^k &= \bar{c}^T u^k \\ &\geq p(Q, \bar{c}) \|u^k\| \\ &= p(Q, \bar{c}) \|v^k\| \end{aligned} \quad (5.22)$$

since $\|v^k\|^2 = u_k^T Q^T Q u_k = \|u_k\|^2$. As $\{c^T x^k\}$ is bounded, from Theorem 5.1.19,

$$\infty > \sum_{k=1}^{\infty} c^T(x^k - x^{k+1}) = \sum_{k=1}^{\infty} \alpha \theta_k c^T v^k \geq p(Q, \bar{c}) \sum_{k=1}^{\infty} \alpha \theta_k \|v^k\| \quad (5.23)$$

and thus $\{x^k\}$ converges, say to x^*, and we are done. ∎

We assume that $\{x^k\}$ converges to x^*. From the previous theorem, this will happen if the dual has a feasible solution. We now show that, with the non-degeneracy assumption (4), x^* is an optimum solution, and that the dual sequences $\{y^k\}$ and $\{s^k\}$ also converge to an optimum solution of the dual.

Define

$$B = \{j : x_j^* > 0\}.$$

5.1.21 Theorem Let the sequence $\{c^T x^k\}$ be bounded, and the assumptions (1) through (4) hold. Then

1. $x^k \longrightarrow x^*$

2. $y^k \longrightarrow y^*$

3. $s^k \longrightarrow s^*$

where x^* is an optimum solution of the primal linear program, (y^*, s^*) is an optimum solution of its dual.

Proof: From Theorem 5.1.11, $\{y^k\}$ is bounded, and thus has a cluster point y^*; i.e., on some subsequence K, $y^{k'} \to y^*$ for $k' \in K$. Let $s^{k'} = c - A^T y^{k'}$

5.1 PRIMAL AFFINE SCALING METHOD

Then $s^{k'} \to s^* = c - A^T y^*$. For $j \in B$, from Theorem 5.1.19, $x^{k'} s^{k'} \to 0$, and $x^{k'} \to x_j^* > 0$. Thus $s_j^* = 0$ and

$$A_B^T y^* = c_B$$

for every cluster point y^*. From Assumption (4), the rank of A_B is m. Thus this equation has at most one solution, and thus the sequence $\{y^k\}$ has exactly one cluster point, and so converges to y^*.

Let $s_j^* < 0$ for some $j \notin B$. Then, there is an $L \geq 1$ such that for all $k \geq L$, $s_j^k < 0$. But

$$\begin{aligned} x_j^{k+1} &= x_j^k(1 - \alpha \frac{s_j^k x_j^k}{\phi(X_k s^k)}) \\ &> x_j^k. \end{aligned}$$

$x_j^k \not\to 0$, a contradiction. Thus $s^* \geq 0$ and $X_* s^* = 0$, we are done. ∎

Convergence Rate of Primal Sequences

The sequences $\{c^T x^k\}$ and $\{x^k\}$ converge without the non-degeneracy assumption (4). We now establish the rates at which these sequences converge. We will first show that the sequence $\{c^T x^k\}$ converges, at least, linearly. We then derive the convergence rate of the other sequences in terms of the rate of $\{c^T x^k\}$.

Let

$$c^T x^k \longrightarrow c^*,$$

$$x^k \longrightarrow x^*,$$

$B = \{j : x_j^* \neq 0\}$ and $N = \{j : x_j^* = 0\}$,

$q = |N|$, the number of elements in $|N|$,

and, for any given vector u, $\phi(u) = \max_j u_j$.

We note that the min ratio test, Step 4, of the large step affine scaling method can be modified to

$$\theta_k = \min \{\frac{\|X_k s^k\|}{x_j^k s_j^k} : s_j^k > 0\} = \frac{\|X_k s^k\|}{\phi(X_k s^k)}. \tag{5.24}$$

We are now ready to derive the convergence rate of $\{c^T x^k\}$.

5.1.22 Linear Convergence of Objective Value
There exists an $L \geq 1$ such that for all $k \geq L$

$$(c^T x^{k+1} - c^*) \leq (1 - \frac{\alpha}{\sqrt{n}})(c^T x^k - c^*).$$

Proof: Consider the problem 5.21. We note that $\frac{X_k^2 s^k}{\|X_k s^k\|}$ solves this problem and that $\frac{(x^k - x^*)}{\|X_k^{-1}(x^k - x^*)\|}$ is feasible for it. Thus, as $c^T X_k^2 s^k = \|X_k s^k\|^2$ and $c^T x^* = c^*$, we have

$$\frac{c^T x^k - c^*}{\|X_k^{-1}(x^k - x^*)\|} \leq \frac{c^T X_k^2 s^k}{\|X_k s^k\|} = \|X_k s^k\|. \tag{5.25}$$

Now, for an n vector of all ones, $e = (1, \cdots, 1)^T$, $\|X_k^{-1}(x^k - x^*)\|^2 = \|e_B - X_{B,k}^{-1} x_B^*\|^2 + \|e_N\|^2$. Since $X_{B,k}^{-1} x_B^* \to e_B$, there is an $L \geq 1$ such that for all $k \geq L$, $\|e_B - X_{B,k}^{-1} x_B^*\|^2 \leq n - q$. Thus, for each $k \geq L$, $\|X_k^{-1}(x^k - x^*)\| \leq \sqrt{n}$. Now, as $c^T x^{k+1} - c^* = c^T x^k - c^* - \alpha \theta_k \|X_k s^k\|$, and $\theta_k \geq 1$ we have

$$\frac{c^T x^{k+1} - c^*}{c^T x^k - c^*} \leq 1 - \alpha \frac{\|X_k s^k\|}{c^T x^k - c^*}$$

and the result follows from the Equation (5.25). ∎

We will now establish the convergence rate of the sequence $\{x^k\}$.

5.1.23 Primal Convergence Rate Theorem
There are constants $\delta > 0$ and $\gamma > 0$ such that for every $k = 1, 2, \cdots$ we have

(a) $\gamma(c^T x^k - c^*) \geq \|x^k - x^*\| \geq \frac{1}{\|c\|}(c^T x^k - c^*)$.

(b) $\frac{c^T x^k - c^*}{\sum_{j \in N} x_j^k} \geq \delta$.

(c) $\frac{c^T x^k - c^*}{\sum_{j \in B} |x_j^k - x_j^*|} \geq \delta$

Proof: Using the result (5.22), and letting $\frac{1}{\gamma} = p(Q, \bar{c})$, we obtain

$$c^T x^k - c^* = \sum_{j=0}^{\infty} c^T (x^{k+j} - x^{k+j+1})$$

$$\geq \frac{1}{\gamma} \sum_{j=0}^{\infty} \|x^{k+j} - x^{k+j+1}\|$$

5.1 PRIMAL AFFINE SCALING METHOD

$$\geq \frac{1}{\gamma}\|\sum_{j=0}^{\infty}(x^{k+j} - x^{k+j+1})\|$$

$$= \frac{1}{\gamma}\|x^k - x^*\|$$

and we have the left inequality of part (a). The right inequality readily follows from the Cauchy-Schwartz inequality. Also, from part (a),

$$c^T x^k - c^* \geq \frac{\|x_N^k\|}{\gamma} \geq \frac{\sum_{j \in N} x_j^k}{\gamma\sqrt{q}},$$

and

$$c^T x^k - c^* \geq \frac{\|x_B^k - x_B^*\|}{\gamma} \geq \frac{\sum_{j \in B} |x_j^k - x_j^*|}{\gamma\sqrt{n-q}}.$$

We have our result with $\delta = \min\{\frac{1}{\gamma\sqrt{q}}, \frac{1}{\gamma\sqrt{n-q}}\}$. ∎

Convergence Rate of Other Sequences

We now investigate the convergence rates of $\{s_B^k\}$ and $\{X_k s^k\}$, both of which converge to 0. In this investigation we will need a result related to the following problem: given a $k \times l$ matrix R, a k vector r with r in the range (or column) space $\mathcal{R}(R)$ of R; and, an arbitrary l vector z, consider the following problem

$$\begin{aligned} \text{minimize} \quad & \|z - x\|^2 \\ & Rx = r \end{aligned} \tag{5.26}$$

and let $x(z)$ be its solution. We now prove the result we need.

5.1.24 Lemma There is a constant $q(R) > 0$ such that for every $z \in \mathbf{R}^l$, the solution $x(z)$ of the above problem satisfies

$$\|z - x(z)\| \leq q(R)\|Rz - r\|.$$

Proof: Since $r \in \mathcal{R}(R)$, there exists $\bar{R}$, a $\bar{k} \times l$ submatrix of R, which has full row rank, with

$$\{x : Rx = r\} = \{x : \bar{R}x = \bar{r}\}$$

for some $\bar{r}$. The smaller system is obtained by dropping all the redundant equations in original system. It is readily seen (by setting up the Lagrangian) that $x(z)$ solves the problem (5.26) if and only if for some y,

$$\begin{aligned} 2(z - x(z)) - \bar{R}^T y &= 0 & (5.27) \\ \bar{R}x(z) &= \bar{r}. & (5.28) \end{aligned}$$

Multiplying the Equation (5.27) by $\bar{R}$ and substituting Equation (5.28), we derive the formula
$$y = 2(\bar{R}\bar{R}^T)^{-1}(\bar{R}z - \bar{r}).$$
Substituting the above formula for y into Equation (5.27), we obtain
$$z - x(z) = \bar{R}^T(\bar{R}\bar{R}^T)^{-1}(\bar{R}z - \bar{r}).$$
For $q(R) = \bar{R}^T(\bar{R}\bar{R}^T)^{-1}$, our result follows from the fact that $(\bar{R}z - \bar{r}) \leq \|Rz - r\|$. ∎

We are now ready to establish the convergence rate of $\{s_B^k\}$. Define the polyhedron
$$\mathcal{D} = \{(y, s) : A_B^T y = c_B, A_N^T y + s_N = c_N, s_B = 0\} \tag{5.29}$$
and note that, because of Theorem 5.1.19, any cluster point of the sequence $\{(y^k, s^k)\}$ belongs to this polyhedron. $\mathcal{D} \neq \emptyset$ since, by Theorems 5.1.11 and 5.1.12, the above sequences are bounded, and thus have cluster points.

We now establish a simple property relating to $\mathcal{D}$.

5.1.25 Lemma For any $(\tilde{y}, \tilde{s}) \in \mathcal{D}$, $c^T x^k - c^* = \tilde{s}_N^T x_N^k$.

Proof: Consider the identities:
$$\begin{aligned} c^T x^k - c^* &= c^T(x^k - x^*) \\ &= (A^T \tilde{y} + \tilde{s})^T(x^k - x^*) \\ &= \tilde{s}^T(x^k - x^*) \\ &= \tilde{s}_N^T x_N^k. \end{aligned}$$

Our result now follows. ∎

We now establish an important relationship between $\mathcal{D}$ and $\{s^k\}$. For each k, define the problem:
$$\begin{aligned} \text{minimize} \quad & \|s^k - s\|^2 \\ (y, s) \in \mathcal{D} & \end{aligned} \tag{5.30}$$
and let its solution be $(\tilde{y}^k, \tilde{s}^k)$. We can then prove:

5.1 PRIMAL AFFINE SCALING METHOD

5.1.26 Theorem If $(\tilde{y}^k, \tilde{s}^k)$ is the solution to the problem (5.30), then there is a $\rho > 0$ such that for each $k = 1, 2, \cdots$

1. $\|\tilde{s}^k - s^k\| \leq \rho \|s_B^k\|$.

2. $\{\tilde{s}^k\}$ is bounded.

Proof: From the Theorem 5.1.12, $\{s^k\}$ is bounded, and thus there is an $M > 0$ such that for each $k = 1, 2, \cdots$, $\|s^k\| \leq M$ and $\|s_B^k\| \leq M$. Thus, part 2 follows from part 1 as

$$\begin{aligned}\|\tilde{s}^k\| &\leq \|\tilde{s}^k - s^k\| + \|s^k\| \\ &\leq \rho\|s_B^k\| + M \\ &= (\rho + 1)M.\end{aligned}$$

Now, using Lemma 5.1.24, we note that there is a $\rho > 0$ such that

$$\|(y^k, s^k) - (\tilde{y}^k, \tilde{s}^k)\| \leq \frac{\rho}{2} \left\| R \begin{bmatrix} y^k \\ s_B^k \\ s_N^k \end{bmatrix} - \begin{bmatrix} c_B \\ c_N \\ 0 \end{bmatrix} \right\|$$

where

$$R = \begin{bmatrix} A_B^T & 0 & 0 \\ A_N^T & 0 & I \\ 0 & I & 0 \end{bmatrix}.$$

But $A_B^T y^k + s_B^k = c_B$ and $A_N^T y^k + s_N^k = c_N$. As

$$R \begin{bmatrix} y^k \\ s_B^k \\ s_N^k \end{bmatrix} = \begin{bmatrix} c_B - s_B^k \\ c_N \\ s_B^k \end{bmatrix},$$

our result follows. ∎

We now investigate the convergence properties of the sequences $\{s_B^k\}$ and $\{X_k s^k\}$, both of which are converging to zero. To facilitate this study, we define

$$w^k = X_k^2 s^k$$

and

$$u^k = \frac{X_k s^k}{c^T x^k - c^*}.$$

We now prove some preliminary lemmas.

5.1.27 Lemma There is a $\rho_1 > 0$ such that for all $k = 1, 2, \cdots$

$$\|X_k s^k\| \leq \rho_1 (c^T x^k - c^*)$$

Proof: Now

$$\begin{aligned}
\|X_k s^k\|^2 &= c^T X_k^2 s^k \\
&= (A^T \tilde{y}^k + \tilde{s}^k)^T w^k \\
&= (\tilde{s}^k)^T w^k \\
&= (\tilde{s}_N^k)^T w_N^k \\
&= (X_{k,N} \tilde{s}_N^k)^T (X_{k,N}^{-1} w_N^k) \\
&\leq \phi(x_N^k) \|\tilde{s}_N^k\| \|X_{k,N} s_N^k\|.
\end{aligned}$$

From Theorem 5.1.23 part (b) we have

$$\phi(x_N^k) \leq \sum_{j \in N} x_N^k \leq \frac{1}{\gamma}(c^T x^k - c^*)$$

and from Lemma 5.1.26 that $\tilde{s}^k$ is bounded, and thus we are done. ∎

5.1.28 Lemma There is a $\rho_2 > 0$ such that for every $k = 1, 2, \cdots$,

$$\|w_B^k\| \leq \rho_2 (c^T x^k - c^*)^2.$$

Proof: Using the result 5.22 and the fact that $v^k = \frac{w^k}{\|X_k s^k\|}$, we note that there is a $\gamma > 0$ such that

$$\|w^k\| \leq \gamma c^T w^k. \tag{5.31}$$

As $\|w_B^k\| \leq \|w^k\|$ and $c^T w^k = \|X_k s^k\|^2$, our result follows from the Lemma 5.1.27 with $\rho_2 = \gamma \rho_1^2$. ∎

We now prove a result showing that the error of the sequence $\{s_B^k\}$ goes as the square of the error of the sequence $\{c^T x^k - c^*\}$. We show this in the next theorem.

5.1.29 Theorem There is a $\rho_3 > 0$ and an $L \geq 1$ such that for every $k \geq L$

$$\|s_B^k\| \leq \rho_3 (c^T x^k - c^*)^2.$$

5.1 PRIMAL AFFINE SCALING METHOD

Proof: Now
$$\|s_B^k\| \leq \|X_{k,B}^{-2}\|\|w_B^k\|.$$
Since $x_B^k \to x_B^*$ and $x_B^* > 0$, $M = \|X_{*,B}^{-2}\| < \infty$ and there is an $L \geq 1$ such that for all $k \geq L$, $\|X_{k,B}^{-2}\| \leq 2M$. Our result now follows from Lemma 5.1.28 with $\rho_3 = 2M\rho_2$. ∎

The Theorem 5.1.29 falls short of establishing the convergence rate of $\{s_B^k\}$. It does establish the fact that the error of the dual sequence is the square of the error of the primal sequence $\{x_N^k\}$.

We now consider the sequence $\{u^k\}$ and establish some of its important properties. This will also enable us to show that the convergence rate of this sequence is asymptotically, linear. We now prove an important theorem.

5.1.30 Properties of u^k

1. $\{u^k\}$ is bounded.
2. There is an $L \geq 1$ and $\theta_1 > 0$, $\theta_2 > 0$ such that for all $k \geq L$
   (a) $e^T u_N^k = 1 + \delta_k$ with $|\delta_k| \leq \theta_1(c^T x^k - c^*)^2$ and $\sum_{k=L}^{\infty} |\delta_k| < \infty$.
   (b) $\|u^k\|^2 = \|u_N^k\|^2 + \gamma_k$, $0 \leq \gamma_k \leq \theta_2(c^T x^k - c^*)^2$ and $\sum_{k=L}^{\infty} \gamma_k < \infty$.
   (c) $\phi(u_N^k) \geq \frac{1}{2q}$.
   (d) $\|u_N^k\|^2 \geq \frac{1}{2q}$.
   (e) $\phi(u^k) = \phi(u_N^k)$.

Proof: Part 1 readily follows readily from Lemma 5.1.27.

To see part 2(a), note that from the definition of u^k, Lemma 5.1.25 and Theorems 5.1.23, 5.1.26 and 5.1.29 we obtain that there is an $L_1 \geq 1$ such that for all $k \geq L_1$,

$$\begin{aligned}
|e^T u_N^k - 1| &= \frac{|(x_N^k)^T s_N^k - (x_N^k)^T \tilde{s}_N^k|}{c^T x^k - c^*} \\
&\leq \frac{\|x_N^k\|}{c^T x^k - c^*} \|s_N^k - \tilde{s}_N^k\| \\
&\leq \gamma\rho\rho_3(c^T x^k - c^*)^2.
\end{aligned}$$

Defining $\theta_1 = \gamma\rho\rho_3$, and $\delta_k = e^T u_N^k - 1$ we have 2(a). Also, from Theorem 5.1.22,

$$\sum_{k=L}^{\infty} |\delta_k| = \theta_1 \sum_{k=L}^{\infty} (c^T x^k - c^*)^2$$

$$= \theta_1(c^T x^L - c^*)^2 \sum_{j=0}^{\infty}(1 - \frac{\alpha}{\sqrt{n}})^{2j}$$
$$< \infty.$$

To see part 2(b), note that from the definition of u^k, Lemma 5.1.28 and the fact that $x_B^k \to x_B^* > 0$ implies there is an $L_2 \geq 1$ such that for all $k \geq L_2$

$$\|u_B^k\| \leq \frac{\|X_{k,B}^{-1}\| \|w_B^k\|}{c^T x^k - c^*}$$
$$\leq M\rho_2(c^T x^k - c^*).$$

Letting $\gamma_k = M^2 \rho_2^2 (c^T x^k - c^*)^2$ and $\theta_2 = M^2 \rho_2^2$ we obtain the remaining part exactly as in the case of part 2(a).

To see Part 2(c), using Part 2(a), we note that

$$\phi(u_N^k) \geq \frac{1+\delta_k}{q}.$$

But, there is an $L_3 \geq 1$ such that for all $k \geq L_3$, $|\delta_k| \leq 0.50$. Thus, we have the result.

To see Part 2(d), consider the following optimization problem:

$$\begin{aligned}\text{minimize} \quad & x^T x \\ & e^T x = 1 + \delta_k.\end{aligned}$$

The solution to this problem is $x = \frac{1+\delta_k}{q} e$. Thus, for $k \geq L_3$ we have our result.

To see Part 2(e), note that from the result of Part 2(b),

$$\phi(u_B^k) < \|u_B^k\| \leq M\rho_2(c^T x^k - c^*)$$

and for $L_4 \geq 1$ and all $k \geq L_4$ we have $M\rho_2(c^T x^k - c^*) \leq \frac{1}{2q}$. And the result now follows from Part 2(c) and the fact that $\phi(u^k) = \max\{\phi(u_B^k), \phi(u_N^k)\}$. We obtain Part 2 by defining $L = \max\{L_1, L_2, L_3, L_4\}$. ∎

We are now ready to show that the convergence rate of the sequence $\{X_k s^k\}$, asymptotically, is the same as that of $\{c^T x^k - c^*\}$, and is thus linear.

5.1.31 Corollary There is an $L \geq 1$ and $\theta_1 > 0$, $\theta_2 > 0$ such that for all $k \geq L$,

$$\theta_1(c^T x^k - c^*) \leq \|X_k s^k\| \leq \theta_2(c^T x^k - c^*).$$

Proof: The upper bound follows readily from 5.1.30 part 1. The lower bound follows readily from 5.1.30 parts 2(b) and (d). ∎

5.1 PRIMAL AFFINE SCALING METHOD

Strict Complementary Solution

An important property of the interior point methods is that they generally converge to the relative interior of the optimum solution set. The simplex method, on the other hand, converges to an extreme point, which is on the relative boundary of the optimum solution set. We now prove a theorem which establishes a condition under which the large step primal affine scaling method generates a strictly complementary solution.

5.1.32 Strict Complementary Theorem Let the primal and the dual sequences generated by the primal affine scaling algorithm converge, and the primal converge to a non-degenerate solution. Then this pair satisfies the strict complementarity condition.

Proof: Let the primal and the dual sequences generated be $\{x^k\}$, $\{y^k\}$ and $\{s^k\}$, and let the sequences converge, respectively, to x^*, y^* and s^*. Indeed, as shown in Theorem 5.1.20 these limit points satisfy the complementary slackness conditions, and are thus optimal for their respective problems. Since x^* is non-degenerate, there is a constant $\gamma_1 > 0$ such that

$$\|(AX_k^2 A^T)^{-1}\| \le \gamma_1 \text{ for all } k = 1, 2, \cdots$$

and thus $\|X_k A^T (AX_k^2 A^T)^{-1} A\| \le \gamma_2$ for some $\gamma_2 > 0$.

Let $N = \{j : s_j^* > 0\}$ and thus $x_j^* = 0$ for all $j \in N$. We now show the contrary, that if $j \notin N$ then $x_j^* > 0$. Thus, let $j \notin N$. Then $\{x_j^k\}$ converges to a positive number x_j^* if $\sum_{k=0}^{\infty} |\frac{x_j^{k+1} - x_j^k}{x_j^k}| < \infty$. This is so because then $a_k = \frac{x_j^{k+1}}{x_j^k} \to 1$ and thus there is an $L \ge 1$ such that for all $k \ge L$,

$$|a_k - 1| \ge |\frac{(a_k - 1)^2}{2} - \frac{(a_k - 1)^3}{3} + \cdots|$$

Thus, using the identity $\log(1 - a) = -a - a^2/2 - a^3/3 - \cdots$ we obtain $|\log(a_k)| = |\log(1 - (1 - a_k))| \le 2|a_k - 1|$. Thus

$$\infty > \sum_{k=L}^{\infty} |\log(\frac{x_j^{k+1}}{x_j^k})|$$

$$\ge |\sum_{k=L}^{\infty} \log(\frac{x_j^{k+1}}{x_j^k})|$$

$$= |\log(x_j^*) - \log(x_j^L)|$$

and so $x_k^* > 0$. To see that $\sum_{k=0}^{\infty} |\frac{x_j^{k+1} - x_j^k}{x_j^k}| < \infty$, consider the following:

$$\begin{aligned} s^k - s^* &= A^T(y^* - y^k) \\ &= A^T(AX_k^2 A^T)^{-1} AX_k^2 (A^T y^* - c) \\ &= -A^T(AX_k^2 A^T)^{-1} AX_k^2 s^*. \end{aligned}$$

Now, for $j \notin N$, and e^j the jth unit vector

$$s_j^k = s_j^k - s_j^* = -(e^j)^T A^T (AX_k^2 A^T)^{-1} AX_k^2 s^*$$

and $x_j^k s_j^k = -(e^j)^T X_k A^T (AX_k^2 A^T)^{-1} AX_k^2 s^*$. Now

$$\begin{aligned} \sum_{k=0}^{\infty} |\frac{x_j^{k+1} - x_j^k}{x_j^k}| &= \alpha \sum_{k=0}^{\infty} \frac{|x_j^k s_j^k|}{\phi(X_k s^k)} \\ &= \alpha \sum_{k=0}^{\infty} |\frac{u_j^T X_k A^T (AX_k^2 A^T)^{-1} AX_k^2 s^*}{\phi(X_k s^k)}| \\ &\leq \gamma_2 \alpha \sum_{l \in N} \sum_{k=0}^{\infty} \frac{(x_l^k)^2 s_l^*}{\phi(X_k s^k)} \end{aligned}$$

Our theorem now follows from Theorems 5.1.23, 5.1.22 and 5.1.30, Part 2 (c). ■

5.1.33 On Convergence of Dual Sequence We have seen in Theorem 5.1.21, that the dual sequence converges when the linear program (5.1) is non-degenerate. Unfortunately, without this assumption, there exist examples for which the dual sequence does not converge, and even has a cluster point which is not dual feasible. Such cluster points exist since the dual sequence is bounded, Theorem 5.1.11. We demonstrate this with the following simple example:

$$\begin{aligned} \text{minimize } & x_1 + x_2 + x_3 \\ & x_1 + x_2 - x_3 - x_4 = 0 \\ & x_1 \geq 0 \quad x_2 \geq 0 \quad x_3 \geq 0 \quad x_4 \geq 0 \end{aligned}$$

whose dual is

$$\begin{aligned} \text{maximize } & 0.y \\ & 0 \leq y \leq 1. \end{aligned}$$

Starting with the interior point $(10, 10, 19, 1)^T$, and the step size $\alpha = 0.995$, the large step primal affine scaling procedure generates the following tentative solutions to the dual:

k	y^k	s^k			
1	-.28648	1.28648	1.28648	.71352	-.28648
2	.33416	.66584	.66584	1.33416	.33416
3	-.28513	1.28513	1.28513	.71487	-.28513
4	.33413	.66587	.66587	1.33413	.33413
5	-.28373	1.28373	1.28373	.71627	-.28373
6	.33411	.66589	.66589	1.33411	.33411
7	-.28227	1.28227	1.28227	.71773	-.28227
8	.33417	.66583	.66583	1.33417	.33417

The corresponding primal solutions generated are:

k	x^k			
1	10.0D-00	10.0D-00	19.0D-00	1.0D-00
2	.5115D-00	.5115D-00	1.9D-03	1.0211D-00
3	9.94D-04	9.94D-04	1.8859D-03	1.0211D-04
4	5.2253D-05	5.2253D-05	1.8859D-07	1.0432D-04
5	9.883D-08	9.8829D-08	1.8723D-07	1.0432D-08
6	5.3404D-09	5.3404D-09	1.8723D-11	1.0662D-08
7	9.8289D-12	9.8289D-12	1.8591D-11	1.0662D-12
8	5.4615D-13	5.4615D-13	1.8591D-15	1.0902D-12

We note that the dual sequence $\{y^k\}$ has two cluster points, one positive and the other negative. Both these cluster points are dual feasible though the negative one takes a large number of iterations to become dual feasible. On the other hand, the primal sequence has converged to the unique optimum solution $(0, 0, 0, 0)^T$. There is now an example for which none of the cluster points of the dual sequence are dual feasible, and the primal does not converge to an optimum solution. This example requires the step size α to be .999, see Mascarenhas [141].

5.2 Degeneracy Resolution by Step-Size Control

We have seen in the previous section that the dual sequence does not converge. We will show here that if the step size, α, implemented in the large step primal affine scaling method is required to be less than $\frac{2}{3}$, the dual sequence indeed converges to the analytic center of the set of optimal dual solutions. For this purpose we introduce the concept of a potential function.

5.2.1 Potential Function

Given sequence $\{c^T x^k\}$ that converges to c^* and the sequence $\{x^k\}$, whenever $c^T x^k > c^*$, we define the potential of x^k as

$$F(x^k) = n\log(c^T x^k - c^*) - \sum_{j=1}^{n} \log(x_j^k); \tag{5.32}$$

and, for a given subset $N \subset \{1, \cdots, n\}$, we define the 'local' potential function as

$$F_N(x^k) = |N|\log(c^T x^k - c^*) - \sum_{j \in N} \log(x_j^k). \tag{5.33}$$

Potential function plays an important role in the proof of convergence of the projective transformation method, while local potential function will allow us to investigate the local behavior of the large step primal affine scaling method when the step size is changed. We need the following two straight forward lemmas related to the natural logarithm.

5.2.2 Lemma

Let $0 \neq a < 1$ be any real number. Then

$$\log(1 - a) < -a.$$

Proof: Follows readily from the concavity of $\log(x)$. ∎

5.2.3 Lemma

Let $w \in \mathbf{R}^q$, and $0 < \lambda < 1$ be such that $w_j \leq \lambda$ for every $j = 1, \cdots, q$. Then

$$\sum_{j=1}^{q} \log(1 - w_j) \geq -e^T w - \frac{\|w\|^2}{2(1-\lambda)}.$$

Proof: The result follows as a consequence of the following facts: for $0 < a \leq \lambda$, we see that

$$\begin{aligned}
\log(1-a) &= -a - \frac{a^2}{2} - \frac{a^3}{3} - \cdots \\
&\geq -a - \frac{a^2}{2(1-a)} \\
&\geq -a - \frac{a^2}{2(1-\lambda)};
\end{aligned}$$

and, for $a < 0$, by considering the function

$$g(x) = \log(1+x) - x + \frac{x^2}{2(1-\lambda)}$$

5.2 DEGENERACY RESOLUTION

we note that $g(0) = 0$ and $g'(x) > 0$ for all $x > 0$. So $g(|a|) > 0$ for all $a < 0$, and as $|a| = -a$ for all $a < 0$, we have

$$\log(1-a) \geq -a - \frac{a^2}{2(1-\lambda)}$$

and we are done. ∎

Decrease of Local Potential along $\{x^k\}$

We now calculate the decrease in the local potential along the primal sequence $\{x^k\}$. For this, let x^* be the limit of $\{x^k\}$, with

$$N = \{j : x_j^* = 0\}, \text{ and } q = |N|.$$

Define F_N as in Equation (5.33), and, because of Theorem 5.1.19, F_N is defined for each $k = 1, 2 \cdots$. Also, let

$$d_N^k = u_N^k - \frac{1}{q}e,$$

and

$$\tilde{\alpha}_k = \frac{\alpha}{\phi(u_N^k)}.$$

We now prove a straightforward lemma:

5.2.4 Lemma Let $0 < \alpha < 1$. There is an $L \geq 1$ and a $\theta > 0$ such that for all $k \geq L$, $q - \tilde{\alpha}_k \geq \theta$.

Proof: From Theorem 5.1.30, Part 2(a), we note that for some $L \geq 1$ and all $k \geq L$,

$$\phi(u_N^k) \geq \frac{1 + \delta_k}{q}.$$

Thus

$$q - \tilde{\alpha}_k \geq q - \frac{q\alpha}{1 + \delta_k}$$
$$= \frac{q}{1 + \delta_k}(1 + \delta_k - \alpha).$$

Since $\alpha < 1$, and $\delta_k \to 0$, there is an $L_1 \geq L$ and a $\theta > 0$ such that for all $k \geq L_1$, $1 + \delta_k - \alpha \geq \theta$, and we are done. ∎

We are now ready to establish a formula for the local potential decrease along the sequence $\{x^k\}$.

5.2.5 Lemma There exists an $L \geq 1$ such that for all $k \geq L$

$$F_N(x^{k+1}) - F_N(x^k) = q\log(1 - \mu_k\|d_N^k\|^2 + \mu_k(\frac{2\delta_k}{q} + \gamma_k)) - \sum_{j \in N} \log(1 - \mu_k d_j^k)$$

where $\mu_k = \frac{q\tilde{\alpha}_k}{q - \tilde{\alpha}_k} > 0$ and $\sum_{k=L}^{\infty} |\delta_k| < \infty$, $\sum_{k=L}^{\infty} |\gamma_k| < \infty$.

Proof: From Step 4, we note that $c^T x^{k+1} - c^* = c^T x^k - c^* - \frac{\alpha}{\phi(X_k s^k)} c^T X_k^2 s^k$. The fact that $c^T X_k^2 s^k = \|X_k s^k\|^2$ and simple algebra we see that

$$\frac{c^T x^{k+1} - c^*}{c^T x^k - c^*} = 1 - \frac{\alpha}{\phi(u^k)} \|u^k\|^2; \qquad (5.34)$$

and, for each $j \in N$

$$\frac{x_j^{k+1}}{x_j^k} = 1 - \frac{\alpha}{\phi(u^k)} u_j^k.$$

From Theorem 5.1.30 part 2(b), there is an $L \geq 1$ such that for all $k \geq L$,

$$\frac{c^T x^{k+1} - c^*}{c^T x^k - c^*} = 1 - \frac{\alpha}{\phi(u_N^k)} \|u_N^k\|^2 - \frac{\alpha \gamma_k}{\phi(u_N^k)}.$$

Now let $k \geq L$, and note that from Theorem 5.1.30 part 2(a),

$$\|d_N^k\|^2 = \|u_N^k\|^2 - \frac{2e^T u_N^k}{q} + \frac{1}{q} = \|u_N^k\|^2 - \frac{1}{q} - \frac{2\delta_k}{q}.$$

We also note that

$$\frac{q}{q - \tilde{\alpha}_k}(1 - \tilde{\alpha}_k \|u_N^k\|^2 - \tilde{\alpha}_k \gamma_k) = 1 - \mu_k \|d_N^k\|^2 - \mu_k(\frac{2\delta_k}{q} + \gamma_k),$$

$$\frac{q}{q - \tilde{\alpha}_k}(1 - \tilde{\alpha}_k u_j^k) = 1 - \mu_k d_j^k;$$

and, from Lemma 5.2.4, the constant $\frac{q}{q - \tilde{\alpha}_k} > 0$. Since

$$F_N(x^{k+1}) - F_N(x^k) = q\log(\frac{c^T x^{k+1} - c^*}{c^T x^k - c^*}) - \sum_{j \in N} \log(\frac{x_j^{k+1}}{x_j^k})$$

we obtain our result by substitution of the appropriate identities established above, and noting that the constant $\frac{q}{q - \tilde{\alpha}_k}$ cancels.

We are now ready to obtain an upper bound on the decrease of the potential function.

5.2.6 Decrease of Local Potential Function

There is an $L \geq 1$ and a $\theta > 0$ such that for all $k \geq L$ with $\phi(d_N^k) > 0$, we have

$$F_N(x^{k+1}) - F_N(x^k) \leq \mu_k \|d_N^k\|^2 \left(-q + \frac{\alpha}{2(1-\alpha)} \frac{q}{1+\phi(d_N^k)} \right) - \mu_k(q\gamma_k + \delta_k)$$

with $\theta > \mu_k > 0$, $\sum_{k=L}^{\infty} |\delta_k| < \infty$ and $\sum_{k=L}^{\infty} |\gamma_k| < \infty$.

Proof: From Lemmas 5.2.2, 5.2.3, 5.2.5 and Theorem 5.1.30 Part 2(a), we obtain an $L \geq 1$ and a $\theta > 0$ such that for each $k \geq L$ $\theta \geq \mu_k > 0$ and there exist γ_k and δ_k such that when $\phi(d_N^k) > 0$,

$$
\begin{aligned}
F_N(x^{k+1}) - F_N(x^k) &\leq -q\mu_k(\|d_N^k\|^2 + \frac{2\delta_k}{q} + \gamma_k) + \\
&\quad \mu_k(e^T d_N^k + \mu_k \frac{\|d_N^k\|^2}{2(1 - \mu_k \phi(d_N^k))}) \\
&= \mu_k \|d_N^k\|^2 \left(-q + \frac{\mu_k}{2(1 - \mu_k \phi(d_N^k))} \right) - \mu_k(q\gamma_k + \delta_k) \\
&= \mu_k \|d_N^k\|^2 \left(-q + \frac{\alpha}{2(1-\alpha)} \frac{1}{\phi(u_N^k)} \right) - \mu_k(q\gamma_k + \delta_k).
\end{aligned}
$$

Now, as $d_N^k = u_N^k - \frac{1}{q}e$, we note that, when $\phi(d_N^k) > 0$, $\phi(u_N^k) \geq \frac{1+\phi(d_N^k)}{q}$, and we have our result by substituting this expression. The required property of δ_k and γ_k follows from Lemma 5.2.5. ∎

Analytic Center of a Dual Face

We will prove that when the step size $\alpha \leq \frac{2}{3}$, the dual sequence converges to the analytic center of the optimal dual face. We derive here the properties of such an analytic center.

Let $N \subset \{1, \cdots, n\}$ and $B = \{1, \cdots, n\} \setminus N$. We define the analytic center of the face of the dual defined by N as the solution to the following optimization problem:

$$
\begin{aligned}
\text{maximize} \quad & \sum_{j \in N} \log(s_j) \\
& A_B^T y = c_B \\
& A_N^T y + s_N = c_N \\
& (s_N > 0).
\end{aligned} \tag{5.35}
$$

Such an analytic center may not exist, as the feasible region defined by the equality constraints may be empty or unbounded. Indeed, when it does exist, it is unique. Writing the Karush-Kuhn-Tucker conditions for this problem, we obtain

$$\begin{aligned} S_N^{-1} e + v_N &= 0 \\ A_B v_B + A_N v_N &= 0 \\ A_B^T y &= c_B \\ A_N^T y + s_N &= c_N \\ (s_N > 0). & \end{aligned} \quad (5.36)$$

where v_B and v_N are the Lagrange multipliers associated with the two equality constraints.

5.2.7 Convergence of the Dual Sequence We are now ready to prove the convergence of the primal affine scaling method without the nondegeneracy assumption. This is proved for the case when the step size $\alpha \leq \frac{2}{3}$. We will now assume that:

1. The matrix A is $m \times n$ and has full rank m.

2. The linear program has an interior solution.

3. The objective function is not constant on the feasible region.

4. The dual has a feasible solution.

We can then prove:

5.2.8 Theorem Let the assumptions (1) - (4) hold, the sequence generated by the affine scaling algorithm be infinite, and the step size $\alpha \leq \frac{2}{3}$. Then there exist vectors x^*, y^* and s^* such that

1. $x^k \to x^*$

2. $y^k \to y^*$

3. $s^k \to s^*$, and $s^* \geq 0$.

The limits are optimal solutions of their respective problems, and satisfy the strict complementarity condition. In addition, the dual solution (y^*, s^*) converges to the analytic center of the optimal dual face, and the primal solution to the relative interior of the optimal primal face.

Proof: For each $\alpha \leq \frac{2}{3}$, $\frac{\alpha}{2(1-\alpha)} \leq 1$. Hence, when $\phi(d_N^k) > 0$, from Theorem 5.2.6, we get

$$F_N(x^{k+1}) - F_N(x^k) \leq -\frac{q\mu_k \phi(d_N^k) \|d_N^k\|^2}{1 + \phi(d_N^k)} - \mu_k(q\gamma_k + \delta_k).$$

5.2 DEGENERACY RESOLUTION

In case $\phi(d_N^k) \leq 0$, $d_N^k \leq 0$ and thus $\log(1 - \mu_k d_j^k) \geq 0$ and from Lemma 5.2.5 and the fact that $\log(1-a) < -a$, we get

$$F_N(x^{k+1}) - F_N(x^k) \leq -q\mu_k \|d_N^k\|^2 - \mu_k(q\gamma_k + 2\delta_k).$$

Note that $\phi(u_N^k)^2 \leq \|u^k\|^2$. Using Equation (5.34) and Theorem 5.1.30 Part 2(e), we obtain $\phi(u_N^k) \leq \frac{1}{\alpha}$. Thus $\mu_k \geq \frac{q\alpha^2}{q-\alpha^2}$. Now, let $M > L+1$. Then, from Theorem 5.1.23 Part (b), there is a $\delta > 0$ such that

$$\sum_{k=L}^{M-1} (F_N(x^{k+1}) - F_N(x^k)) = F_N(x^M) - F_N(x^L)$$
$$\geq q\log(\delta) - \sum_{j \in N} \log \frac{c^T x^L - c^*}{x_j^L}$$
$$> -\infty.$$

Thus, $\sum_{k=L}^{\infty} (F_N(x^{k+1}) - F_N(x^k)) > -\infty$. Also, from Theorem 5.2.6, $0 \leq \sum_{k=L}^{\infty} (|\delta_k| + |\gamma_k|) < \infty$. Thus, when the sequence $\{\phi(d_N^k)\}$ has either a strictly positive or a strictly negative cluster point, we must have $\|d_N^k\| \to 0$. Also, if $\phi(d_N^k) \longrightarrow 0$, since $e^T d_N^k = \delta_k$ and $\delta_k \longrightarrow 0$, $d_N^k \longrightarrow 0$. Thus, we get

$$\lim_{k \to \infty} u_N^k = \frac{1}{q} e. \tag{5.37}$$

From Corollary 5.1.12, the sequence $\{s^k\}$ has a cluster point s^*. Thus, let $s^k \to s^*$ on some subsequence $K' \subset \{1, 2, \cdots\}$. Consider the sequences $\{y^k\}$, $\{\frac{x_j^k}{c^T x^k - c^*}\}$ for each $j \in N$ and $\{\frac{x_j^k - x_j^*}{c^T x^k - c^*}\}$ for each $j \in B$. From Theorems 5.1.11 and 5.1.23 Parts (b) and (c), these sequences are bounded. Thus, on some common subsequence $K \subset K'$

$$y^k \to y^*, \ s^k \to s^*$$

$$\frac{qx_j^k}{c^T x^k - c^*} \to a_j \text{ for each } j \in N$$

and

$$\frac{q(x_j^k - x_j^*)}{c^T x^k - c^*} \to b_j \text{ for each } j \in B.$$

Since $u_j^k = \frac{x_j^k s_j^k}{c^T x^k - c^*}$, in view of (5.37), $a_j > 0$ for each $j \in N$. Also, since $qu_j^k \to 1$, we note that $s_j^* = \lim_{k \in K} s_j^k = a_j^{-1}$.

Since $A_N x_N^k + A_B x_B^k = A_B x_B^*$, we see that

$$A_N a + A_B b = 0$$

and thus $s_j = a_j^{-1}$ for each $j \in N$ and $v_N = -a$, $v_B = -b$ and $s_B = 0$, $s_N = s_N^*$, $y = y^*$ solve the conditions 5.36 for the analytic center of the dual face generated by N.

Now, each cluster point of the sequence $\{s_N^k\}$ converges to the analytic center, a^{-1}, and thus parts (2) and (3) of the theorem follow. We finish the proof by observing that x^* is primal feasible, (y^*, s^*) is dual feasible and the pair satisfies the complementary slackness theorem (by Theorem 5.1.19) and thus are optimal solutions for the respective problems. The strict complementarity holds as $s_N^* > 0$. ∎

We now prove convergence for a larger step size than $\frac{2}{3}$, namely $\alpha = \frac{2q}{3q-1}$. This is that largest step size known for the convergence of the affine scaling method, and is conjectured to be so.

We now establish a lemma, before proving the main result.

5.2.9 Lemma Assume that on some subsequence K, $\|d_N^k\| \not\to 0$. Then there is an $L \geq 1$ such that for every $k \geq L$ and $k \in K$

$$\phi(u_N^k) \geq \frac{1}{q} + \frac{\|d_N^k\|}{\sqrt{q(q-1)}} - \frac{2|\delta_k|}{q}$$

Proof: Consider the optimization problem:

$$\begin{array}{lrcl}
\text{minimize} & \rho & & \\
\sum_{i=1}^{q-1} x_i & + \rho & = & 1 + \delta_k \\
x_i & - \rho & \leq & 0 \\
\sum_{i=1}^{q-1} (x_i - \frac{1}{q})^2 & + (\rho - \frac{1}{q})^2 & = & \|d_N^k\|^2
\end{array}$$

It is readily confirmed that the solution of this optimization problem is:

$$\rho - \frac{1}{q} = \sqrt{\frac{\|d_N^k\|^2 - \frac{\delta_k^2}{q}}{q(q-1)}} + \frac{\delta_k}{q}$$

and thus we obtain our result by observing that $\phi(u_N^k) \geq \rho$ for every u_N^k with $e^T u_N^k = 1 + \delta_k$ and $u_N^k - \frac{1}{q}e = d_N^k$; and for every $a > b > 0$, $\sqrt{a^2 - b^2} \geq a - b$. ∎

We are now ready to prove the main theorem:

5.2.10 Convergence For Larger Step Size Let $\frac{2}{3} < \alpha \leq \frac{2q}{3q-1}$. Then, there exists a subsequence $K \subset \{1, 2, \cdots\}$ and vectors x^*, y^* and s^* such that

5.2 DEGENERACY RESOLUTION

1. $\lim_{k \in K} x^k \to x^*$
2. $\lim_{k \in K} y^k \to y^*$
3. $\lim_{k \in K} s^k \to s^*$, and $s^* \geq 0$.

Thus x^* and (y^*, s^*) are optimum solutions of the dual pair. Also, when $\alpha < \frac{2q}{3q-1}$ they satisfy the strict complementarity condition.

Proof: As a result of 5.1.23, Part (b) either on some subsequence K', $F(x^{k+1}) - F(x^k) \geq 0$ for each $k \in K'$ or $\lim(F(x^{k+1}) - F(x^k)) = 0$. Thus, on some subsequence, from 5.2.6

$$\liminf_{k \in K'} (-q + \frac{\alpha}{2(1-\alpha)} \frac{1}{\phi(u_N^k)}) \geq 0.$$

Let $K \subset K'$ be such that as $k \in K$ goes to infinity, $x^k \to x^*$, $y^k \to y^*$, $s^k \to s^*$, $u_N^k \to u_N^*$ and $\|d_N^k\| \to \theta \geq 0$. If $\theta = 0$ then we are done since $u_N^* = \frac{1}{q} e > 0$. Otherwise, from Lemma 5.2.9

$$-q + \frac{\alpha}{2(1-\alpha)} \frac{q}{1 + \theta\sqrt{\frac{q}{q-1}}} \geq 0.$$

Letting $\alpha \leq \frac{2q}{3q-1}$, it is readily confirmed that

$$\theta \leq \frac{1}{\sqrt{q(q-1)}}.$$

Since $\|u_N^* - \frac{1}{q}e\| = \theta$ and $e^T u_N^* = 1$, we can conclude that $u_N^* \geq 0$ since it lies in the largest inscribed hypersphere in the simplex $\{x : e^T x = 1, x \geq 0\}$. The strict complementarity follows since under the hypothesis, u_N^* lies strictly inside the hypersphere, and is thus strictly positive, and we are done. ∎

The following theorem gives a sharp bound on the convergence of $\{c^T x^k - c^*\}$.

5.2.11 Theorem Let $\alpha \leq \frac{2}{3}$. Then

$$\lim_{k \to \infty} \frac{c^T x^{k+1} - c^*}{c^T x^k - c^*} = 1 - \alpha.$$

Proof: From Equation (5.34) and Theorem 5.1.30 Parts 2(b) and 2(e) we obtain an $L \geq 1$ and a $\theta > 0$ such that for all $k \geq L$,

$$\frac{c^T x^{k+1} - c^*}{c^T x^k - c^*} = 1 - \frac{\alpha}{\phi(u_N^k)} \|u_N^k\|^2 + \delta_k$$

where $|\delta_k| \leq \theta(c^T x^k - c^*)^2$. Taking limits, the result follows from Theorem 5.2.8 and Equation (5.37). ∎

5.3 Accelerated Affine Scaling Method

By making a connection between the step of Newton's method applied to finding an analytic center of a certain polyhedron; and, the affine scaling step, it is possible to make the method converge quadratically to the solution. This is achieved by simply controlling the step size at each iteration. We now explore this relationship.

Consider the primal sequence $\{x^k\}$ generated by the algorithm. Then this converges, say to x^*. Define the sets B and N of positive and zero variables in x^*, respectively, and the sequence

$$v^k = \frac{x^k - x^*}{c^T x^k - c^*}.$$

Let $(\bar{y}, \bar{s}) \in \mathcal{D}$ be arbitrary, but fixed, where $\mathcal{D}$ is defined by (5.29). Then we can prove:

5.3.1 Lemma For every $k = 1, 2, \cdots$,

$$\bar{s}_N^T v_N^k = 1.$$

Proof: Readily follows from Lemma 5.1.25. ∎

An Analytic Center

We now define the polyhedron whose analytic center we will find using Newton's method. This polyhedron is:

$$\mathcal{P} = \{v : A_N v_N + A_B v_B = 0, \bar{s}_N^T v_N = 1, v_N \geq 0\}.$$

We can then prove:

5.3.2 Lemma For each $k = 1, 2, \cdots$, $v^k \in \mathcal{P}$.

Proof: Follows readily from the definition of v^k and the Lemma 5.3.1. ∎

$\mathcal{P}$ is non-empty, but may be unbounded, and thus may have no analytic center. In case it does have one, let v^* be its analytic center. Thus v^* is the solution to the following optimization problem:

$$\begin{aligned}
\text{maximize} \quad & \sum_{j \in N} \log(v_j) \\
& A_N v_N + A_B v_B = 0 \\
& \bar{s}_N^T v_N = 1 \\
& v_N > 0,
\end{aligned} \quad (5.38)$$

5.3 ACCELERATED AFFINE SCALING METHOD

when it exists. Writing the Karush - Kuhn - Tucker conditions for this problem we get:
$$\begin{aligned} V_N^{-1}e - A_N^T y - \bar{s}_N \theta &= 0 \\ -A_B^T y &= 0 \\ A_N v_N + A_B v_B &= 0 \\ \bar{s}_N^T v_N &= 1 \end{aligned} \tag{5.39}$$

where y and θ are the Lagrange multipliers on the equality constraints of the problem (5.38).

Relationship Between Two Analytic Centers

We will now explore the relationship between the two centers defined by the problems (5.38) and (5.35). It turns out that one has an analytic center if and only if the other has. We establish this in the next theorem:

5.3.3 Theorem The analytic center (y^*, s^*) with $s_N^* > 0$ of (5.35) exists if and only if the analytic center v^* with $v_N^* > 0$ of (5.38) exists. Also, for some $\rho^* > 0$
$$v_N^* = \rho^* S_{*,N}^{-1} e.$$

Proof: Let v^* be the analytic center of $\mathcal{P}$. Then, for some θ^* and y^*, the conditions (5.39) are satisfied. Now, $\theta^* > 0$, since it is readily seen that if the right hand side of the equation $\bar{s}_N^T v_N = 1$ is increased, to say $\rho > 1$, then ρv^* remains feasible, and the objective function strictly increases by at least $q \log(\rho) > 0$. Now, define
$$\begin{aligned} s_N &= \frac{1}{\theta^*} A_N^T y^* + \bar{s}_N, (= V_{*,N}^{-1} e > 0) \\ v &= \theta^* v^*, (v_N > 0) \\ y &= \bar{y} - \frac{1}{\theta^*} y^* \end{aligned} \tag{5.40}$$

and note that these vectors satisfy the conditions (5.36).

Also, if (y^*, s^*), $s_N^* > 0$ is the analytic center of the (5.35), then (5.36) is satisfied for some v^*. Then
$$\begin{aligned} \theta &= \bar{s}_N^T v_N^* \\ v &= \frac{1}{\theta} v^* \\ y &= \theta(\bar{y} - y^*) \end{aligned}$$

satisfy the conditions (5.39). The remaining part follows. ∎

Newton Method and Analytic Center of $\mathcal{P}$

We will now investigate the application of Newton's method to compute the analytic center of $\mathcal{P}$. This is done by applying the method to the 'zero' finding problem represented by the Karush - Kuhn - Tucker conditions (5.39), starting at some arbitrary point (v, y, θ) with $v \in \mathcal{P}$. Thus this investigation is possible even when the analytic center does not exist; but this method will not converge in that case.

Let $(\Delta v, \Delta y, \Delta \theta)$ be the Newton step at (v, y, θ). Then, it is readily seen that this is obtained by solving the following system:

$$\begin{aligned} -V_N^{-2}\Delta v_N - A_N^T\Delta y - \bar{s}_N\Delta\theta &= -V_N^{-1}e + A_N^T y + \bar{s}_N\theta \\ -A_B^T\Delta y &= A_B^T y \\ A_N\Delta v_N + A_B\Delta v_B &= 0 \\ \bar{s}_N^T\Delta v_N &= 0. \end{aligned} \qquad (5.41)$$

By the simple substitution,

$$\begin{aligned} \hat{y} &= y + \Delta y \\ \hat{\theta} &= \theta + \Delta\theta \\ t_N &= V_N^{-1}\Delta v_N \\ \hat{A}_N &= A_N V_N \\ \hat{s}_N &= V_N \bar{s}_N \end{aligned}$$

and rearrangement, we can rewrite the system (5.41) as the following:

$$\begin{aligned} t_N + \hat{A}_N^T\hat{y} + \hat{s}_N\hat{\theta} &= e \\ A_B^T\hat{y} &= 0 \\ \hat{A}_N t_N + A_B\Delta v_B &= 0 \\ \hat{s}_N^T t_N &= 0. \end{aligned} \qquad (5.42)$$

We can then prove:

5.3.4 Theorem Let A have full row rank, and $v_N > 0$. There exists a unique vector $(t_N, \hat{y}, \hat{\theta})$ such that for some Δv_B, the System (5.42) is satisfied. In case A_B has full column rank, the vector Δv_B is unique.

Proof: Considering t_N, $\hat{y}$, $\hat{\theta}$ and Δv_B as variables, the System (5.42) is linear with the underlying matrix:

$$M(v_N) = \begin{bmatrix} I & \hat{A}_N^T & \hat{s}_N & 0 \\ 0 & A_B^T & 0 & 0 \\ \hat{A}_N & 0 & 0 & A_B \\ \hat{s}_N^T & 0 & 0 & 0 \end{bmatrix}. \qquad (5.43)$$

5.3 ACCELERATED AFFINE SCALING METHOD

In case A_B has full column rank, this matrix can easily be seen as non-singular. Thus, the second conclusion follows.

Now let $A_B = (A_C, A_D)$ where A_C has full column rank, and spans the column space $\mathcal{R}(A_B)$ of A_B. Then, for some matrix Λ, $A_D = A_C \Lambda$. Substituting A_C for A_B in the matrix (5.43), we generate a new system with the underlying matrix non-singular. This new system has a unique solution. By setting $\Delta v_D = 0$ and so $\Delta v_B = (\Delta v_C, \Delta v_D)$ we obtain a solution for the System (5.42). To show that each solution to (5.42) has the same value for the variables $(t_N, \hat{y}, \hat{\theta})$, note that for a given solution $(t_N, \hat{y}, \hat{\theta}, \Delta v_B)$ (to (5.42)), we generate the unique solution $(t_N, \hat{y}, \hat{\theta}, \Delta v_C + \Lambda \Delta v_D)$, to the modified system. Thus we are done, since only the vector Δv_B is modified. ■

We will now show that the Newton stepin $\mathcal{P}$ is bounded.

5.3.5 Theorem For some given $v_N > 0$, let t_N solve the System (5.42). Then

1. $e^T t_N = t_N^T t_N$.

2. $\|t_N\| \leq \sqrt{q}$, where q is the number of elements in N.

Proof: Part 1 is obtained by multiplying the first relation by t_N^T and substituting $t_N^T \hat{A}_N^T \hat{y} = 0$ obtained from second and third relations of the System (5.42).

Part 2 can be obtained by considering the problem

$$\text{maximize} \quad z^T z$$
$$e^T z - z^T z = 0$$

which has the solution $z = e$. ■

More on Affine Scaling Step

We now investigate the primal affine scaling step with a view to relating it to the Newton step investigated in the previous section. For this purpose we will adopt the ellipsoidal approximating problem (5.2) generated by some given interior point x^0. Substituting $p = x^0 - x$, where p is the affine scaling step at x^0, (as $Ax^0 = b$) we obtain the following equivalent problem:

$$\begin{aligned}\text{maximize} \quad & c^T p \\ Ap &= 0 \\ \|X_0^{-1}p\|^2 &\leq 1\end{aligned} \qquad (5.44)$$

whose unique solution, by Theorem 5.2, is

$$p = \frac{X_0^2 s^0}{\|X_0 s^0\|}.$$

Let $(\bar{y}, \bar{s}) \in \mathcal{D}$ be arbitrary, but fixed. Then using Lemma 5.1.25, we obtain the following equivalent problem:

$$\begin{aligned}\text{maximize} \quad & \bar{s}_N^T p_N \\ A_N p_N + A_B p_B &= 0 \\ p_N^T X_{0,N}^{-2} p_N + p_B^T X_{0,B}^{-2} p_B &\leq 1.\end{aligned} \qquad (5.45)$$

By noting that the solution of this problem is on the boundary of the ellipsoid (i.e., the second constraint is at equality), and setting up the Lagrangian, we obtain the Karush - Kuhn - Tucker conditions for this optimization problem:

$$\begin{aligned}\bar{s}_N - A_N^T y - 2\theta X_{0,N}^{-2} p_N &= 0 \\ -A_B^T y - 2\theta X_{0,B}^{-2} p_B &= 0 \\ A_N p_N + A_B p_B &= 0 \\ \|X_0^{-1} p\| &= 1.\end{aligned} \qquad (5.46)$$

Now, let

$$\begin{aligned}v &= \frac{x^0 - x^*}{c^T x^0 - c^*} \\ u &= \frac{X_0 s^0}{c^T x^0 - c^*} \\ \hat{A}_N &= A_N V_N \\ \hat{s}_N &= V_N \bar{s}_N\end{aligned}$$

and consider the following system of equations:

$$\begin{aligned}\frac{u_N}{\|u\|^2} - \hat{A}_N^T \tilde{y} - \frac{\rho}{\|u\|^2} \hat{s}_N &= 0 \\ -A_B^T \tilde{y} &= -\frac{s_B^0}{\|u\|^2} \\ \hat{A}_N \frac{u_N}{\|u\|^2} + A_B p'_B &= 0 \\ \frac{\hat{s}_N^T u_N}{\|u\|^2} &= 1.\end{aligned} \qquad (5.47)$$

We can then prove:

5.3 ACCELERATED AFFINE SCALING METHOD

5.3.6 Theorem Consider the conditions represented by the Systems (5.46) and (5.47).

1. (5.46) has a unique solution which also solves (5.47).

2. The solution to (5.47) is unique up to a choice of p'_B; and, there is a value for p'_B such that its solution also solves (5.46).

3. When A_B has full column rank, the two systems are equivalent.

Proof: Since the System (5.46) represents the solution to the ellipsoidal approximating problem (5.2), it has a unique solution. Using this solution, define the vectors

$$\tilde{y} = \frac{-y(c^T x^0 - c^*)}{2\theta \|u\|}$$

$$\rho = \frac{(c^T x^0 - c^*)\|u\|}{2\theta}$$

$$p'_B = \frac{p_B}{(c^T x^0 - c^*)\|u\|}.$$

It is now readily confirmed, by simple algebra, that u, $\tilde{y}$ ρ and p'_B solve the System (5.47). Thus we have proved Part 1.

From Part 1, it follows that (5.47) has a solution. Considering $q_N = \frac{u_N}{\|u\|^2}$, $\tilde{\rho} = \frac{\rho}{\|u\|^2}$, $\tilde{y}$ and p'_B as variables, this system is linear in these variables, with the underlying matrix given by (5.43). If A_B has full column rank, the solution to (5.47) is unique, and Part 3 follows. Otherwise, since (5.47) can have a solution only if s_B lies in the row space $\mathcal{R}(A_B^T)$ of A_B, the third condition must have redundant constraints readily identified by choosing any full column rank submatrix of A_B. Using an argument identical to that of the proof of Theorem 5.3.4, we can establish that the values of the variables q_N, $\tilde{\rho}$ and $\tilde{y}$ are the same for every solution of (5.47). The values of these variables are thus uniquely determined by the solution of (5.46). This then establishes Part 2 with the required $p'_B = \frac{p_B}{(c^T x^0 - c^*)\|u\|}$. ∎

Affine Scaling and Newton Steps in $\mathcal{P}$

Consider the sequence $\{v^k\}$ with $v^k \in \mathcal{P}$ generated by the affine scaling method, and let $\{t_N^k\}$ be the sequence of Newton steps, with $t_N^k = V_N^{-1} \Delta v_N^k$. Then we can show that

5.3.7 Theorem There is an $L \geq 1$ and a $\rho > 0$ such that for all $k \geq L$

$$e - t_N^k - \frac{u_N^k}{\|u\|^2} = \Delta^k$$

with $\|\Delta^k\| \leq \rho(c^T x^k - c^*)^2$.

Proof:

By selecting possibly a submatrix of A_B, we consider Systems (5.42) and (5.47) with the columns of A_B linearly independent. Using Theorems 5.3.4 and 5.3.6, the vectors t_N and $\frac{u_N}{\|u\|^2}$ are unaffected by this choice, and any result independent of p'_B and Δv_B will carry over, since these variables are the only ones effected by this choice.

For some k, let $v = v^k$, and consider the System (5.42). In this system, make the change of variables $t'_N = e - t_N$ and $v'_B = v_B - \Delta v_B$. Using Theorem 5.1.25, by substitution in (5.42), obtain the following equivalent system:

$$\begin{aligned}
t'_N - \hat{A}_N^T \hat{y} - \hat{s}_N \hat{\theta} &= 0 \\
A_B^T \hat{y} &= 0 \\
\hat{A}_N t'_N + A_B v'_B &= 0 \\
\hat{s}_N^T t'_N &= 1.
\end{aligned} \qquad (5.48)$$

Defining

$$\begin{aligned}
V_N^{-1} a_1 &= \frac{u_N}{\|u\|^2} - t'_N \\
a_2 &= \tilde{y} - \hat{y} \\
a_3 &= \frac{\rho}{\|u\|^2} - \hat{\theta} \\
a_4 &= p'_B - v'_B \\
s'_B &= \frac{s_B}{\|u\|^2}
\end{aligned}$$

we can generate, using Systems (5.48) and (5.47)

$$\begin{bmatrix} V_N^{-2} & A_N^T & \bar{s}_N & 0 \\ 0 & A_B^T & 0 & 0 \\ A_N & 0 & 0 & A_B \\ \bar{S}_N^T & 0 & 0 & 0 \end{bmatrix} \begin{bmatrix} a_1 \\ a_2 \\ a_3 \\ a_4 \end{bmatrix} = \begin{bmatrix} 0 \\ s'_B \\ 0 \\ 0 \end{bmatrix}.$$

This system is readily seen as the Karush - Kuhn - Tucker conditions of the following quadratic programming problem:

$$\begin{aligned}
\text{minimize} \quad & \tfrac{1}{2} a_1^T V_N^{-2} a_1 + (s'_B)^T a_4 \\
& A_N a_1 + A_B a_4 = 0 \\
& \bar{s}_N^T a_1 = 0
\end{aligned}$$

5.3 ACCELERATED AFFINE SCALING METHOD

with a_2 and a_3, respectively, the Lagrange multipliers of the two constraints. Since A_B has full column rank, by using the change of variables

$$a_4 = -(A_B^T A_B)^{-1} A_B^T A_N a_1$$

we can eliminate a_4 to generate the following equivalent quadratic program:

$$\begin{aligned} \text{maximize} \quad & \tfrac{1}{2} a_1^T V_N^{-2} a_1 + (s_B^*)^T a_1 \\ & A_N^* a_1 = 0 \\ & \bar{s}_N^T a_1 = 0 \end{aligned}$$

where $s_B^* = A_N^T A_B (A_B^T A_B)^{-1} s_B'$ and $A_N^* = A_N - A_B (A_B^T A_B)^{-1} A_B^T A_N$. From Theorems 5.1.16 and 5.1.11, the Lagrange multipliers (a_2, a_3) of this problem are bounded by a function of the form $q(A_N^*, \bar{s}_N) \|s_B^*\|$, independent of the diagonal matrix V_N, where $q(A_N^*, \bar{s}_N) > 0$ is some constant determined by A_N^* and $\bar{s}_N$. Thus

$$\|(a_2, a_3)\| \le \frac{q(A_N^*, \bar{s}_N) \|A_N^T A_B (A_B^T A_B)^{-1}\| \|s_B\|}{\|u\|^2}.$$

Since $V_N^{-2} a_1 = A_N^T a_2 + \bar{s}_N a_3$, independent of a_4, we conclude that

$$\begin{aligned} \|e - t_N - \frac{u_N}{\|u\|^2}\| &= \|V_N^{-1} a_1\| \\ &= \|V_N (A_N^T a_2 + \bar{s}_N a_3)\| \end{aligned}$$

we obtain our result from Theorem 5.1.29 and Theorem 5.1.30 parts 1, 2(b) and 2(d). ∎

We are now ready to establish a relationship between Newton's direction and the affine scaling direction at a given point $v^k \in \mathcal{P}$. We will actually compare the two directions as they apply to points in the projected polyhedron:

$$\mathcal{P}_N = \{v_N : v \in \mathcal{P}\}.$$

In case v^* is the analytic center of $\mathcal{P}$, v_N^* is the analytic center of $\mathcal{P}_N$.

At step k of the large step affine scaling method, let $\alpha_k > 0$ be the implemented step size. We now explore this relationship:

5.3.8 Theorem *The affine scaling direction at v_N^k in $\mathcal{P}_N$ is*

$$v_N^{k+1} - v_N^k = \frac{\alpha_k \delta(u^k)}{1 - \alpha_k \delta(u^k)} (v_N^k - V_{k,N} \frac{u_N^k}{\|u^k\|^2})$$

where $\delta(u^k) = \frac{\|u^k\|^2}{\phi(u^k)}$. Also, the Newton direction Δv_N^k at v_N^k in $\mathcal{P}_N$ is:

$$\Delta v_N^k = v_N^k - V_{k,N}\frac{u_N^k}{\|u^k\|^2} + V_{k,N}\Delta^k$$

where Δ^k is as in Theorem 5.3.7.

Proof: Since $t_N^k = V_{k,N}^{-1}\Delta v_N^k$ the formula for the Newton direction follows from Theorem 5.3.7. To see the affine direction note that:

$$v_N^{k+1} - v_N^k = \frac{x_N^{k+1}}{c^T x_N^{k+1} - c^*} - \frac{x_N^k}{c^T x_N^k - c^*}.$$

Now, from Step 5 and substitution of the formula for θ_k, we have

$$x^{k+1} = x^k - \alpha_k \frac{X_k^2 s^k}{\phi(X_k s^k)}$$

from Theorem 5.1.19 and substitution of the formula for θ_k

$$c^T x^{k+1} - c^* = c^T x^k - c^* - \alpha_k \frac{\|X_k s^k\|^2}{\phi(X_k s^k)}.$$

Thus we have,

$$\begin{aligned} v_N^{k+1} - v_N^k &= \frac{v_N^k - \frac{\alpha_k V_{k,N} u_N^k}{\phi(u^k)}}{1 - \frac{\alpha_k \|u^k\|^2}{\phi(u^k)}} - v_N^k \\ &= \frac{\frac{\alpha_k \|u^k\|^2}{\phi(u^k)}}{1 - \frac{\alpha_k \|u^k\|^2}{\phi(u^k)}}(v_N^k - V_{N,k}\frac{u_N^k}{\|u^k\|^2}) \end{aligned}$$

and we are done. ∎

Large Step Accelerated Affine Scaling Method

In this section we present a step-size selection strategy that asymptotically integrates a predictor-corrector strategy for computing the analytic center of $\mathcal{P}_N$. As we will subsequently see, this method has the advantage of attaining a quadratic rate of convergence. The accelerated method follows:

Step 0 Let x^0 be an interior point solution, $0 < \alpha < 1$ (normally between 0.95 and 0.99) and let $k = 0$.

5.3 ACCELERATED AFFINE SCALING METHOD

Step 1 Tentative Solution to the Dual:
$$y^k = (AX_k^2 A^T)^{-1} AX_k^2 c$$

Step 2 Tentative Dual Slack:
$$s^k = c - A^T y^k$$

If $s^k \leq 0$ then STOP. The solution is unbounded. Otherwise, define
$$d^k = X_k s^k$$

Step 3 Min-Ratio test:
$$\theta_k = \min\{\frac{\|d^k\|}{d_j^k} : s_j^k > 0\}$$
$$= \frac{\|d^k\|}{\phi(d^k)}.$$

Also, if $\theta_k = 1$ set $\alpha_k = 1$, and go to Step 7.

Step 4 Tentative Non-Basic set:
$$N_k = \{j : x_j^k \leq \sqrt{\phi(d^k)}\}$$
$$\gamma_k = |e^T d_{N_k}^k|$$

Step 5 Estimate of Newton Step:
$$h_{N_k}^k = \frac{x_{N_k}^k}{\gamma_k} - X_{k,N_k} \frac{d_{N_k}^k}{\|d^k\|^2}$$
$$\epsilon_k = \|h_{N_k}^k\|$$

Step 6 Step-size calculation:

1. Normal Step: If $\gamma_k \geq 1$ then set
$$\alpha_k = \alpha$$

2. Predictor Step: If $\gamma_k < 1$ and $\epsilon_k^{0.75} \leq \gamma_k$ then set:
$$\alpha_k = 1 - \max\{\gamma_k, \epsilon_k^{0.50}\}$$

3. Corrector Step: Otherwise, set

$$\alpha_k = \min\{\frac{\gamma_k}{2\theta_k\|d^k\|}, \frac{2}{3}\}$$

Step 7 Next Interior Point:

$$x^{k+1} = x^k - \alpha_k\theta_k \frac{X_k^2 s^k}{\|X_k s^k\|}$$

Step 8 Iterative Step: If $x_j^{k+1} = 0$ for some j, then STOP. x^{k+1} is an optimal solution to the primal, y^k the optimum solution to the dual. Otherwise set $k = k + 1$ and go to step 1.

Some comments are in order here. From Theorem 5.1.19, $d^k \to 0$. As $x_j^* > 0$ for each $j \in B$, and $x_j^* = 0$ for each $j \in N$, there is a sufficiently large $L \geq 1$ such that, in Step 4, $N_k = N$, for all $k \geq L$.

In this acceleration procedure, an estimate h_N^k of the Newton step at $v_N^k \in \mathcal{P}_N$ is used to conclude how close it is to the analytic center v_N^*. In case it is very close to the center, more precisely, if $\|h_N^k\| = O(c^T x^k - c^*)^2$, α_k is assigned a value close to 1. Another requirement on this value is that the next iterate v_N^{k+1} be not *too far* from v_N^*, thus equivalently, not *too close* to the boundary of $\mathcal{P}_N$; more precisely $\|v_N^{k+1} - v_N^*\| = O(c^T x^k - c^*)^{\frac{1}{2}}$. This we call the *predictor* step. After a predictor step, the connection with Newton's method is used to bring the iterates close to the center, which we call the *corrector* step. This is done by assigning the step size the moderate value defined in Step 6, part 3. By assuring that $\alpha_k \leq \frac{2}{3}$, global convergence is maintained even if no predictor steps are taken. In addition

$$\frac{\gamma_k}{2\theta_k\|d^k\|} \to \frac{1}{2},$$

and, as we will see later, this makes the affine scaling step sufficiently close to the Newton step to make the information of Newton's step valuable within the affine scaling framework.

We now prove the important properties of the accelerated procedure.

5.3.9 Lemma There exists an $L \geq 1$ such that for all $k \geq L$, $N_k = N$.

Proof: From Theorem 5.1.19, $\phi(d^k) \to 0$, and Theorem 5.1.23 and Corollary 5.1.31 x_N^k and $X_k s^k$ go to zero at the same rate as $c^T x^k - c^*$. Thus, there is an $L \geq 1$ such that for all $k \geq L$, $x_j^k \geq \phi(d^k)$ for each $j \in B$, and $x_j^k \leq \sqrt{\|X_k s^k\|}$ for each $j \in N$. Thus we are done. ∎

We now prove a technical lemma related to the Newton step.

5.3 ACCELERATED AFFINE SCALING METHOD

5.3.10 Lemma Let L be as in Lemma 5.3.9. There is an $\hat{L} \geq L$ and a $\beta > 0$ such that for all $k \geq \hat{L}$

1. $\frac{c^T x^{k+1} - c^*}{c^T x^k - c^*} = 1 - \alpha_k \delta(u_N^k)$ where $\delta(u_N^k) = \frac{\|u^k\|^2}{\phi(u_N^k)}$.

2. $1 - \|t_N^k\| - \gamma_k \leq \delta(u_N^k) \leq 1$ where $0 \leq \gamma_k \leq \beta (c^T x^k - c^*)^2$.

Proof: From Equation (5.34) and Theorem 5.1.30 Part 2(e), there is an $\hat{L} \geq L$ such that Part 1 holds. Since Part 1 holds for every $\alpha_k \leq 1$, we obtain the upper bound of Part 2 by considering $\alpha_k = 1$. Also, from Theorem 5.3.7,

$$\frac{\phi(u_N^k)}{\|u^k\|^2} = \phi\left(\frac{u_N^k}{\|u^k\|^2}\right)$$
$$= \phi(e + t_N^k - \Delta^k)$$
$$\leq 1 + \|t_N^k\| + \gamma_k$$

Setting $0 \leq \gamma_k = \|\Delta^k\|$, and noting that for every real number a, $\frac{1}{1+a} \geq 1 - a$, we obtain

$$\delta(u_N^k) = \frac{\|u^k\|^2}{\phi(u_N^k)}$$
$$\geq \frac{1}{1 + \|t_N^k\| + \gamma_k}$$
$$\geq 1 - \|t_N^k\| - \gamma_k$$

and we are done by Theorem 5.3.7. ∎

The next result establishes the quality of the estimator h_N^k at Step 5, used in the estimation of the Newton step Δv_N^k.

5.3.11 Theorem Let L be as in Lemma 5.3.9. There is a $\hat{L} \geq L$ and a $\rho > 0$ such that for all $k \geq \hat{L}$

$$\|h_N^k - \Delta v_N^k\| \leq \rho (c^T x^k - c^*)^2$$

Proof: From Theorem 5.1.30 Part 2(a) there is a $\hat{L} \geq L$ such that for some $\theta > 0$ and all $k \geq \hat{L}$

$$e^T d_N^k = (c^T x^k - c^*)(1 + \delta_k)$$

where $|\delta_k| \leq \theta (c^T x^k - c^*)^2$. Thus, there is an $\hat{L} \geq L$ such that $e^T d_N^k > 0$ for all $k \geq \hat{L}$. From Step 5 and definitions of u^k and v^k, we obtain

$$h_N^k = \frac{v_N^k}{e^T u_N^k} - V_{k,N} \frac{u_N^k}{\|u^k\|^2}.$$

Theorem 5.1.30 Part2(a) and Theorem 5.3.8 give

$$h_N^k - \Delta v_N^k = \frac{\delta_k v_N^k}{1+\delta_k} - V_{k,N}\Delta^k$$

and the result now follows from Theorem 5.1.30 Part 1 and Theorem 5.3.7. ∎

5.3.12 Theorem x^* lies in the relative interior of the optimal primal face.

Proof: Let L be as in Lemma 5.3.9. In case the predictor step is only taken a finite number of times, there is an $\hat{L} \geq L$ such that for all $k \geq \hat{L}$, $\alpha_k \leq \frac{2}{3}$. Thus the result follows from Theorem 5.2.8.

Now, assume the predictor step is taken an infinite number of times, and let K be the subsequence of such iterates. Let $k \in K$. Then, from Steps 4, 5 and 6 Part 2,

$$\|h_N^k\| \leq (e^T X_{k,N} s_N^k)^{\frac{4}{3}} \leq (\|X_k s^k\|)^{\frac{4}{3}}$$

and, from Theorem 5.1.19

$$h_N^k \longrightarrow 0 \text{ for } k \in K.$$

Also, from Theorem 5.3.11, $\Delta v_N^k \to 0$, for $k \in K$ as

$$\|\Delta v_N^k\| \leq \|h_N^k\| + \|\Delta v_N^k - h_N^k\|.$$

But, $v_N^* \in \mathcal{P}_N$ is the only vector for which $\Delta v_N^* = 0$. Thus

$$v_N^k \longrightarrow v_N^* \text{ for } k \in K.$$

As $\Delta v_N^k \to 0$, $t_N^k \to 0$ for $k \in K$. Thus from Theorem 5.3.7

$$\frac{u_N^k}{\|u^k\|^2} \longrightarrow e \text{ for } k \in K,$$

and, from Theorem 5.1.30 Part 2(b),

$$\|u^k\|^2 \longrightarrow \frac{1}{q} \text{ for } k \in K.$$

Thus

$$u_N^k \longrightarrow \frac{1}{q} e \text{ for } k \in K.$$

5.3 ACCELERATED AFFINE SCALING METHOD

The sequences $\{y^k\}$ and $\{s^k\}$ are bounded and so, on some subsequence $K' \subset K$, they converge, say to (y^*, s^*). So

$$V_{*,N} s_N^* = \frac{1}{q} e.$$

Also, as v_N^* is the analytic center of $\mathcal{P}_N$, from Theorem 5.3.3, (y^*, s^*) is an analytic center of $\mathcal{D} \cap \{s : s_N \geq 0\}$, and we are done. ∎

5.3.13 Lemma Let $t_N^k \to 0$ on some subsequence K. Then, on K

$$\rho_k = \frac{|e^T u_N^k| \phi(u_N^k)}{2\|u^k\|^2} \to \frac{1}{2}$$

Proof: From Theorem 5.1.30 Part 2(a) and Lemma 5.3.10, $2\rho_k \geq e^T u_N^k = 1 - |\delta_k|$. Again, from Lemma 5.3.10

$$\rho_k \leq \frac{1 + |\delta_k|}{2} (1 + \|t_N^k\| + \|\Delta^k\|)$$

and we get the result as δ_k and $\|\Delta^k\|$ go to zero. ∎

On Predictor and Corrector Steps

In this section we will establish the asymptotic rate of convergence of the sequence $\{c^T x^k - c^*\}$ generated by the accelerated affine scaling method. We will show that this sequence converges three-step quadratically to 0. Because of Theorem 5.3.12, the analytic center v_N^* of $\mathcal{P}_N$ exists, and thus Newton's method, when initiated close to the center, will converge to it. Thus, there is a sufficiently large $\hat{L} \geq 1$ for which the following properties hold:

For all $k \geq \hat{L}$,

1. $v_j^* \geq (c^T x^k - c^*)^{0.25}$ for all $j \in N$.

2. If θ is the constant of Theorem 2.7.4 related to Newton's method, then $(c^T x^k - c^*)^{0.25} \leq \theta$.

3. The conditions of Theorem 2.7.3 related to Newton's method; and, Lemma 5.3.9, are satisfied.

Such an $\hat{L}$ exists as $v_j^* > 0$ for each $j \in N$, $\theta > 0$ and $c^T x^k - c^* \to 0$. We can then prove the following straight forward lemmas:

5.3.14 Lemma There is an $L \geq \hat{L}$ such that for all $k \geq L$
$$0.50(c^T x^k - c^*) \leq e^T d_N^k \leq 1.5(c^T x^k - x^*).$$

Proof: From Theorem 5.1.30 Part 2(a), there is a $L' \geq \hat{L}$ and a $\theta > 0$ such that for all $k \geq L'$
$$e^T d_N^k = (1 + \delta_k)(c^T x^k - c^*)$$
where $|\delta_k| \leq \theta(c^T x^k - c^*)^2$. Thus, there is a $L \geq L'$ such that $|\delta_k| \leq 0.50$ and we are done. ∎

5.3.15 Lemma Let $\|v_N^k - v_N^*\| \leq \beta(c^T x^k - c^*)^p$ for some $\beta > 0$ and $p > 0.25$. There is an $L \geq \hat{L}$ and a $\theta > 0$ such that for all $k \geq L$
$$\|t_N^k\| \leq \theta(c^T x^k - c^*)^{p-0.25}.$$

Proof: Consider
$$v_j^k \geq v_j^* - |v_j^k - v_j^*| \geq (c^T x^k - c^*)^{0.25} - \beta(c^T x^k - c^*)^p$$
and thus there is a $L \geq \hat{L}$ such that for all $k \geq L$
$$v_j^k \geq 0.5(c^T x^k - c^*)^{0.25}.$$
Thus, $\|V_{N,k}^{-1}\| \leq \frac{2}{(c^T x^k - c^*)^{0.25}}$, and since $t_N^k = V_{N,k}^{-1} \Delta v_N^k$, we obtain our result from Theorem 2.7.4. ∎

We are now ready to investigate the predictor step and the corrector step of the accelerated method. The next theorem investigates the predictor step, Step 6, part 2 of the accelerated method.

5.3.16 Theorem Assume that for some $k \geq \hat{L}$ and $\beta > 0$, $\|v_N^k - v_N^*\| \leq \beta(c^T x^k - c^*)^{2p}$ for some $0.50 < p \leq 1$ and the predictor step is taken. Then

1. There is a $\tilde{\theta} > 0$ such that
$$0.50(c^T x^k - c^*)^2 \leq c^T x^{k+1} - c^* \leq \tilde{\theta}(c^T x^k - c^*)^{2p}.$$

2. There is a $\hat{\theta} > 0$ such that
$$\|v_N^{k+1} - v_N^*\| \leq \hat{\theta}(c^T x^{k+1} - c^*)^{\frac{p}{2}}.$$

5.3 ACCELERATED AFFINE SCALING METHOD

Proof: To see (1), note that from Step 6, part 2; Lemmas 5.3.14, 5.3.15, and 5.3.10, we get

$$\begin{aligned}
0.50(c^T x^k - c^*) &\leq e^T d_N^k \\
&\leq 1 - \alpha_k \\
&\leq 1 - \alpha_k \frac{\|u^k\|^2}{\phi(u_N^k)} \\
&= \frac{c^T x^{k+1} - c^*}{c^T x^k - c^*} \\
&\leq 1 - (1 - \max\{\epsilon_k^{0.50}, \gamma_k\})(1 - 2\|t_N^k\|) \\
&\leq \max\{\epsilon_k^{0.50}, \gamma_k\} + 2\|t_N^k\| \\
&\leq \tilde{\theta}(c^T x^k - c^*)^{1+p}.
\end{aligned}$$

To see (2), from Theorem 5.3.8

$$v_N^{k+1} - v_N^* = v_N^k + \Delta v_N^k - v_N^* + \frac{2\alpha_k \delta(u^k) - 1}{1 - \alpha_k \delta(u^k)}(v_N^k - V_{N,k}\frac{u_N^k}{\|u^k\|^2}) - V_{N,k}\Delta^k.$$

From Lemma 5.3.10, Theorems 5.3.11, 5.3.7, and some simple algebra, we obtain

$$\|v_N^{k+1} - v_N^*\| \leq \|v_N^k + \Delta v_N^k - v_N^*\| + \frac{2\alpha_k - 1}{1 - \alpha_k}\|h_N^k\| + \beta(c^T x^k - c^*)^2.$$

Choosing $k \geq \hat{L}$ and substituting the formula for α_k from Step 6, Part 2; Part 1 of this theorem, and the Theorem 2.7.3

$$\begin{aligned}
\|v_N^{k+1} - v_N^*\| &\leq \delta_1 \|v_N^k - v_N^*\|^2 + \theta(c^T x^k - c^*)^p \\
&\leq \hat{\beta}(c^T x^k - c^*)^p \\
&\leq \hat{\theta}(c^T x^{k+1} - c^*)^{\frac{p}{2}}
\end{aligned}$$

where $\delta_1 > 0$, $\hat{\theta} > 0$, and $\hat{\beta} > 0$ are appropriate constants. We have used

$$\begin{aligned}
\frac{2\alpha_k - 1}{1 - \alpha_k}\epsilon_k &\leq \frac{1 - 2\max\{\epsilon_k^{0.50}, \gamma_k\}}{\max\{\epsilon_k^{0.50}, \gamma_k\}}\epsilon_k \\
&\leq \epsilon_k^{0.50}.
\end{aligned}$$

∎

We now investigate the corrector step.

5.3.17 Theorem There is an $L \geq \hat{L}$ and $\alpha > 0$ such that for some $k \geq L$, assume $\|v_N^k - v_N^*\| \leq \alpha(c^T x^k - c^*)^p$ where $0.25 < p \leq 0.50$.

1. Let $p = 0.50$. Then two iterations of the corrector step, Step 6, Part 3, can be taken, and for some $\theta > 0$,
$$\|v_N^{k+2} - v_N^*\| \leq \theta(c^T x^{k+2} - c^*)^2.$$

2. Let $0.25 < p < 0.50$. Then three iterations of the corrector step can be taken, and for some $\theta > 0$,
$$\|v_N^{k+3} - v_N^*\| \leq \theta(c^T x^{k+3} - c^*)^2.$$

Proof: From Lemma 5.3.15, we note that for all $k \geq \hat{L}$
$$\|t_N^k\| \leq \theta(c^T x^k - c^*)^{p-0.25}.$$

Let $L \geq \hat{L}$ be such that for all $k \geq L$, $\theta(c^T x^k - c^*)^{p-0.25} \leq 0.25$. Now, consider $k \geq L$. Then from Lemma 5.3.10

$$\rho_k = \frac{\gamma_k}{2\theta_k \|d^k\|} = \frac{|e^T u_N^k| \phi(u_N^k)}{2\|u^k\|^2}$$
$$\leq \frac{1}{2}(1 + \delta_k)(1 + \|t_N^k\| + \gamma_k)$$
$$< \frac{2}{3}$$

where δ_k and γ_k are positive and of the order $O(c^T x^k - c^*)^2$.

Thus, during the corrector step, $\alpha_k = \rho_k$, and thus, from Theorem 5.3.10 Part 1 we get
$$\frac{c^T x^{k+1} - c^*}{c^T x^k - c^*} = 1 - \frac{e^T u_N^k}{2}$$

and from Theorem 5.1.30, Part 2(a), for all $k \geq L$

$$0.25(c^T x^k - c^*) \leq c^T x^{k+1} - c^* \leq 0.75(c^T x^k - c^*). \tag{5.49}$$

For the first corrector step, after substituting α_k Theorem 5.3.8, we get

$$v_N^{k+1} = v_N^k + \frac{e^T u_N^k}{2 - e^T u_N^k}(v_N^k - V_{k,N} \frac{u_N^k}{\|u^k\|^2}).$$

5.3 ACCELERATED AFFINE SCALING METHOD

Thus, by simple algebra and substitutions we get:

$$\begin{aligned}
\|v_N^{k+1} - v_N^*\| &\leq \|v_N^k + \Delta v_N^k - v_N^*\| + \frac{2|\delta_k|}{1-|\delta_k|}\|V_{k,N} t_N^k\| + \\
&\qquad \frac{1+|\delta_k|}{1-|\delta_k|}\|V_{k,N}\Delta^k\| \\
&\leq \rho\|v_N^k - v_N^*\|^2 + \beta(c^T x^k - c^*)^2 \\
&\leq \rho\alpha^2(c^T x^k - c^*)^{2p} + \beta(c^T x^k - c^*)^2 \qquad (5.50) \\
&\leq \rho_1(c^T x^k - c^*)^{2p} \\
&\leq (4)^{2p}\rho(c^T x^{k+1} - c^*)^{2p}.
\end{aligned}$$

Applying the same argument as above to one more corrector step, we obtain

$$\|v_N^{k+2} - v_N^*\| \leq \tilde{\rho}(c^T x^{k+2} - c^*)^{4p} \qquad (5.51)$$

Now, if $p = 0.50$, then $4p = 2$ and we have part (1). Otherwise after one more corrector step, we obtain

$$\|v_N^{k+3} - v_N^*\| \leq \hat{\rho}(c^T x^{k+3} - c^*)^{8p} + \beta(c^T x^{k+2} - c^*)^2. \qquad (5.52)$$

Since $8p > 2$ for $0.25 < p < 0.50$, we obtain part (2) and we are done. ∎

Rate of Convergence of the Accelerated Method

We are now ready to prove the main theorem.

5.3.18 Theorem Let the sequences $\{x^k\}$, $\{y^k\}$ and $\{s^k\}$ be generated by the accelerated affine scaling method. Then

1. $x^k \longrightarrow x^*$
2. $y^k \longrightarrow y^*$
3. $s^k \longrightarrow s^*$

where x^* lies in the relative interior of the optimal face of the primal, (y^*, s^*) is the analytic center of the optimal face of the dual. In addition, asymptotically, the sequence $\{c^T x^k - c^T x^*\}$ converges three-step quadratically to zero; i.e., there is an $L \geq 1$, $\theta > 0$ and $K = \{L, L+3, L+6, \cdots\}$ such that the subsequence $\{c^T x^k - c^T x^*\}_{k \in K}$ converges quadratically to zero, or

$$c^T x^{(k+1)'} - c^T x^* \leq \theta(c^T x^{k'} - c^T x^*)^2 \text{ for } k' \in K.$$

Proof: Using Equation (5.49) and Lemma 5.3.14 we note that, during corrector steps, asymptotically the sequence $\{\gamma_k\}$ decreases linearly in $c^T x^k - c^*$. Also, using equations (5.51), (5.52), Theorems 2.7.4 and 5.3.11 asymptotically, if no predictor step is taken, the sequence $\{\epsilon_k\}$ decreases at least as fast as $O(c^T x^k - c^*)^2$. Thus, for some large k, we must have

$$\epsilon_k^{0.75} \leq \gamma_k \tag{5.53}$$

and a predictor step must be taken. The first time the condition of equation (5.53) occurs, if $\epsilon_k > \gamma_k^2$ then we enter the predictor step with $1 > p > 0.50$; and, after the predictor step we enter the corrector step with $p < 0.50$ and thus three steps may be required to guarantee, by Lemma 2.7.4 and proposition 5.3.11 that we enter the next predictor step with $p = 1$. Then every subsequent corrector step will be entered with $p = 0.50$, and will thus require at most two corrector iterations to satisfy the conditions for the predictor step, i.e, enter the predictor step with $p = 1$. Thus, if during the first corrector step, three corrector iterations are taken, from Theorem 5.3.17, we guarantee that the predictor step is initiated with $p = 1$, and the three step quadratic convergence of $\{c^T x^k - c^*\}$ follows.

It is readily confirmed from Theorems 5.3.16 and 5.3.17 that

$$\|v_N^k - v_N^*\| \longrightarrow 0$$

and thus $v_N^k \longrightarrow v_N^*$. But this can only happen, by Theorem 5.3.3, if (y^k, s^k) converges to the analytic center of $D \cap \{s : s_N \geq 0\}$ and we are done. ∎

5.4 Primal Power Affine Scaling Method

In this section we consider a variant of the primal affine scaling method based on the following modification of ellipsoidal approximating problem, 5.1.2: let $r > \frac{1}{2}$ an arbitrary real number. Given an interior point $x^0 > 0$, define the following approximating set

$$E = \{x : \|X_0^{-r}(x - x^0)\| \leq \rho\}$$

Note that when $\rho \leq 1$, solution of this problem is interior to the non-negative orthant. Variant thus generated is called the large step power affine scaling method:

5.4.1 Large Step Primal Power Method

Step 0 Let x^0 be an interior point solution, $0 < \alpha < 1$. Let $k = 0$.

5.4 POWER AFFINE SCALING METHOD

Step 1 Tentative Solution to the Dual:

$$y^k = (AX_k^{2r}A^T)^{-1}AX_k^{2r}c$$

Step 2 Tentative Dual Slack:

$$s^k = c - A^T y^k$$

If $s^k \leq 0$ then stop. The solution is unbounded.

Step 3 Min Ratio Test:

$$\begin{aligned}\theta_k &= \min\{\frac{\|X_k^{2r-1}s^k\|}{(x_j^k)^{2r-1}s_j^k} : s_j^k > 0\} \\ &= \frac{\|X_k^{2r-1}s^k\|}{\phi(X_k^{2r-1}s^k)}.\end{aligned}$$

If $\theta_k = 1$, set $\alpha = 1$.

Step 4 Next Interior Point:

$$x^{k+1} = x^k - \alpha\theta_k \frac{X_k^{2r}s^k}{\|X_k^r s^k\|}$$

Step 4 Iterative Step: If $x_j^{k+1} = 0$ for some j, then STOP. x^{k+1} is an optimal solution. Otherwise set $k = k+1$ and go to step 1.

When $r = 1$, the above method reduces to the primal affine scaling method of Section 5.1.9. We will analyze the power affine method when $r > \frac{1}{2}$. It is defined for $0 < r \leq 0.50$, but analysis of its convergence is difficult. It appears that the method fails when $r < \frac{1}{2}$ and its convergence behavior is not known when $r = \frac{1}{2}$.ere we will analysis the method for $r \neq 1$, and $r > \frac{1}{2}$. Although the power method is a generalization of affine scaling method of Section 5.1.9, its analysis does not specialize to that method when applied to the case when $r = 1$.

We assume the following about the given linear program:

1. The matrix A is $m \times n$ and has full rank m.

2. The linear program has an interior solution.

3. The objective function is not constant on the feasible region.

Finite Convergence of the Power Method

We show here that if the method stops at steps 2 and 5, then the linear program is solved in finite number of iterations, and objective function decreases monotonically along the generated sequence. We do this in the next theorem:

5.4.2 Theorem $\{c^T x^k\}$ is strictly decreasing. Also exactly one of the following holds:

(a) The algorithm stops at Step 2. Then the linear program has an unbounded solution, i.e., its dual is infeasible.

(b) The algorithm stops at Step 5. Then x^{k+1} is an optimal solution of the primal and y^k is an optimal solution of the dual. These solutions also satisfy the strict complementarity condition.

(c) The sequence $\{x^k\}$ is infinite and $\{c^T x^k\}$ is not bounded below. Then the linear program has an unbounded solution.

(d) The sequence $\{x^k\}$ is infinite and $\{c^T x^k\}$ is bounded below. Then $\{c^T x^k\}$ converges to, say c^*.

Proof: To see the first part, from Step 4, we note that

$$c^T x^{k+1} = c^T x^k - \alpha \theta_k \frac{c^T X_k^{2r} s^k}{\|X_k^{2r-1} s^k\|}.$$

As can be readily established from the definitions, $x^k > 0$ and $\theta_k \geq 1$. Also,

$$\begin{aligned} c^T X_k^{2r} s^k &= c^T X_k^{2r}(c - A^T y^k) \\ &= c^T X_k^r (I - X_k^r A^T (A X_k^{2r} A^T)^{-1} A X_k^r) X_k^r c \\ &= \|P_k X_k^r c\|^2 \end{aligned}$$

where $P_k = I - X_k^r A^T (A X_k^{2r} A^T)^{-1} A X_k^r$ is the projection matrix into the null space $\mathcal{N}(A X_k^r)$ of the matrix $A X_k^r$. Now, by a simple calculation, we see that $\|P_k X_k^r c\| = \|X_k^r s^k\|$ and thus we have

$$c^T x^{k+1} = c^T x^k - \alpha \theta_k \frac{\|X_k^r s^k\|^2}{\|X_k^{2r-1} s^k\|}. \tag{5.54}$$

Since the objective function is not constant on the feasible region, the subtracted term in the above formula is non-zero.

5.4 POWER AFFINE SCALING METHOD

To see Part (a), we note that for $s^k \leq 0$, x^{k+1} remains strictly positive for every $\alpha > 0$, and thus $c^T x^{k+1} \to -\infty$ as $\alpha \to \infty$.

To see Part (b), let $x_l^{k+1} = 0$. Then, from Step 4 we see that

$$0 = x_l^{k+1} = x_l^k - \alpha \frac{(x_l^k)^{2r} s_l^k}{\phi(X_k^{2r-1} s^k)}$$

and thus $1 = \alpha \theta_k \frac{(x_l^k)^{2r-1} s_l^k}{\phi(X_k^{2r-1} s^k)}$. Since $x^{k+1} > 0$ for each $\alpha < 0$, $\alpha = 1$. But, from Step 3, $\alpha = 1$ if and only if $\theta_k = 1$, and thus $(x_l^k)^{2r-1} s_l^k = \phi(X_k^{2r-1} s^k) > 0$ so $s_l^k > 0$. Thus, from Step 3, $(x_l^k)^{2r-1} s_l^k = \|X_k^{2r-1} s^k\|$. Hence, for every $j \neq l$ $(x_j^k)^{2r-1} s_j^k = 0$, and so $s_j^k = 0$ and $x_j^{k+1} = x_j^k > 0$. Thus $s^k \geq 0$ and $x^{k+1} \geq 0$ satisfy the conditions of the complementary slackness theorem.

Part (c) follows from the monotonicity of $\{c^T x^k\}$, and Part (d) from the fact that every bounded monotone sequence converges. ∎

A consequence of Theorem 5.4.2 is that we need to analyze the case when sequences generated are infinite, and $\{c^T x^k\}$ is bounded below and converges. We will henceforth assume that this is the case, and now establish some important properties of the sequences $\{x^k\}$, $\{y^k\}$, $\{s^k\}$ and $\{X_k s^k\}$.

Convergence Properties of Sequences

We will now show convergence of sequences $\{x^k\}$, $\{y^k\}$, $\{s^k\}$ and $\{X_k s^k\}$, and establish some of their important properties. For this purpose consider the transformed ellipsoidal approximating problem (like the problem (5.21)), by setting $p = x^k - x$:

$$\begin{aligned} \text{maximize} \quad & c^T p \\ Ap &= 0 \\ \|X_k^{-r} p\| &\leq 1. \end{aligned} \quad (5.55)$$

and it is readily seen, as in the case of the affine scaling method, that the solution to this problem is

$$\hat{p}^k = \frac{X_k^{2r} s^k}{\|X_k^r s^k\|}. \quad (5.56)$$

We are now ready to prove the main convergence result about these sequences.

5.4.3 Convergence of Primal Sequence

Let $\{c^T x^k\}$ be bounded below, and converge to c^*. Then

(a) The sequence $\{x^k\}$ converges, to say x^*.

(b) There is a $\rho > 0$ such that for every $k = 1, 2, \cdots$
$$\frac{c^T x^k - c^*}{\|x^k - x^*\|} \geq \rho.$$

(c) The sequences $\{y^k\}$ and $\{s^k\}$ are bounded.

(d) For each $r \geq \frac{1}{2}$, $X_k s^k \longrightarrow 0$.

Proof: Using the result of Theorem 5.1.17 (as in Theorem 5.1.20), there is a $\rho_1 > 0$ such that $c^T \hat{p}^k \geq \rho_1 \|\hat{p}^k\|$, where $\hat{p}^k$ is the solution to the problem (5.55). To see Part (a), define $\gamma_k > 0$ such that $x^{k+1} - x^k = \gamma_k \hat{p}^k$. Thus, we obtain

$$\infty > c^T x^1 - c^* = \sum_{k=1}^{\infty} c^T(x^k - x^{k+1}) \geq \rho_1 \sum_{k=1}^{\infty} \gamma_k \|\hat{p}^k\| = \rho_1 \sum_{k=1}^{\infty} \|x^{k+1} - x^k\|,$$

and thus the sequence is Cauchy, and converges, say to x^*. To see Part (b), using the triangular inequality we obtain the following:

$$c^T x^k - c^* = \sum_{j=0}^{\infty} c^T(x^{k+j} - x^{k+j+1}) \geq \rho_1 \sum_{j=0}^{\infty} \|x^{k+j+1} - x^{k+j}\| \geq \rho_1 \|x^k - x^*\|.$$

Part (c) follows from the Corollary 5.1.12. To see Part (d), note that $\{c^T x^k\}$ converges, and thus from relation (5.54), we note that

$$\alpha \theta_k \frac{\|X_k^r s^k\|^2}{\|X_k^{2r-1} s^k\|} \longrightarrow 0,$$

where $\alpha > 0$ and $\theta_k \geq 1$. Since $\{x^k\}$ converges and $\{y^k\}$ and $\{s^k\}$ are bounded, when $r \geq \frac{1}{2}$, the denominator of the above expression is bounded, thus

$$X_k^r s^k \longrightarrow 0.$$

But this is only possible if $X_k s^k \longrightarrow 0$, and we are done. ∎

We know that the sequence $\{x^k\}$ converges to a solution x^*. It is clear that this limit point belongs to the boundary of the linear programming polyhedron. Define the sets:

$$\begin{aligned} B &= \{j : x_j^* > 0\} \\ N &= \{j : x_j^* = 0\}. \end{aligned} \qquad (5.57)$$

5.4 POWER AFFINE SCALING METHOD

We now show with the additional assumption that primal linear program is non-degenerate, the sequences $\{y^k\}$ and $\{s^k\}$ also converge, and that the limit points of primal and dual sequences are optimal for their respective problems.

5.4.4 Convergence with Primal Non-Degeneracy Let assumptions (1) - (3) hold, primal be non-degenerate and $0 < \alpha < 1$. Then, there exist vectors x^*, y^* and s^* such that

(a) $x^k \longrightarrow x^*$,

(b) $y^k \longrightarrow y^*$,

(c) $s^k \longrightarrow s^*$,

where x^* is an optimal solution of primal, (y^*, s^*) is an optimal solution of the dual linear program.

Proof: Using Part (c) of Theorem 5.4.3, let (y^*, s^*) be a cluster point of the sequence $\{(y^k, s^k)\}$. From Part (d) of Theorem 5.4.3, $s_B^* = 0$. Thus

$$A_B^T y^* = c_B.$$

From the non-degeneracy assumption, A_B has full row rank m, and thus the above system has at most one solution. But each cluster point y^* of $\{y^k\}$ solves this system, thus the sequence has only one cluster point y^*, and so

$$y^k \longrightarrow y^*$$

and thus $s^k \longrightarrow s^*$. Now assume that for some $j \in N$, $s_j^* < 0$. Then there is an $L \geq 1$ such that for all $k \geq L$, $s_j^k < 0$. Thus, from Step 4

$$\begin{aligned} x_j^{k+1} &= x_j^k - \alpha \frac{(x_j^k)^{2r} s_j^k}{\phi(X_k^{2r-1} s^k)} \\ &> x_j^k \end{aligned}$$

so $x_j^k \not\to 0$, and we have a contradiction. Hence $s^* \geq 0$ and so (y^*, s^*) is dual feasible, and the theorem follows from the complementary slackness theorem. ∎

Convergence without Non-Degeneracy Assumption

To investigate convergence to optimality of sequences under degeneracy, we will exploit a connection between Newton step for computing the 'power' center of a polyhedron, and the affine scaling step in that polyhedron. This allows analysis of choice of step size using a merit function. This function replaces the local potential function used in the analysis of affine scaling method. Before we do this, we introduce several important sequences:

$$\begin{aligned} v^k &= \frac{x^k - x^*}{c^T x^k - c^*} \\ \tilde{u}^k &= \frac{X_k s^k}{c^T x^k - c^*} \\ u^k &= \frac{X_k^r s^k}{(c^T x^k - c^*)^r} \\ p^k &= X_k^{2r} s^k. \end{aligned} \qquad (5.58)$$

Define the polyhedron

$$F_\mathcal{D} = \{(y, s) : A^T y + s = c, s_B = 0\}. \qquad (5.59)$$

It is easily seen that every cluster point of the sequence $\{(y^k, s^k)\}$ belongs to this polyhedron. Thus, it is non-empty. Let $(\bar{y}, \bar{s}) \in F_\mathcal{D}$ be arbitrary, but fixed.

5.4.5 Properties of Sequences The sequences $\{v^k\}$ and $\{u_N^k\}$ are bounded. Also:

(a) There exist $\rho_1 > 0$, $\rho_2 > 0$, $\rho_3 > 0$ and $\rho_4 > 0$ such that for every $k = 1, 2, \cdots$

  (i) $Ap^k = 0$.
  (ii) $c^T x^k - c^* = \bar{s}_N x_N^k$.
  (iii) $\rho_1(c^T x^k - c^*) \leq \|x_N^k\| \leq \rho_2(c^T x^k - c^*)$.
  (iv) $\|X_k^r s^k\| \leq \rho_3 \phi(x_N^k)^r$.
  (v) $\|p_B^k\| \leq \rho_4 \phi(x_N^k)^r \|X_{k,N}^r s_N^k\|$.

(b) There exist constants $\rho_5 > 0$ and $\rho_6 > 0$ and an $L \geq 1$ such that for every $k \geq L$

  (i) $\|u_B^k\| \leq \rho_5 (c^T x^k - c^*)^r \|u_N^k\|$.
  (ii) $\|s_B^k\| \leq \rho_6 (c^T x^k - c^*)^{2r}$.
  (iii) $\phi(V_k^{r-1} u^k) = \phi(V_{k,N}^{r-1} u_N^k)$.

5.4 POWER AFFINE SCALING METHOD

(iv) $\frac{c^T x^{k+1} - c^*}{c^T x^k - c^*} = 1 - \alpha\delta(u_N^k)$ where $\delta(u_N^k) = \frac{\|u^k\|^2}{\phi(V_{k,N}^{r-1} u_N^k)}$.

Proof: The boundedness of sequences follows from definitions and 5.4.3 Parts (b) and (c). a(i) and (ii) follow readily from definitions. The lower bound of a(iii) follows from 5.4.3 Part (b) and a(ii).

To see a(iv), using argument from the proof of 5.4.2 and a(i), we obtain:

$$\begin{aligned}
\|X_k^r s^k\|^2 &= c^T p^k \\
&= (A^T \bar{y} + \bar{s})^T p^k \\
&= \bar{s}_N^T p_N^k \\
&= (X_{k,N}^r \bar{s}_N)^T (X_{k,N}^{-r} p_N^k) \\
&\leq \|\bar{s}_N\| \phi(x_N^k)^r \|X_{k,N}^r s_N^k\|
\end{aligned}$$

and the part follows with $\rho_3 = \|\bar{s}_N\|$. To see a(v), we note that for some $\rho > 0$

$$\|p_N^k\| \leq \|p^k\| \leq \rho c^T p^k = \rho(\bar{s}_N)^T p_N^k \leq \rho \|\bar{s}_N\| \phi(x_N^k)^r \|X_{k,N}^r s_N^k\|$$

and the result follows from a(iv) with $\rho_4 = \rho\rho_3$.

Since x^* is the limit of $\{x^k\}$, and $x_B^* > 0$, there is a $\rho > 0$ and an $L \geq 1$ such that for all $k \geq l$, $\|X_{k,B}^{-r}\| \leq \rho$. b(i) follows from 5.4.3 Part (b), a(v) and

$$\|u_B^k\| \leq \frac{\|X_{k,B}^{-r}\| \|p_B^k\|}{(c^T x^k - c^*)^r}.$$

To see b(ii), note that

$$\|s_B^k\| \leq \|X_{k,B}^{-r}\| \frac{\|X_{k,B}^r s_B^k\|}{(c^T x^k - c^*)^r} (c^T x^k - c^*)^r.$$

b(ii) follows by substituting b(i). b(iii) follows from the boundedness of v^k and b(i). b(iv) follows readily by substitution, boundedness of u_N^k and b(i). ∎

Two Power Centers

Consider the polyhedron,

$$\mathcal{V} = \{v : A_N v_N + A_B v_B = 0, \bar{s}_N^T v_N = 1, v_N \geq 0\}. \tag{5.60}$$

From the results of 5.4.5, specifically a(ii), we see that sequence $\{v^k\}$ generated by the power method lies in this polyhedron. Also, the power affine scaling step interpreted in polyhedron $\mathcal{V}$ has a close connection to Newton step for computing its power center, defined as the unique solution (if it exists) of the concave maximization problem:

$$\begin{aligned}
\text{maximize} \quad & \tfrac{-1}{2(r-1)} \sum_{j \in N} v_j^{-2(r-1)} \\
A_N v_N + A_B v_B &= 0 \\
\bar{s}_N^T v_N &= 1 \\
v_N &> 0.
\end{aligned}$$

K. K. T. conditions defining this center are:

$$\begin{aligned}
V_N^{-(2r-1)} e - A_N^T z - \theta \bar{s}_N &= 0 \\
-A_B^T z &= 0 \\
A_N v_N + A_B v_B &= 0 \\
\bar{s}_N^T v_N &= 1.
\end{aligned} \quad (5.61)$$

We will explore this in the next section. Here we will investigate the connection between power centers of $\mathcal{V}$ and $F_\mathcal{D} \cap \{s : s_N \geq 0\}$ (see (5.59) for definition of $F_\mathcal{D}$). The power center of $F_\mathcal{D} \cap \{s : s_N \geq 0\}$ is defined by following concave maximization problem:

$$\begin{aligned}
\text{maximize} \quad & \tfrac{2r-1}{2(r-1)} \sum_{j \in N} s_j^{\frac{2(r-1)}{2r-1}} \\
A_N^T y + s_N &= c_N \\
A_B^T y &= c_B \\
s_N &> 0
\end{aligned}$$

whose K. K. T. conditions are:

$$\begin{aligned}
S_N^{-\frac{1}{(2r-1)}} e - u_N &= 0 \\
A_N u_N + A_B u_B &= 0 \\
A_N^T(y - \bar{y}) + (s_N - \bar{s}_N) &= 0 \\
A_B^T(y - \bar{y}) &= 0.
\end{aligned} \quad (5.62)$$

The next result establishes an intimate relationship between the power centers of the above problems:

5.4.6 Relationship of power centers (v^*, z^*, θ^*) solves (5.61) if and only if (s^*, u^*, y^*) solves (5.62), and

$$s_N^* = \frac{1}{\theta^*} V_{*,N}^{-2r+1} e.$$

Proof: Let (v^*, z^*, θ^*) be the solution to system (5.61). Then it can be

5.4 POWER AFFINE SCALING METHOD

verified that

$$s_N = \frac{1}{\theta^*} V_{*,N}^{-2r+1} e$$
$$y - \bar{y} = -\frac{1}{\theta^*} z^*$$
$$u = (\frac{1}{\theta^*})^{-\frac{1}{2r-1}} v^*$$

will solve the system (5.62).

Now let the center of $F_D \cap \{s : s_N \geq 0\}$ exist and the system (5.62) have the solution (s^*, u^*, y^*). Then it can be verified that the transformation

$$\beta^{-1} = e^T S_{*,N}^{\frac{2(r-1)}{2r-1}} e$$
$$\theta = \beta^{-(2r-1)}$$
$$z = -\theta(y^* - \bar{y})$$
$$v = \beta u^*$$

will solve the system (5.61). Here we have used $\bar{s}_N^T u_N = s_N^T u_N$, which is readily established using last three relations of system (5.62). ∎

Newton Step in $\mathcal{V}$

Starting at the point v, z, θ, with $v \in \mathcal{V}$, the Newton direction $\Delta v, \Delta z, \Delta \theta$ obtained when applying Newton's method (see section 2.7.1) to find the zero of the system (5.61), is obtained by solving the following system:

$$(2r-1)V_N^{-2r}\Delta v_N - A_N^T \Delta z - \Delta \theta \bar{s}_N = -V_N^{-2r+1}e + A_N^T z + \theta \bar{s}_N$$
$$-A_B^T \Delta z = A_B^T z$$
$$A_N \Delta v_N + A_B \Delta v_B = 0$$
$$\bar{s}_N^T \Delta v_N = 0.$$

Consider the change of variables:

$$\hat{z} = z + \Delta z$$
$$\hat{\theta} = \theta + \Delta \theta$$
$$\Delta v_B' = (2r-1)\Delta v_B \quad (5.63)$$
$$w_N = (2r-1)V_N^{-r}\Delta v_N.$$

Substituting this change in the above system, we can derive the following equivalent system with $\hat{A}_N = A_N V_N^r$ and $\hat{s}_N = V_N^r \bar{s}_N$,

$$\begin{aligned} w_N + \hat{A}_N^T \hat{z} + \hat{s}_N \hat{\theta} &= V_N^{-(r-1)} e \\ -A_B^T \hat{z} &= 0 \\ \hat{A}_N w_N + A_B \Delta v_B' &= 0 \\ \hat{s}_N^T w_N &= 0 \end{aligned} \quad (5.64)$$

We now prove a property of w_N:

5.4.7 Property of w_N

Up to a choice of $\Delta v_B'$, the solution to the system (5.64) is unique. Also,

(a) $e^T V_N^{-(r-1)} w_N = w_N^T w_N$.

(b) $\|w_N\|^2 \leq e^T V_N^{-2(r-1)} e$.

Proof: The uniqueness is easy to see when the columns of the matrix A_B are linearly independent. Otherwise, a unique solution results when A_B is replaced with a submatrix spanning the column space $\mathcal{R}(A_B^T)$ of A_B. This only changes the value of $\Delta v_B'$.

Multiplying the first equation of (5.64) by w_N^T we obtain

$$w_N^T w_N + w_N^T \hat{A}_N^T \hat{z} + w_N^T \hat{s}_N \hat{\theta} = w_N^T V_N^{-(r-1)} e.$$

The second and the third terms of left hand side of the above expression, from second, third and fourth equations of (5.64), are seen as zero, and we have (a). (b) follows from (a) and the Cauchy-Schwartz inequality. ∎

Affine and Newton steps in $\mathcal{V}_N$

We now investigate the relationship between affine scaling step interpreted in the polyhedron $\mathcal{V}$, and Newton step for computing the power center of this polyhedron. We will show that there is a close connection between these steps, which will be used later to establish convergence and convergence rate. We first investigate the affine scaling step.

Affine scaling step is defined by solving optimization problem (5.55). Using a(ii) of Lemma 5.4.5 we can restate problem (5.55) as:

$$\begin{aligned} \text{minimize} \quad & \bar{s}_N^T p_N \\ & A_N p_N + A_B p_B = 0 \\ & p_N^T X_{k,N}^{-2r} p_N + p_B^T X_{k,B}^{-2r} p_B \leq 1 \end{aligned}$$

5.4 POWER AFFINE SCALING METHOD

whose K.K.T. conditions are:

$$\begin{aligned}
\bar{s}_N - A_N^T y - 2\theta X_{k,N}^{-2r} p_N &= 0 \\
-A_B^T y - 2\theta X_{k,B}^{-2r} p_B &= 0 \\
A_N p_N + A_B p_B &= 0 \\
\|X_k^{-r} p\| &= 1 \\
(\theta > 0).
\end{aligned} \qquad (5.65)$$

Using the definitions of u^k and v^k given in (5.58), setting $u = u^k$, $v = v^k$ and

$$\begin{aligned}
p'_B &= \frac{p_B^k}{(c^T x^k - c^*)^r \|u\|} \\
\tilde{y} &= \frac{-(c^T x^k - c^*)^r y^k}{2\theta \|u\|} \\
\tilde{\theta} &= \frac{(c^T x^k - c^*)^r \|u\|}{2\theta} \\
&= 1
\end{aligned} \qquad (5.66)$$

we can rewrite (5.65) as:

$$\begin{aligned}
\frac{u_N}{\|u\|^2} - \hat{A}_N^T \tilde{y} - \hat{s}_N \frac{\tilde{\theta}}{\|u\|^2} &= 0 \\
-A_B^T \tilde{y} &= \frac{-s_B^k}{\|u\|^2} \\
\hat{A}_N \frac{u_N}{\|u\|^2} + A_B p'_B &= 0 \\
\hat{s}_N^T \frac{u_N}{\|u\|^2} &= 1
\end{aligned} \qquad (5.67)$$

where $\hat{A}_N = A_N V_N^r$ and $\hat{s}_N = V_N^r \bar{s}_N$. The next result explores the relationship between (5.65) and (5.67).

5.4.8 Proposition Consider systems represented by equations (5.65) and (5.67).

(a) (5.65) has a unique solution which generates a solution to (5.67).

(b) The solution to (5.67) is unique up to a choice of p'_B; and, there is a value for p'_B for which the resulting solution also solves (5.65).

(c) When A_B has full column rank, two systems are equivalent.

Proof: Since system (5.65) represents a solution to strictly convex problem, it has a unique solution. By simple algebra, it is seen that u, $\tilde{y}$ ρ and p'_B, defined by the change of variables (5.66), solve system (5.67). Thus (a) follows.

From (a), it follows that (5.67) has a solution. Considering $q_N = \frac{u_N}{\|u\|^2}$, $\tilde{\rho} = \frac{\rho}{\|u\|^2}$, $\tilde{y}$ and p'_B as variables, this system is linear in these variables. If A_B

has full column rank, the solution to (5.67) is unique, and (c) follows. Otherwise, since (5.67) can have a solution only if s_B lies in the row space $\mathcal{R}(A_B^T)$ of A_B, third condition must have redundant constraints readily identified by choosing any full column rank submatrix of A_B.

To see (b), let $A_B = (A_C, A_D)$ where A_C has full column rank and spans the range (or column space) $\mathcal{R}(A_B)$ of A_B. Thus, for some unique matrix Λ, $A_D = A_C \Lambda$. Replacing second and third equations of (5.67) by

$$-A_C^T \tilde{y} = \frac{-s_C^k}{\|u\|^2} \qquad (5.68)$$

and

$$\hat{A}_N \frac{u_N}{\|u\|^2} + A_C p_C' = 0 \qquad (5.69)$$

respectively, we obtain a new system that has a unique solution. By setting $p_B' = (p_C', p_D')$, and letting $p_D' = 0$, the solution to (5.69) generates a solution to the third equation of (5.67). Now, let $(q_N, \tilde{y}, \tilde{\rho}, p_B')$ be any solution to (5.67). This then generates the unique solution $(q_N, \tilde{y}, \tilde{\rho}, p_C' - \Lambda p_D')$ to the first and fourth equations of (5.67) and (5.68) - (5.69). Since only the vector p_B' is modified in any solution to (5.67), (b) is established with the required $p_B' = \frac{p_B}{(c^T x^k - c^*)\|u\|}$. ∎

As a byproduct of the representation of affine scaling step by system (5.67), we prove the following important property of the sequence $\{\tilde{u}_N^k\}$:

5.4.9 Lemma There exists a $\rho > 0$ and an $L \geq 1$ such that for all $k \geq L$

$$e^T \tilde{u}_N^k = 1 + \delta_k$$

where $|\delta_k| \leq \rho(c^T x^k - c^*)^{2r}$.

Proof: Multiplying first equation of system (5.67) by $e^T V_{k,N}^{-(r-1)}$, we obtain

$$\frac{e^T \tilde{u}_N^k}{\|u^k\|^2} - e^T V_{k,N} A_N^T \tilde{y} - \frac{e^T V_{k,N} \bar{s}_N}{\|u^k\|^2} = 0.$$

Substituting $A_N v_N = -A_B v_B$, and the second equation, we obtain

$$e^T \tilde{u}_N^k + v_B^T s_B^k - 1 = 0.$$

The lemma now follows from 5.4.5, b(ii). ∎

Now consider the system (5.64), defining the Newton step in $\mathcal{V}$. By making the change of variables

$$w_N' = V_N^{-(r-1)} e - w_N$$
$$\Delta \hat{v}_B = v_B - \Delta v_B'$$

5.4 POWER AFFINE SCALING METHOD

we obtain an equivalent system:

$$\begin{aligned}
w'_N - \hat{A}_N^T \hat{z} - \hat{s}_N \hat{\theta} &= 0 \\
-A_B^T \hat{z} &= 0 \\
\hat{A}_N w'_N + A_B \Delta \hat{v}_B &= 0 \\
\hat{s}_N^T w'_N &= 1.
\end{aligned} \quad (5.70)$$

Comparing systems (5.70) and (5.67), we note that for the 4-vectors f and g defined by

$$\begin{aligned}
f_1 &= V_{k,N}^r (w'_N - \frac{u_N}{\|u\|^2}) \\
f_2 &= \hat{z} - \tilde{y} \\
f_3 &= \hat{\theta} - \frac{1}{\|u\|^2} \\
f_4 &= \Delta \hat{v}_B - p'_B
\end{aligned}$$

and $g = (0, \frac{-s_B^k}{\|u\|^2}, 0, 0)^T$, we obtain

$$M(v_N^k) f = g \quad (5.71)$$

where

$$M(v_N^k) = \begin{bmatrix} V_{k,N}^{-2r} & -A_N^T & \bar{s}_N & 0 \\ 0 & -A_B^T & 0 & 0 \\ A_N & 0 & 0 & A_B \\ \bar{s}_N^T & 0 & 0 & 0 \end{bmatrix}.$$

We now prove a connection between the two systems (5.67) and (5.70).

5.4.10 Proposition There is a $\beta > 0$ such that for every $k = 1, 2, \cdots$

$$\frac{u_N^k}{\|u^k\|^2} = V_{k,N}^{-(r-1)} e - w_N^k + V_{k,N}^r \Delta^k$$

with $\|\Delta^k\| \leq \beta \frac{\|s_B^k\|}{\|u^k\|^2}$.

Proof: It is readily confirmed that system (5.71) represents K.K.T. conditions of the following quadratic programming problem:

$$\begin{aligned}
\text{minimize} \quad & \tfrac{1}{2} f_1^T V_{k,N}^r f_1 + f_4^T \tfrac{s_B^k}{\|u\|^2} \\
& A_N f_1 + A_B f_4 = 0 \\
& \bar{s}_N^T f_1 = 0.
\end{aligned}$$

where f_2 and f_3 are Lagrange multipliers on the two equality constraints. We can assume, because of 5.4.7 and Proposition 5.4.8, without of loss of generality that the columns of A_B are linearly independent. Thus, after the substitution

$$f_4 = -(A_B^T A_B)^{-1} A_B^T A_N f_1$$

in the above optimization problem, obtain the equivalent problem

$$\begin{aligned}
\text{minimize} \quad & \tfrac{1}{2} f_1^T V_{k,N}^r f_1 - f_1^T \tilde{s}_B^k \\
& \tilde{A}_N f_1 = 0 \\
& \tilde{s}_N f_1 = 0
\end{aligned}$$

where $\tilde{A}_N = A_N - A_B(A_B^T A_B)^{-1} A_B^T A_N$ and $\tilde{s}_B^k = A_N^T A_B (A_B^T A_B)^{-1} \frac{s_B^k}{\|u\|^2}$.

Above problem is a quadratic programming problem, and from 5.14 and Theorem 5.1.11, its Lagrange multipliers are bounded by

$$\|(f_2, f_3)\| \leq q(A, \bar{s}_N) \|\tilde{s}_B^k\|. \tag{5.72}$$

where $q(A, \bar{s}_N)$ is a positive constant determined only by A and $\bar{s}_N$. Since

$$V_{k,N}^{-r}(V_{k,N}^{-(r-1)} e - w_N^k - \frac{u_N^k}{\|u^k\|^2}) = A_N^T f_2 + \bar{s}_N f_3$$

by defining

$$\Delta^k = -(A_N^T f_2 + \bar{s}_N f_3)$$

our result follows. ∎

We now investigate Newton and affine scaling directions at v_N in the projected polyhedron $\mathcal{V}_N = \{v_N : v \in \mathcal{V}\}$.

Consider the sequence $\{v_N^k\} \in \mathcal{V}_N$ generated by affine scaling method. The following corollary to Proposition 5.4.10 establishes a connection between affine scaling and Newton directions.

5.4.11 Affine and Newton Steps
Let $v_N^k \in \mathcal{V}_N$. Then

(a) Affine scaling direction at v_N^k is

$$v_N^{k+1} - v_N^k = \frac{\tilde{\alpha}_k}{1 - \tilde{\alpha}_k}(v_N^k - V_{k,N} \frac{V_{k,N}^{r-1} u_N^k}{\|u^k\|^2})$$

where $\tilde{\alpha}_k = \alpha \frac{\|u^k\|^2}{\phi(V_k^{r-1} u^k)}$.

5.4 POWER AFFINE SCALING METHOD

(b) Newton direction at v_N^k is

$$\Delta v_N^k = \frac{1}{2r-1}(v_N^k - \frac{V_{k,N}^r u_N^k}{\|u^k\|^2} + V_{k,N}^{2r}\Delta^k)$$

where Δ^k is as in Proposition 5.4.10.

Proof: To see the proof of (a), by simple substitutions, we note that

$$v_N^{k+1} - v_N^k = \frac{x_N^{k+1}}{c^T x^{k+1} - c^*} - \frac{x_N^k}{c^T x^k - c^*}$$

$$= \frac{v_N^k - \frac{\alpha V_{k,N}^r u_N^k}{\phi(V_k^r u^k)}}{1 - \frac{\alpha\|u^k\|^2}{\phi(V_k^{r-1} u^k)}} - v_N^k$$

$$= \frac{\frac{\alpha\|u^k\|^2}{\phi(V_k^{r-1} u^k)}}{1 - \frac{\alpha\|u^k\|^2}{\phi(V_k^{r-1} u^k)}}(v_N^k - \frac{V_{k,N}^r u_N^k}{\|u^k\|^2}).$$

(b) follows from the change of variables (5.63), and Proposition 5.4.10. ∎

Degeneracy Resolution by Step Size Control

In the affine scaling method, we resolved degeneracy by controlling the step size, Section 5.2. Like the case of affine scaling method, methodology for investigating the power method uses a merit function (in place of a local potential function) to achieve this goal. It is shown that dual sequence converges to the power center of the optimal dual face (instead of analytic center). Also, the established connection between the affine step and Newton step in $\mathcal{V}$ is used to achieve this step size control.

5.4.12   The Merit Function   When $r = 1$, convergence was proved by using a local potential function, 5.2.1. We assume here that $r \neq 1$ and $r > \frac{1}{2}$. Define the following merit function

$$F_N(x) = \frac{1}{2(r-1)}\sum_{j \in N}(v_j)^{-2(r-1)}$$

and note that it is negative of the objective function of power center problem on $\mathcal{V}$. We now prove a simple lemma related to this function.

5.4.13 Lemma Let $w > -e$ and $v > 0$ be arbitrary p vectors with V the diagonal matrix generated by v.

(a) Let $\phi(-w) \leq 0$. Then

$$\frac{1}{2(r-1)} \sum_{j=1}^{p} v_j((1+w_j)^{-2(r-1)} - 1) \leq -e^T Vw + \frac{(2r-1)}{2} w^T Vw.$$

(b) Let $1 > \phi(-w) > 0$. Then

$$\frac{1}{2(r-1)} \sum_{j=1}^{p} v_j((1+w_j)^{-2(r-1)} - 1) \leq -e^T Vw + \frac{(2r-1)}{2(1-\phi(-w))^{2r}} w^T Vw.$$

Proof: (a) follows from the following identity obtained by using mean value version of Taylor's formula, and noting that as $w_j \geq 0$, $\hat{w}_j \geq 0$. This implies that the last term in the expression below is non-negative.

$$\frac{1}{2(r-1)}(1+w_j)^{-2(r-1)} = \frac{1}{2(r-1)} - w_j + \frac{(2r-1)}{2} w_j^2 - \frac{(2r-1)2r}{3!(1+\hat{w}_j)^{2r+1}} w_j^3$$

(b) follows from the following relations, obtained by again applying the mean value version of Taylor's formula, where $\hat{w}_j \geq -\phi(-w)$:

$$\frac{1}{2(r-1)}(1+w_j)^{-2(r-1)} = \frac{1}{2(r-1)} - w_j + \frac{(2r-1)}{2(1+\hat{w}_j)^{2r}} w_j^2$$

$$\leq \frac{1}{2(r-1)} - w_j + \frac{(2r-1)}{2(1-\phi(-w))^{2r}} w_j^2$$

and we have our result. ∎

Decrease in Merit Function along x^k

After proving a simple lemma, we will establish that the merit function decreases along the sequence.

5.4.14 Lemma There is an $L \geq 1$ $\beta > 0$, $\gamma > 0$, $\delta > 0$ and $\delta_1 > 0$ such that for every $k \geq L$

(a) $\|u_N^k\| \geq \delta_1$.

(b) $1 - \|\hat{w}_N^k\| - |\delta_k| \leq \delta(u_N^k) \leq 1$ with $\hat{w}_N^k = V_{k,N}^{r-1} w_N^k$ and $|\delta_k| \leq \beta(c^T x^k - c^*)^{2r}$ where $\delta(u_N^k) = \frac{\|u^k\|^2}{\phi(V_{k,N}^{r-1} u_N^k)}$.

5.4 POWER AFFINE SCALING METHOD

(c) $\delta(u_N^k) \geq \gamma > 0$.

(d) $(c^T x^{k+1} - c^*) \leq \delta(c^T x^k - c^*)$ and thus $\sum_{k=0}^{\infty}(c^T x^k - c^*) < \infty$

Proof: From Lemma 5.4.9 and definitions, $u_N^k = V_{k,N}^{r-1} \tilde{u}_N^k$, and for some $L' \geq 1$ and all $k \geq L'$, $e^T \tilde{u}_N^k = 1 + \tilde{\delta}_k$ with $\tilde{\delta}_k \leq \rho(c^T x^k - c^*)^{2r}$. Thus, there is an $L \geq L'$ such that for all $k \geq L$

$$\phi(\tilde{u}_N^k) \geq \frac{1+\tilde{\delta}_k}{q} \geq \frac{1}{2q}.$$

Let $s_{j_k}^k v_{j_k}^k = \tilde{u}_{j_k}^k = \phi(\tilde{u}_N^k)$. As $\|s_N^k\|$ is bounded, by say $M > 0$, $v_{j_k}^k \geq \frac{1}{2qM}$. Part (a) follows with $\delta_1 = \frac{1}{(2q)^r M^{r-1}} > 0$.

To see (b), from Proposition 5.4.10,

$$\frac{\phi(V_{k,N}^{r-1} u_N^k)}{\|u^k\|^2} = \phi\left(\frac{V_{k,N}^{r-1} u_N^k}{\|u^k\|^2}\right)$$

$$= \phi(e - \hat{w}_N^k + V_{k,N}^{2r-1} \Delta^k)$$

$$\leq 1 + \|\hat{w}_N^k\| + \delta_k \qquad (5.73)$$

where $\|\Delta^k\| = V_{k,N}^{2r-1} \delta_k$; and the lower bound follows from Lemma 5.4.9, (a) and the fact that for each real $a > -1$, $\frac{1}{1+a} \geq 1-a$. The upper bound follows since 5.4.5 b(iv) holds for every $\alpha \leq 1$. (c) follows from (a) and 5.4.5. (d) follows from (c) and 5.4.5 b(vi). ∎

We are now ready to prove the main result related to decrease in merit function.

5.4.15 Decrease in Merit Function Let $r \neq 1$, $r > \frac{1}{2}$, $\tilde{\alpha} = \alpha \frac{\|u^k\|^2}{\phi(V_k^{r-1} u^k)}$ and $\hat{w}_N^k = (2r-1) V_{k,N}^{-1} \Delta v_N^k$. There is an $L \geq 1$ and a $\beta > 0$ such that for every $k \geq L$

$$\Delta F_N(x^k) \leq -\frac{\tilde{\alpha}}{1-\tilde{\alpha}}\left(\left(1 - \frac{(2r-1)\tilde{\alpha}}{2(1-\alpha)^{2r}}\right)((\hat{w}_N^k)^T V_{k,N}^{-2(r-1)} \hat{w}_N^k) + \mu_k\right).$$

where $|\mu_k| \leq \beta(c^T x^k - c^*)^{2r}$, and $\Delta F_N(x^k) = F_N(x^{k+1}) - F_N(x^k)$.

Proof: From Step 4 and Proposition 5.4.10, for each $j \in N$,

$$\frac{x_j^{k+1}}{x_j^k} = 1 - \alpha \frac{\|u^k\|^2}{\phi(V_k^{r-1} u^k)} \frac{(v_j^k)^{r-1} u_j^k}{\|u^k\|^2}$$

$$= 1 - \tilde{\alpha} \frac{(v_j^k)^{r-1} u_j^k}{\|u^k\|^2}$$

$$= 1 - \tilde{\alpha}(1 - \hat{w}_j^k + \hat{\Delta}_j^k),$$

where $\hat{\Delta}^k = V_{k,N}^{2r-1}\Delta^k$. From the above, definitions and 5.4.5 b(iii) and b(iv) we obtain

$$\begin{aligned}
\frac{v_j^{k+1}}{v_j^k} &= \frac{x_j^{k+1}}{x_j^k}\frac{c^T x^k - c^*}{c^T x^{k+1} - c^*} \\
&= 1 + \frac{\tilde{\alpha}}{1-\tilde{\alpha}}(\hat{w}_j^k - \hat{\Delta}_j^k) \\
&> 0.
\end{aligned}$$

Let $\theta_j^k = \frac{v_j^{k+1}}{v_j^k} - 1 = \frac{\tilde{\alpha}}{1-\tilde{\alpha}}(\hat{w}_j^k - \hat{\Delta}_j^k)$, and note that $\theta_j^k > -1$. First, consider the case when $1 > \phi(-\theta^k) > 0$. Then, from Lemma 5.4.13, (b) we obtain

$$\begin{aligned}
\Delta F_N(x^k) &= \frac{1}{2(r-1)}\sum_{j\in N}((v_j^{k+1})^{-2(r-1)} - (v_j^k)^{-2(r-1)}) - \frac{(2r-1)\tilde{\alpha}}{2(1-\tilde{\alpha}}\\
&\leq \frac{\tilde{\alpha}}{1-\tilde{\alpha}}(e^T V_{k,N}^{-2(r-1)}(\hat{w}_N^k - \hat{\Delta}^k) \\
&\quad - \frac{1}{(1-\phi(-\theta^k))^{2r}}(\hat{w}_N^k - \hat{\Delta}^k)^T V_{k,N}^{-2(r-1)}(\hat{w}_N^k - \hat{\Delta}^k)).
\end{aligned}$$

Using Proposition 5.4.10 and Lemma 5.4.5 b(ii), we obtain

$$\begin{aligned}
1 - \phi(-\theta^k) &= 1 - \frac{\tilde{\alpha}}{1-\tilde{\alpha}}\phi(-\hat{w}_N^k + \hat{\Delta}^k) \\
&= \frac{1}{1-\tilde{\alpha}}(1 - \tilde{\alpha}\phi(e - \hat{w}_N^k + \hat{\Delta}^k)) \\
&= \frac{1}{1-\tilde{\alpha}}(1 - \tilde{\alpha}\frac{\phi(V_{k,N}^{r-1}u_N^k)}{\|u^k\|^2}) \\
&= \frac{1-\alpha}{1-\tilde{\alpha}}.
\end{aligned}$$

Thus, from Lemma 5.4.7, as $1 - \tilde{\alpha} < 1$, substituting above we obtain

$$\begin{aligned}
\Delta F_N(x^k) &\leq \frac{-\tilde{\alpha}}{1-\tilde{\alpha}}((1-\frac{(2r-1)\tilde{\alpha}}{2(1-\alpha)^{2r}})((\hat{w}_N^k)^T V_{k,N}^{-2(r-1)}\hat{w}_N^k) \\
&\quad -\mu_k)
\end{aligned} \tag{5.74}$$

where

$$\begin{aligned}
\mu_k &= e^T V_{k,N}^{-2(r-1)}\hat{\Delta}^k - \frac{(2r-1)\tilde{\alpha}(1-\tilde{\alpha})^{2r-1}}{2(1-\alpha)^{2r}} \\
&\quad (2(\hat{w}_N^k)^T V_{k,N}^{-2(r-1)}\hat{\Delta}^k - (\hat{\Delta}^k)^T V_{k,N}^{-2(r-1)}\hat{\Delta}^k) \\
&= e^T V_{k,N}\Delta^k - \frac{(2r-1)\tilde{\alpha}(1-\tilde{\alpha})^{2r-1}}{2(1-\alpha)^{2r}}(2(w_N^k)^T V_{k,N}^r\Delta^k - (\Delta^k)^T V_{k,N}^{2r}\Delta^k).
\end{aligned}$$

5.4 POWER AFFINE SCALING METHOD

Now consider the case when $\phi(-\theta^k) \leq 0$. Then Lemma 5.4.13 (a) must be used to calculate this change. By an identical analysis as above, in this case we obtain

$$\Delta F_N(x^k) \leq -\frac{\tilde{\alpha}}{1-\tilde{\alpha}}((1-\frac{(2r-1)\tilde{\alpha}}{2(1-\tilde{\alpha})})((\hat{w}_N^k)^T V_{k,N}^{-2(r-1)}\hat{w}_N^k) - \mu'_k)$$

where

$$\mu'_k = e^T V_{k,N} \Delta^k - \frac{(2r-1)\tilde{\alpha}}{2(1-\tilde{\alpha})}(2(w_N^k)^T V_{k,N}^r \Delta^k - (\Delta^k)^T V_{k,N}^{2r} \Delta^k).$$

As $\alpha < 1$, $\tilde{\alpha} < \alpha$ it is readily seen that the bound obtained by (b) of Lemma 5.4.13 is larger. Using the first equation of (5.64), we can derive

$$V_{k,N}^r w_N^k + V_{k,N}^{2r} A_N^T \hat{z} + V_{k,N}^{2r} \bar{s}_N \hat{\theta} = v_N^k.$$

From definitions $f_2 = \hat{z} - \tilde{y}$, $f_3 = \hat{\theta} - \frac{1}{\|u\|^2}$, $\tilde{y} = -y^k$, 5.4.3 (c), Lemma 5.4.14 (a) and (5.72), we see that $\hat{z}$ and $\hat{\theta}$ are bounded. Thus $V_{k,N}^r w_N^k$ is bounded, and for some $\beta_1 > 0$ and $\beta_2 > 0$ we have

$$|e^T V_{k,N} \Delta^k| \leq \beta_1 \|\Delta^k\|$$
$$|(w_N^k)^T V_{k,N}^r \Delta^k| \leq \beta_2 \|\Delta^k\|$$

and our result follows from Proposition 5.4.10 and Lemma 5.4.14 (a). ∎

We are now ready to prove the main convergence theorem.

5.4.16 Convergence with Fixed Step Let $\frac{\alpha}{(1-\alpha)^{2r}} < \frac{2}{2r-1}$, and assumptions (1) - (3) hold. Then, there exist vectors x^*, y^* and s^* such that

1. $x^k \longrightarrow x^*$

2. $y^k \longrightarrow y^*$

3. $s^k \longrightarrow s^*$

where x^* is an optimum solution of the primal, and (y^*, s^*) is an optimum solution of the dual. In addition, (y^*, s^*) is the power center of the optimal dual face, and thus strict complementarity holds between this pair of optimum solutions.

Proof: From 5.4.14 (c) we see that there is a $\gamma > 0$ such that $\tilde{\alpha}_k \geq \gamma \alpha$. Thus

$$\frac{2(r-1)\tilde{\alpha}_k}{1-\tilde{\alpha}_k} > \frac{2(r-1)\alpha\gamma}{1-\alpha\gamma} = \theta_1 > 0$$

and from 5.4.14 upper bound in (b), $\alpha \geq \tilde{\alpha}_k$. Thus

$$1 - \frac{(2r-1)\tilde{\alpha}_k}{2(1-\alpha)^{2r}} \geq 1 - \frac{(2r-1)\alpha}{2(1-\alpha)^{2r}} = \theta_2 > 0.$$

Now, $(\hat{w}_N^k)^T V_{k,N}^{-2(r-1)} \hat{w}_N^k = (2r-1)^2 (\Delta v_N^k)^T V_{k,N}^{-2r} \Delta v_N^k \geq (2r-1)^2 \|\Delta v_N^k\|^2 \rho^{2r}$.
From 5.4.15, there is a $\beta > 0$ such that $|\mu_k| \leq \beta(c^T x^k - c^*)^{2r}$ and

$$\Delta F_N(x^k) \leq -\theta_1((2r-1)^2 \rho^{2r} \theta_2 \|\Delta v_N^k\|^2 - |\mu_k|).$$

Thus

$$\sum_{k=L}^{\infty}(\Delta F_N(x^k)) \leq -\theta_1\theta_2(2r-1)^2 \rho^{2r} \sum_{k=L}^{\infty} \sigma_k \|\Delta v_N^k\|^2 + \theta_1 \sum_{k=L}^{\infty} |\mu_k|.$$

From 5.4.14 (d),

$$\sum_{k=L}^{\infty} |\mu_k| \leq \sum_{k=L}^{\infty} \beta(c^T x^k - c^*)^{2r} < \infty.$$

When $r > 1$, $F_N(x^k)$ is bounded below by zero and from 5.4.3 (b), when $\frac{1}{2} < r < 1$, it is bounded away from $-\infty$ on this sequence. Thus $\|\Delta v_N^k\| \longrightarrow 0$.
From Proposition 5.4.10

$$\frac{V_{k,N}^{r-1} u_N^k}{\|u^k\|^2} \longrightarrow e.$$

Since the only vector in $\mathcal{V}_N$ for which the Newton step $\Delta v_N = 0$ is its power center v_N^*, we must have,

$$v_N^k \longrightarrow v_N^*.$$

Let K be such that for $k \in K$ (it exists since all these sequences are bounded) for some vectors x^*, y^* and s^*,

$$s^k \longrightarrow s^*$$
$$y^k \longrightarrow y^*$$
$$u^k \longrightarrow u^*,$$

and, so on K

$$\frac{V_{k,N}^{r-1} u_N^k}{\|u^k\|^2} \longrightarrow \frac{V_{*,N}^{r-1} u_N^*}{\|u^*\|^2} = e. \tag{5.75}$$

Using 5.4.6, it is readily seen that (y^*, s^*) is the power center of the optimal dual face; and our result follows from the complementary slackness theorem. ∎

We get the following sharp bound on the linear convergence rate of $\{c^T x^k\}$.

5.4 POWER AFFINE SCALING METHOD

5.4.17 Proposition Let $\frac{\alpha}{(1-\alpha)^{2r}} < \frac{2}{2r-1}$. Then

$$\lim_{k\to\infty} \frac{c^T x^{k+1} - c^*}{c^T x^k - c^*} = 1 - \alpha.$$

Proof: Follows readily from Proposition 5.4.10 5.4.14 (b) and Equation (5.75). ■

5.4.16 proves the convergence to optimality for a constant step size $\alpha > 0$ determined such that

$$\frac{\alpha}{(1-\alpha)^{2r}} < \frac{1}{2r-1}.$$

For $r = 1$ this requires $\alpha < \frac{1}{2}$. To obtain a result analogous to the $\frac{2}{3}$rd for the standard affine scaling method, we introduce a variable step strategy, i.e., we will allow the stepsize to vary, and in iteration k, we will choose the stepsize α_k by rules described below. This increases the step size implemented, which will be shown, asymptotically to be given by the formula

$$\frac{\alpha_k}{1-\alpha_k} < \frac{2}{2r-1}.$$

which for $r = 1$ gives the required $\frac{2}{3}$rd. We obtain this increase by using the following estimate for $\delta(u^k)$ (of 5.4.14)

$$\tau_k = \frac{\|X_k^r s^k\|^2}{\phi(X_k^{2r-1} s^k)(x^k)^T s^k}. \quad (5.76)$$

We now establish the goodness of the estimate (5.76).

5.4.18 Lemma Let τ_k be defined by Equation (5.76). There is an $L \geq 1$ and a $\beta > 0$ such that for all $k \geq L$

$$\delta(u_N^k) = \tau_k + \delta_k$$

where $|\delta_k| \leq \beta(c^T x^k - c^*)^{2r}$ and $\delta(u_N^k) = \frac{\|u^k\|^2}{\phi(V_{k,N}^{r-1} u_N^k)}$.

Proof: This result follows readily from 5.4.5 (b), Lemma 5.4.9 and the upper bound of (b) of 5.4.14. ■

Since τ_k is, asymptotically, a very good estimate of $\delta(u_N^k)$, in place of using $\frac{\alpha}{(1-\alpha)^{2r}}$ as an estimate of $\frac{\tilde{\alpha}(1-\tilde{\alpha})^{2r-1}}{(1-\alpha)^{2r}}$ in the relations (5.74), we will use the estimate $\tau_k \alpha$ of $\tilde{\alpha}$ in the above formula. Also, from 5.4.14, (b) whenever $\|w_N^k\| \longrightarrow 0$, $\delta(u_N^k) \longrightarrow 1$. Thus, in this case $\tau_k \longrightarrow 1$, and the new estimate approaches $\frac{\alpha}{1-\alpha}$. We now present this step selection strategy.

Define α^* such that

$$\frac{\alpha^*}{(1-\alpha^*)^{2r}} < \frac{2}{2r-1}.$$

Let $\theta > 0$ be very small and at iteration k choose the step size α_k by the following strategy:

Step 1 Let α' be such that

$$\frac{\tau_k \alpha'(1 - \tau_k \alpha')^{2r-1}}{(1-\alpha')^{2r}} = \frac{2}{(2r-1)} - \theta. \tag{5.77}$$

Step 2 Define

$$\alpha_k = \begin{cases} \alpha' & \text{if } \alpha' \geq \alpha^* \\ \alpha^* & \text{otherwise} \end{cases} \tag{5.78}$$

The above choice guarantees that the estimate is not smaller than one obtained from Theorem 5.4.16. The next lemma establishes a relationship between the α' computed in (5.77) and the related expression in system (5.74). Note that the non-linear system (5.77) has to be solved to obtain α'.

5.4.19 Lemma Let α' be computed by (5.77) and let $\tilde{\alpha} = \delta(u_N^k)\alpha'$. Then, there is an $L \geq 1$ and a $\beta > 0$ such that for all $k \geq L$

$$\tilde{\alpha}(1-\tilde{\alpha})^{2r-1} = \tau_k \alpha'(1-\tau_k \alpha')^{2r-1} + \delta_k$$

where $|\delta_k| \leq \beta(c^T x^k - c^*)^{2r}$.

Proof: Follows readily from Lemma 5.4.18, and the fact that for small $\epsilon > 0$, $(1+\epsilon)^{2r-1} < 1 + 4^r \epsilon$. ∎

We are now ready to prove the main theorem with variable step size.

5.4.20 Convergence with Variable Step Size For each k, let α_k, be generated by the above rules and let the assumptions (1) - (3) hold. Then, there exist vectors x^*, y^* and s^* such that

1. $x^k \longrightarrow x^*$

2. $y^k \longrightarrow y^*$

3. $s^k \longrightarrow s^*$

5.4 POWER AFFINE SCALING METHOD

where x^* is an optimum solution of the primal, and (y^*, s^*) is an optimum solution of the dual. In addition, (y^*, s^*) is the power center of the optimal dual face, and thus strict complementarity holds between this pair of optimum solutions.

Proof: Assume that $\alpha_k = \alpha^*$ for each k. Then this theorem follows from 5.4.16. Otherwise for each k for which $\alpha_k = \alpha' > \alpha^*$, using the same argument of 5.4.16, it is readily shown that for some $\beta > 0$

$$\Delta F_N(x^k) \leq -\beta \|\Delta v_N^k\|^2 + |\mu_k|$$

where $|\mu_k| \leq \gamma(c^T x^k - c^*)^{2r}$. The proof is completed in the same way as in the 5.4.16. ∎

We now obtain the asymptotic behavior of α_k.

5.4.21 Corollary $\alpha_k \longrightarrow \hat{\alpha}$ where

$$\frac{\hat{\alpha}}{1-\hat{\alpha}} = \frac{2}{2r-1} - \theta.$$

Proof: As a consequence of 5.4.20, we obtain the fact that $\|w_N^k\| \longrightarrow 0$. From 5.4.14, $\delta(u_N^k) \longrightarrow 1$, from Lemma 5.4.18, $\tau_k \longrightarrow 1$, and we have our result. ∎

Accelerated Power Method

We will now use the connection between Newton and affine scaling step in $\mathcal{V}$, developed earlier, to accelerate the convergence of the method. We first present the accelerated version of the method and then investigate its convergence and convergence rate. This accelerated primal power affine scaling method is generated by replacing steps 3 and 4 by the following three steps, where $0 < \delta < 1$ is a constant whose value will be specified later:

Step 3.1 Min-Ratio Test:

$$\theta_k = \min\left\{\frac{\|X_k^{2r-1} s^k\|}{(x_j^k)^{2r-1} s_j^k} : s_j^k > 0\right\}$$

$$= \frac{\|X_k^{2r-1} s^k\|}{\phi(X_k^{2r-1} s^k)}.$$

If $\theta_k = 1$ set $\alpha_k = 1$, and go to Step 5.

Step 3.2 Step Size Selection: If $e^T X_k s^k \geq 1$, set $\alpha_k = \alpha$ and go to Step 4. Otherwise, define

$$\alpha'_k \quad \text{defined by (5.78)}$$
$$N_k = \{j : x^k_j \leq \sqrt{(x^k)^T s^k}\}$$
$$\gamma_k = e^T X_{k,N_k} s^k_{N_k}$$
$$h^k_{N_k} = \frac{1}{2r-1}\left(\frac{X_{k,N_k}}{\gamma_k} - \frac{X^{2r-1}_{k,N_k} s^k_{N_k}}{\|X^r_k s^k\|^2}\right)$$
$$\epsilon_k = \|h^k_{N_k}\|$$
$$\rho_k = \min\{3r+2, \frac{\log(\epsilon_k)}{\log(\gamma_k)}\}.$$

1. Predictor Step: If $\rho_k \geq 1.5r$ then

$$\alpha_k = 1 - \max\{\epsilon^\delta_k, \gamma^{\delta \rho_k}_k\}.$$

2. Corrector Step: Otherwise,

$$\alpha_k = \min\{\frac{\gamma_k \phi(X^{2r-1}_{k,N_k} s^k_{N_k})}{2r\|X^r_k s^k\|^2}, \alpha'_k\}.$$

Step 4 Next Interior Point:

$$x^{k+1} = x^k - \alpha_k \theta_k \frac{X^{2r}_k s^k}{\|X^{2r-1}_k s^k\|}$$

Some comments are in order here. h^k_N computed in step 4 is a very good estimate of the Newton step Δv^k_N, and its magnitude ϵ_k is used to estimate the distance to the power center v^*_N of $\mathcal{V}_N \cap \{v_N : v_N \geq 0\}$. Asymptotically, we apply a predictor step when the size, ϵ_k of the Newton step is of the order $O(c^T x^k - c^*)^{2r}$; and the corrector step otherwise. As is well known about the Newton step, $\|\Delta v^k_N\|$ is a very good estimate of $\|v^k_N - v^*_N\|$. During the corrector step, α_k is chosen so that

$$v^{k+1}_N - v^k_N = \Delta v^k_N + O(c^T x^k - c^*)^{2r}$$

and thus the affine scaling step behaves, asymptotically, like a Newton step. We now establish two simple lemmas and then prove the convergence of this accelerated method.

5.4 POWER AFFINE SCALING METHOD

5.4.22 Lemma Let N be defined by (5.57). There is an $\hat{L} \geq 1$ such that for all $k \geq \hat{L}$, $N_k = N$, where N_k is defined in step 3.2.

Proof: From 5.4.5 b(ii) and Lemma 5.4.9, there are $\rho_1 > 0$, $\rho_2 > 0$ and $L \geq 1$ such that for all $k \geq L$ we have $\|\tilde{u}_B^k\| = \frac{\|X_{k,B} s_B^k\|}{c^T x^k - c^*} \leq \rho_1 (c^T x^k - c^*)^{2r-1}$ and $e^T \tilde{u}_N^k = 1 + \delta_k$ with $|\delta_k| \leq \rho_2 (c^T x^k - c^*)^{2r}$. Thus for some $\delta_3 > 0$, $\hat{L} \geq L$ and all $k \geq \hat{L}$ we have

$$\begin{aligned} e^T X_k s^k &= (c^T x^k - c^*)(e^T \tilde{u}_B^k + e^T \tilde{u}_N^k) \\ &\geq 0.50(c^T x^k - c^*) \\ e^T x_N^k &\leq \rho_3 (c^T x^k - c^*) \\ x_j^k &> \sqrt{e^T X_k s^k} \text{ for all } j \in B \\ x_j^k &< \sqrt{e^T X_k s^k} \text{ for all } j \in N \end{aligned}$$

where the second inequality follows from 5.4.5 a(iii). The third inequality follows from the fact that $x_j^* > 0$ for each $j \in B$ and the fourth from first and second inequalities. ∎

We now establish another lemma.

5.4.23 Lemma Let $\hat{L}$ be generated by Lemma 5.4.22. There is an $\bar{L} \geq \hat{L}$ and a $\theta > 0$ such that for all $k \geq \bar{L}$

(a) $\|h_N^k - \Delta v_N^k\| \leq \theta (c^T x^k - c^*)^{2r}$.

(b) $0.50(c^T x^k - c^*) \leq \gamma_k \leq 1.50(c^T x^k - c^*)$.

Proof: We note that by substitution of the results of Lemma 5.4.9 and 5.4.11 (b), and definitions, we get

$$\begin{aligned} h_N^k &= \frac{1}{2r-1} \left(\frac{v_N^k}{e^T \tilde{u}_N^k} - \frac{V_{k,N}^r u_N^k}{\|u^k\|^2} \right) \\ &= \frac{1}{2r-1} \left(v_N^k - \frac{V_{k,N}^r u_N^k}{\|u^k\|^2} + V_{k,N}^r \Delta^k \right) \\ &\quad - \frac{1}{2r-1} \left(V_{k,N}^r \Delta^k + \frac{\delta_k}{1+\delta_k} v_N^k \right) \\ &= \Delta v_N^k - t^k. \end{aligned}$$

As v_N^k is bounded, 5.4.5, (a) follows. (b) readily follows from Lemma 5.4.9. ∎

We are now ready to prove the convergence to optimality of a cluster point of the primal sequence generated by the accelerated algorithm. This then establishes the existence of power centers of $F_\mathcal{D}$ and $\mathcal{V}_N$. This fact is used to obtain the convergence rate of this accelerated power primal affine scaling method.

5.4.24 Convergence of Accelerated Method Let assumptions (1) - (3) hold. There exists a subsequence K and vectors x^*, y^* and s^* such that for $k \in K$

1. $x^k \longrightarrow x^*$

2. $y^k \longrightarrow y^*$

3. $s^k \longrightarrow s^*$,

where x^* and (y^*, s^*) are optimal solutions of the primal and the dual problems respectively. Also, (y^*, s^*) is the power center of the optimal dual face.

Proof: Let $\hat{L}$ be as in Lemma 5.4.22. In case the predictor step is taken only a finite number of times, then α_k is selected by the variable step size selection strategy, and our result follows from 5.4.20. Thus, let there exist a subsequence K such that for each $k \in K$, a predictor step is taken. Then, for $k \in K$ and $k \geq \hat{L}$,

$$\|h_N^k\| \leq \gamma_k^{1.5r}.$$

But, from 5.4.3 (d), $\gamma_k \longrightarrow 0$. Thus, as

$$\|\Delta v_N^k\| \leq \|h_N^k\| + \|\Delta v_N^k - h_N^k\|,$$

from Lemma 5.4.23, $\Delta v_N^k \longrightarrow 0$; and the result follows by an argument identical to the proof of 5.4.16. ▶

Existence of the power center v_N^* of $\mathcal{V}_N$ has been established by 5.4.24. We now use the connection of Newton step for computing this center and affine step to establish the convergence rate of the accelerated algorithm. Two variants of the method will be considered, the two step and the three step method. The two step method is generated by taking one corrector step (using a small value of step size α) between each pair of predictor steps (using a step size α close to 1), The three step method is generated by taking two corrector steps between each pair of predictor steps. We now prove a result before investigating predictor and corrector steps.

5.4 POWER AFFINE SCALING METHOD

5.4.25 Lemma Let $\hat{L}$ be as in Lemma 5.4.22. There is a $\bar{\theta} > 0$ and an $\bar{L} \geq \hat{L}$ such that whenever $k \geq \bar{L}$, $\|v_N^k - v_N^*\| \leq \beta(c^T x^k - c^*)^r$, with $M > \beta > 0$ and $p \geq 1.5r$, then

$$\|\hat{w}_N^k\| \leq \bar{\theta}(c^T x^k - c^*)^p.$$

Proof: Since $v_N^* > 0$, there is an $\bar{L} \geq \hat{L}$ such that for all $k \geq \bar{L}$,

$$a = \frac{\min_j v_j^*}{2} \geq M(c^T x^k - c^*)^{1.5r}.$$

Thus, under the hypothesis of the lemma,

$$v_j^k \geq v_j^* - |v_j^* - v_j^k| \geq a.$$

and our result now follows by observing from the change of variables (5.63), that $\|\hat{w}_N^k\| \leq (2r-1)(\min_j v_j^k)^{-1}\|\Delta v_N^k\|$, and Theorem 2.7.4. ∎

We are now ready to investigate the predictor step.

5.4.26 Predictor Step Let $\bar{L}$ be as in Lemma 5.4.25. Assume that $0 < \delta < 1$, $k \geq \bar{L}$, $M > \beta > 0$, $1.5r \leq p$ and $\|v_N^k - v_N^*\| \leq \beta(c^T x^k - c^*)^p$. Then

(a) There are $0 < \theta_1 < \theta_2$ such that

$$\theta_1(c^T x^k - c^*)^{1+\delta\rho_k} \leq (c^T x^{k+1} - c^*) \leq \theta_2(c^T x^k - c^*)^{\rho_k^*}.$$

(b) There is a $\theta_3 > 0$ such that

$$\|v_N^{k+1} - v_N^*\| \leq \theta_3(c^T x^{k+1} - c^*)^{\frac{\rho_k'}{1+\delta\rho_k}}.$$

where $\rho_k^* = 1 + \min\{p, \delta\rho_k\}$ and $\rho_k' = \min\{2p, (1-\delta)\rho_k\}$.

Proof: From Theorem 2.7.4, $\|\Delta v_N^k\| \leq 1.5\beta(c^T x^k - c^*)^p$. Also, by Lemma 5.4.23

$$\begin{aligned}
\|h_N^k\| &\leq \|\Delta v_N^k\| + \|h_N^k - \Delta v_N^k\| \\
&\leq 1.5\beta(c^T x^k - c^*)^p + \theta(c^T x^k - c^*)^{2r} \\
&\leq \tilde{\theta}(c^T x^k - c^*)^p.
\end{aligned}$$

We note the following sequence of inequalities which follow from 5.4.5 (b), Lemmas 5.4.14, 5.4.23 (b), 5.4.25 and some $\bar{\theta} > 0$:

$$
\begin{aligned}
(0.50)^{\delta\rho_k}(c^T x^k - c^*)^{\delta\rho_k} &\leq \gamma_k^{\delta\rho_k} \\
&\leq 1 - \alpha_k \\
&\leq 1 - \alpha_k \frac{\|u^k\|^2}{\phi(V_k^{r-1} u^k)} \\
&= \frac{(c^T x^{k+1} - c^*)}{(c^T x^k - c^*)} \\
&\leq 1 - (1 - \max\{\epsilon_k^\delta, \gamma_k^{\delta\rho_k}\})(1 - \|w_N^k\| - |\delta_k|) \\
&\leq \max\{\epsilon_k^\delta, \gamma_k^{\delta\rho_k}\} + \|\hat{w}_N^k\| + |\delta_k| \\
&\leq \bar{\theta}(c^T x^k - c^*)^{\min\{p, \delta\rho_k\}},
\end{aligned}
$$

and we have (a). From 5.4.11 (a) and (b), and some simple algebra, we obtain

$$v_N^{k+1} - v_N^k = \Delta v_N^k + t^k$$

where

$$t^k = \frac{2r\alpha_k \delta(u_N^k) - 1}{1 - \alpha_k \delta(u_N^k)} \frac{1}{2r-1}(h_N^k + \frac{1}{2r-1}\frac{\delta_k}{1+\delta_k} v_N^k) - \frac{1}{2r-1} V_{k,N}^r \Delta^k.$$

Using the facts that $\delta(u_N^k) \leq 1$, $\alpha_k \leq 1 - \epsilon_k^\delta$, we obtain, for some $\hat{\theta} > 0$

$$\|t^k\| \leq \hat{\theta}(c^T x^k - c^*)^{(1-\delta)\rho_k}.$$

Now for appropriate positive constants, from Theorem 2.7.4 we see that

$$
\begin{aligned}
\|v_N^{k+1} - v_N^*\| &\leq \|v_N^k + \Delta v_N^k - v_N^*\| + \|t^k\| \\
&\leq \rho_2 \|v_N^k - v_N^*\|^2 + \hat{\theta}(c^T x^k - c^*)^{(1-\delta)\rho_k} \\
&\leq \rho_2'(c^T x^k - c^*)^{2p} + \hat{\theta}(c^T x^k - c^*)^{(1-\delta)\rho_k}
\end{aligned}
$$

and (b) follows from the lower bound of (a). ∎

We are now ready to investigate the corrector step of the algorithm.

5.4.27 Corrector Step Assume that for some $k \geq L_\delta$ and $\beta > 0$, $\|v_N^k - v_N^*\| \leq \beta(c^T x^k - c^*)^p$. Then

(a) $\frac{1.5r}{2} \leq p \leq r$. One iteration of the corrector step will be taken, and for some $\theta_1 > 0$
$$\|v_N^{k+1} - v_N^*\| \leq \theta_1(c^T x^{k+1} - c^*)^{2p}.$$

5.4 POWER AFFINE SCALING METHOD

(b) $\frac{1.5r}{4} \leq p \leq \frac{r}{2}$. Then at least one corrector step will be taken, and after at most two steps, for some $\theta_2 > 0$

$$\|v_N^{k+2} - v_N^*\| \leq \theta_1 (c^T x^{k+2} - c^*)^{4p}.$$

Proof: Let

$$\mu_k = \frac{\gamma_k \phi(X_{k,N}^{2r-1} s_N^k)}{2r \|X_k^r s^k\|^2}$$

From Lemmas 5.4.9 and 5.4.14 (b), $2r\mu_k \geq e^T \tilde{u}_k = 1 + \delta_k$. From Equation (5.73), $2r\mu_k \leq (1 + \delta_k)(1 + \|w_N^k\| + \|\Delta_k\|)$. Thus

$$\frac{1}{2} - |\delta_k'| \leq \mu_k \leq \frac{1}{2r} + |\rho_k'|$$

where, using Lemma 5.4.25, we note that $|\delta_k'| \leq \theta^*(c^T x^k - c^*)^{2r}$ and $|\rho_k'| \leq \theta'(c^T x^k - c^*)^p$. From Corollary 5.4.21, as α_k' approaches $\frac{2}{2r+1} - \epsilon$ for some small $\epsilon > 0$, for all sufficiently large k, at step 3.2,

$$\alpha_k = \frac{\gamma_k \phi(X_{k,N}^{2r-1} s_N^k)}{2r \|X_k^r s^k\|^2}.$$

Thus, from 5.4.5 b(i), Corollary 5.4.11 and simple algebra we see that

$$\begin{aligned} v_N^{k+1} - v_N^k &= \frac{1+\delta_k}{2r-1-\delta_k}(v_N^k - V_{k,N}^r \frac{u_N^k}{\|u^k\|^2}) \\ &= \Delta v_N^k - t^k \end{aligned}$$

where

$$t^k = -\frac{2(r-1)\delta_k + \delta_k^2}{(2r-1)(2r-1-\delta_k)}(v_N^k - V_{k,N}^r \frac{u_N^k}{\|u^k\|^2}) - \frac{1}{2r-1} V_{k,N}^r \Delta^k.$$

Using 5.4.5 b(i), and Lemma 5.4.14 (b) we can show that

$$\|t^k\| \leq \beta'(c^T x^k - c^*)^{2r}.$$

Thus, after one step with α_k, we see that (using Theorem 2.7.4),

$$\begin{aligned} \|v_N^{k+1} - v_N^*\| &\leq \|v_N^k + \Delta v_N^k - v_N^*\| + \|t^k\| \\ &\leq \rho' \|v_N^k - v_N^*\|^2 + \beta'(c^T x^k - c^*)^{2r} \quad (5.79) \\ &\leq \beta^*(c^T x^k - c^*)^{2p}. \end{aligned}$$

From, Lemma 5.4.5 b(iv),

$$\frac{c^T x^{k+1} - c^*}{c^T x^k - c^*} = 1 - \alpha_k \delta(u_N^k)$$

$$= 1 - \frac{e^T \tilde{u}_N^k}{2r}.$$

and using Lemma 5.4.9 we see that for all sufficiently large k

$$1 - \frac{1.5}{2r} \leq \frac{c^T x^{k+1} - c^*}{c^T x^k - c^*} \leq 1 - \frac{0.5}{2r}. \tag{5.80}$$

Substituting the above inequality, we obtain (a) with $\theta_1 = (1 - \frac{1.5}{2r})^{2p} \beta^*$.

To see (b), we note that after one corrector step, either $2p$ becomes greater than $1.5r$ and we stop the corrector iterates and go to the predictor step; or, after one more corrector step the desired result is obtained. ∎

We are now ready to prove the main convergence theorem.

5.4.28   Accelerated Convergence Rate Let the sequences $\{x^k\}$, $\{y^k\}$ and $\{s^k\}$ be generated by the accelerated method with $r > 1$, and let assumptions (1) - (3) hold. Then, there exist vectors x^*, y^* and s^* such that

1. $x^k \longrightarrow x^*$

2. $y^k \longrightarrow y^*$

3. $s^k \longrightarrow s^*$

where x^* lies in the relative interior of the optimal face of the primal, and (y^*, s^*) is the power center of the optimal face of the dual. In addition, asymptotically, the sequence $\{c^T x^k - c^T x^*\}$ converges to zero as follows:

(a) For $\delta = \frac{1}{2(r+1)}$, the convergence is two-step superlinear at the rate $1 + \frac{r}{r+1}$.

(b) For $\delta = \frac{3}{2(r+2)}$, the convergence is three-step superlinear at the rate $1 + \frac{3r}{r+2}$.

Proof: Assume that only a finite number of predictor steps are taken. Then there is an $L \geq 1$ such that for every $l \geq L$, a corrector step is taken. As in the proof of 5.4.27, from the Equations (5.79) and (5.80), we obtain

$$\|v_N^{k+1} - v_N^*\| \leq \rho \|v_N^k - v_N^*\|^2 + \hat{\beta}(c^T x^{k+1} - c^*)^{2r}.$$

5.4 POWER AFFINE SCALING METHOD

Thus, after several such corrector iterations, $\|v_N^{k+l} - v_N^*\| \leq \rho'(c^T x^{k+l} - c^*)^p$ for $p \geq 1.5r$ and $l \geq 1$. From Lemma 5.4.23, $\rho_{k+l} \geq 1.5r$ and a corrector step must be taken, and we have a contradiction. We note that the constant $\rho' > 0$ is independent of k, and thus the required M in 5.4.26 exists.

Let k be an index, sufficiently large at which a predictor step is performed. To investigate the convergence rate of the two-step method, assume $\rho_k \geq 1.5r$, and let $\delta = \frac{\rho_k - r}{\rho_k(r+1)}$. By the choice of ρ_k, at Step 3.2, for $p \geq 1.5r$, $\rho'_k = \min\{2p, (1-\delta)\rho_k\} = (1-\delta)\rho_k$ after one corrector step, from Propositions 5.4.26 and 5.4.27, we obtain

$$\|v_N^{k+2} - v_N^*\| \leq \theta_3 (c^T x^{k+2} - c^*)^{\frac{2(1-\delta)\rho_k}{1+\delta\rho_k}}.$$

where $\frac{2(1-\delta)\rho_k}{1+\delta\rho_k} = 2r$. Thus (a) follows as

$$\delta = \frac{1}{2(r+1)} \quad \text{when } \rho_k = 2r$$

and the convergence rate obtained is

$$\rho^* = 1 + \min\{2r, \delta\rho_k\} = 1 + \delta\rho_k = 1 + \frac{r}{r+1}.$$

For the three step method, let $\delta = \frac{2\rho_k - r}{\rho_k(r+2)}$. After two corrector steps

$$\|v_N^{k+3} - v_N^*\| \leq \theta_4 (c^T x^{k+3} - c^*)^{\frac{4(1-\delta)\rho_k}{1+\delta\rho_k}}$$

where $\frac{4(1-\delta)\rho_k}{1+\delta\rho_k} = 2r$. (b) now follows since $\delta = \frac{3}{2(r+2)}$ if $\rho_k = 2r$ and the convergence rate $1 + \delta\rho_k = 1 + \frac{3r}{r+2}$. ∎

We now investigate the efficiency of the convergence rates obtained, and thus get some measure of the relative effectiveness of the acceleration. For this purpose, we will use the measure introduced by Ostrowski to compare algorithms achieving different rates of convergence, and requiring different amounts of work per iteration. He suggested the following measure

$$\frac{\log(\rho)}{w}$$

where ρ is the convergence rate of the algorithm, and w is a measure of the work per iteration. The larger this measure, the more efficient the algorithm.

The convergence of the accelerated primal affine scaling method depends on the choice of r and the two-step or three-step method. The table below shows these calculations for several choices.

	Two - Step		Three - Step	
	rate	efficiency	rate	efficiency
$r = 1.0$	1.5^1	$\frac{0.2027}{w}$	2.0^2	$\frac{0.2310}{w}$
$r = 1.5$	1.6	$\frac{0.2350}{w}$	2.2857	$\frac{0.2756}{w}$
$r = 2.0$	1.67	$\frac{0.2564}{w}$	2.50	$\frac{0.3054}{w}$
$r = 4.0$	1.80	$\frac{0.2939}{w}$	3.0^3	$\frac{0.3662}{w}$

5.5 Obtaining an Initial Interior Point

To start the primal affine scaling method, we need an initial interior point x^0. This is generally obtained by a Big M strategy we describe now.

5.5.1 Big M Strategy Consider the following linear program with the added artificial variable x_{n+1}

$$\begin{aligned} \text{minimize} \quad & c^T x + M x_{n+1} \\ & Ax + A_{.n+1} x_{n+1} = b \\ & x \geq 0 \quad\quad x_{n+1} \geq 0 \end{aligned}$$

where $A_{.n+1} = b - Ax^0$ for some arbitrary vector $x^0 > 0$. It is general practice to choose $x^0 = (1, \cdots, 1)^T$, an n vector of all ones. It is readily confirmed that

$$x = x^0, \quad x_{n+1} = 1$$

is an interior point of the above problem. We call this the artificial primal problem. The dual to this problem is

$$\begin{aligned} \text{maximize} \quad & b^T y \\ & A^T y \leq c \\ & A_{.n+1}^T y \leq M. \end{aligned}$$

In case M is so large that the last constraint of the dual has a positive value of the slack for all solutions of the dual, then the value of the artificial value x_{n+1} at the optimum solution to the primal will be zero (from the Complementary Slackness Theorem 3.3.3) and the problem will be solved. The next theorem gives an idea of how large M should be.

5.5.2 Theorem Let the dual have an extreme point optimum solution and the data (A, b, c) be integral. Then for $M = 4^L$, where L is the length of the string needed to code all the data of the problem, $x_{n+1} = 0$ for all optimum solutions of the artificial primal problem.

5.5 INITIAL INTERIOR POINT

Proof: Let y be an extreme point of the dual. Then, from 2.6.29, $|y_i| < 2^L$. Thus

$$\begin{aligned}
|A_{.n+1}y| &< 2^L \sum_{i=1}^{m} |a_{i,n+1}| \\
&\leq 2^L \sum_{i=1}^{m} |b_i - \sum_{j=1}^{n} a_{ij} x_j^0| \\
&\leq 2^L (\|b\|_1 + \sum_{j=1}^{n} \|A_{.j}\|_1 \|x^0\|_1) \\
&\leq 4^L
\end{aligned}$$

Thus choosing $M = 4^L$ will maintain a positive slack at each vertex of the dual, and we are done. ∎

There is a simple test that may allow the elimination of the artificial variable before the problem is solved. This is given in the next result.

5.5.3 Theorem Let the vector (x^k, x_{n+1}^k) be generated at step k. If

$$\hat{x}^k = \frac{x^k - x_{n+1}^k x^0}{1 - x_{n+1}^k} > 0$$

the artificial variable can be dropped, and the affine scaling method can be continued with the vector $\hat{x}^k$.

Proof: It can be readily confirmed that the vector $\hat{x}^k$ is an interior feasible solution for the problem. ∎

In case the matrix A of the linear program is sparse (which is the case when the problem is large), adding a column $A_{.n+1}$ as we have done for the Big M method, will generally, make the matrix $AX_k A^T$ dense, even when AA^T is sparse. This then is a great disadvantage of this approach. We leave further discussion of this to the implementation chapter.

5.5.4 Phase 1 Phase 2 Strategy We have seen in Theorems 5.2.8 and 5.3.18 that by choosing an appropriate step selection strategy, we can guarantee that the primal sequence converges to the relative interior of the optimal face. This result can be used to generate a starting interior point by a Phase 1 - Phase 2 strategy. An advantage of this method over Big M

method is that the question of how large M should be does not arise, as well as the sparsity pattern of the system to be solved is not effected.

Let $x^0 > 0$ be such that $b - Ax^0 \neq 0$. Define a diagonal matrix D such that $D_{ii} \in \{-1, +1\}$ for each $i = 1, \cdots, m$, and $u^0 = D(b - Ax^0) > 0$. Now define the Phase 1 problem:

$$\begin{array}{rl} \text{minimize} & e^T u \\ Ax + Du & = b \\ x \geq 0 \quad u & \geq 0. \end{array}$$

Note that $x^0 > 0$ and $u^0 > 0$ is an interior point solution for this problem. In case $b_i - A_i x^0 = 0$, we do not add an artificial variable to this constraint, and this can be reflected in the definition of D, where rows corresponding to such constraints are all zero in D, and the corresponding columns are removed from D.

It is clear that the objective function is constant for all values of feasible solutions x, if they exist. As in the simplex method, if the optimum value of the objective function is 0, then an interior point of the primal problem x^* would have been found and Phase 2 can be started with x^*. In case $x_j^* = 0$ for some j, from the strict complementary solution property, $s_j^* > 0$, and we can conclude that $x_j = 0$ for all feasible solutions of the primal, and this variable can be removed from the Phase 2 problem. The sparsity pattern of AA^T is inherited by both the Phase 1 and Phase 2 problems. This is because $(A, I)(A, I)^T = AA^T + I$, and during Phase 1, entries are added to non-zero diagonal elements of AA^T.

5.6 Bounded Variable Affine Scaling Method

Like the simplex method, the primal affine scaling method also specializes for the bounded variable linear program. For such linear programs, we now develop a variant of the primal affine scaling method. Thus, we consider the linear program:

$$\begin{array}{rl} \text{minimize} & c^T x \\ Ax & = b \\ l \leq x & \leq u \end{array}$$

where $l < u$. It is generally convenient to make the lower bound zero, done readily by the change of variables $x' = x - l$, to generate the equivalent problem:

$$\begin{array}{rl} \text{minimize} & c^T x \\ Ax & = b - Al \\ 0 \leq x & \leq u - l \end{array}$$

5.6 BOUNDED VARIABLE METHOD

For the ease of notation, we will use the same symbols b and u to represent the changed vectors in the transformed problem.

This variant is based on the following simple lemma:

5.6.1 Lemma Consider the symmetric positive definite matrix

$$B = \begin{bmatrix} AD_1 A^T & AD_1 \\ D_1 A^T & D_2 \end{bmatrix}$$

where D_1 and D_2 are diagonal matrices with positive diagonal entries. Then there exists a decomposition

$$B = \begin{bmatrix} I & AD_1 D_2^{-\frac{1}{2}} \\ 0 & D_2^{\frac{1}{2}} \end{bmatrix} \begin{bmatrix} AD_3 A^T & 0 \\ 0 & I \end{bmatrix} \begin{bmatrix} I & 0 \\ D_2^{-\frac{1}{2}} D_1 A^T & D_2^{\frac{1}{2}} \end{bmatrix}$$

where D_3 is a diagonal matrix and $D_3 = D_1 - D_1^2 D_2^{-1}$.

Proof: Can be readily confirmed by multiplication of the factors, and equating the two sides of the identity. ∎

By adding slack variables t, convert the linear program into the standard form:

$$\begin{aligned} \text{minimize} \quad & c^T x \\ Ax & = b \\ x + t & = u \\ x \geq 0 \quad & t \geq 0 \end{aligned}$$

and note that the underlying constraint matrix is

$$\bar{A} = \begin{bmatrix} A & 0 \\ I & I \end{bmatrix}$$

and the vector of variables is $d^T = (x^T, t^T)$. Given an interior point $x > 0$ and $t > 0$, the basic operation of the primal affine scaling method is to compute the dual estimate y which requires the inversion of the matrix

$$B = \bar{A} D^2 \bar{A}^T = \begin{bmatrix} AX^2 A^T & AX^2 \\ X^2 A^T & X^2 + T^2 \end{bmatrix}$$

or, equivalently, by solving the system

$$By = \bar{A} D^2 \bar{c}$$

where $\bar{c}^T = (c^T, 0)$. Since $X^2 + T^2$ is diagonal, we can use the result of the Lemma 5.6.1 and define

$$D_3 = X^2 - X^4(X^2 + T^2)^{-1}$$

which gives the jth diagonal element of D_3 as

$$\frac{x_j^2 t_j^2}{x_j^2 + t_j^2}.$$

Setting $D_1 = X^2$ and $D_2 = X^2 + T^2$ we note, from Lemma 5.6.1, that

$$B^{-1} = \begin{bmatrix} I & 0 \\ -D_2^{-1}D_1A^T & D_2^{-\frac{1}{2}} \end{bmatrix} \begin{bmatrix} (AD_3A^T)^{-1} & 0 \\ 0 & I \end{bmatrix} \begin{bmatrix} I & -AD_1D_2^{-1} \\ 0 & D_2^{-\frac{1}{2}} \end{bmatrix}.$$

Also,

$$\bar{A}D^2\bar{c} = \begin{bmatrix} AD_1c \\ D_1c \end{bmatrix},$$

and thus, for $y^T = (y_1^T, y_2^T)$, we have

$$\begin{aligned} y &= (\bar{A}D^2\bar{A}^T)^{-1}\bar{A}D^2\bar{c} \\ &= \begin{bmatrix} (AD_3A^T)^{-1}AD_3c \\ D_1D_2^{-1}(I - A^T(AD_3A^T)^{-1}AD_3)c \end{bmatrix} \\ &= \begin{bmatrix} y_1 \\ D_1D_2^{-1}(c - A^Ty_1) \end{bmatrix}. \end{aligned}$$

Also,

$$\begin{aligned} D^2s &= D^2(\bar{c} - \bar{A}^Ty) \\ &= \begin{bmatrix} D_1(c - A^Ty_1 - y_2) \\ -T_1^2D_1D_2^{-1}(c - A^Ty_1) \end{bmatrix} \\ &= \begin{bmatrix} D_3(c - A^Ty_1) \\ -D_3(c - A^Ty_1) \end{bmatrix} \end{aligned}$$

and

$$\begin{aligned} Ds &= D(\bar{c} - \bar{A}^Ty) \\ &= \begin{bmatrix} X^{-1}D_3(c - A^Ty_1) \\ -T^{-1}D_3(c - A^Ty_1) \end{bmatrix} \end{aligned}$$

We now define the bounded variable variant of the primal affine scaling method.

Bounded Variable Primal Affine Scaling Method

Step 0 Let $u > x^0 > 0$ be an interior point. Define $t^0 = u - x^0$. Also let $0 < \alpha < 1$. Typically $\alpha = 0.95$. Set $k = 0$.

Step 1 Define the diagonal matrix D_3 with the jth diagonal element
$$\frac{(x_j^k)^2 (t_j^k)^2}{(x_j^k)^2 + (t_j^k)^2}.$$

Step 2 Define
$$\begin{aligned} y^k &= (AD_3 A^T)^{-1} AD_3 c \\ w^k &= c - A^T y^k \\ \bar{w}^k &= D_3 w^k. \end{aligned}$$

Step 3 Min Ratio Test: Define
$$\theta_{1,k} = \min \{\frac{x_j^k}{\bar{w}_j^k} : \bar{w}_j^k > 0\}$$
$$\theta_{2,k} = \min \{\frac{t_j^k}{-\bar{w}_j^k} : \bar{w}_j^k < 0\}$$

and
$$\theta_k = \min \{\theta_{1,k}, \theta_{2,k}\}.$$

Step 4
$$\begin{aligned} x^{k+1} &= x^k - \alpha \theta_k D_3 w^k \\ t^{k+1} &= t^k + \alpha \theta_k D_3 w^k \end{aligned}$$

If, for some j, either $x_j^{k+1} = 0$ or $t_j^{k+1} = 0$ STOP. Otherwise, set $k = k + 1$ and go to step 1.

5.7 Affine Scaling and Unrestricted Variables

Allowing unrestricted variables in the Primal problem unifies the treatment of the primal and the dual affine scaling methods (which we will study in the next section). In this case, only the non-negative orthant of the restricted variables is approximated by an ellipsoid; resulting, geometrically in a hyper-cylinder with an ellipsoidal section. This problem has a closed form solution, which is more complicated but readily obtainable. Thus, we consider the following linear program:

$$\begin{align} \text{minimize} \quad & c^T x + d^T y \\ & Ax + By = b \\ & x \geq 0. \end{align}$$

In this linear program, the variables y are unrestricted in sign. We assume that

1. A is an $m \times n$ matrix and has rank m.

2. B is an $m \times k$ matrix and has rank k.

3. The linear program has a interior solution.

4. The objective function is not constant on the feasible region.

Note that when $d \neq 0$, Assumption (2) is necessary to make the problem interesting, since otherwise the problem has an unbounded objective function value. Let $(\bar{x}, \bar{y})$ with $\bar{x} > 0$ be an interior feasible solution with $A\bar{x} + B\bar{y} = b$.

The Approximating Problem

Given an interior point $(\bar{x}, \bar{y}) > 0$ of the linear program, we will use the strategy of the primal affine scaling method to approximate the set $\{x : x \geq 0\}$ by an ellipsoid, and generate the following approximating problem:

$$\begin{align} \text{minimize} \quad & c^T x + d^T y \\ & Ax + By = b \\ & \|\bar{X}^{-1}(x - \bar{x})\| \leq \rho \end{align}$$

for some $0 < \rho \leq 1$.

By the usual affine transformation, let the problem on the hyper-spherical cylinder be:

$$\begin{align} \text{minimize} \quad & c^T \bar{X} x + d^T y \\ & A\bar{X} x + By = b \\ & \|(x - e)\| \leq \rho \end{align}$$

Since the solution to the problem lies on the boundary of the second constraint, using Lagrangian theory, we can derive the following necessary

5.7 WITH UNRESTRICTED VARIABLES

conditions for the solution:

$$\bar{X}c + \bar{X}A^T z + \theta(x - e) = 0 \quad (5.81)$$
$$d + B^T z = 0 \quad (5.82)$$
$$A\bar{X}(x - e) + B(y - \bar{y}) = 0 \quad (5.83)$$
$$\|x - e\| = \rho \quad (5.84)$$
$$\theta > 0. \quad (5.85)$$

Here, θ and z are the Lagrange multipliers on the constraints. We now show that if x^*, y^*, θ^* and z^* solve the system (5.81) - (5.85), then x^* and y^* solve the transformed approximating problem, and thus $\bar{X}x^*$ and y^* solve the approximating problem. We first prove a simple lemma:

5.7.1 Lemma For every x and $\bar{x}$,

$$\|x - e\|^2 \geq \|\bar{x} - e\|^2 + 2(x - \bar{x})^T(\bar{x} - e).$$

Proof: Follows trivially from the identity $(x_j - 1)^2 = (\bar{x}_j - 1)^2 + 2(x_j - \bar{x}_j)(\bar{x}_j - 1) + (x_j - \bar{x}_j)^2$. ∎

5.7.2 Theorem Let x^*, y^*, θ^* and z^* satisfy conditions (5.81) - (5.85). Then $\bar{X}x^*, y^*$ solve the approximating problem.

Proof: Let x and y be any pair of vectors satisfying

$$A\bar{X}x + By = b$$
$$\|x - e\| \leq \rho$$

and consider the following:

$$c^T \bar{X} x + d^T y = c^T \bar{X} x^* + d^T y^* + c^T \bar{X}(x - x^*) + d^T(y - y^*) \quad (5.86)$$
$$0 = A\bar{X}(x - x^*) + B(y - y^*) \quad (5.87)$$
$$\|x - e\|^2 - \rho^2 \geq \|x^* - e\|^2 - \rho^2 + 2(x^* - e)^T(x - x^*) \quad (5.88)$$

where (5.88) is from the Lemma 5.7.1. Now, multiply (5.86) by 1, (5.87) by $(z^*)^T$ and (5.88) by $\theta^*/2$ to obtain

$$c^T \bar{X} x + d^T y + \frac{\theta^*}{2}(\|x - e\|^2 - \rho^2) \geq c^T \bar{X} x^* + d^T y^* +$$
$$(x - x^*)^T(\bar{X}(c + A^T z^*) + \theta^*(x^* - e))$$
$$(y - y^*)^T(d + B^T z^*)$$
$$= c^T \bar{X} x^* + d^T y^*.$$

Since $\|x - e\|^2 - \rho^2 \leq 0$ and $\theta^* > 0$ we note that x^* and y^* solve the transformed problem. Thus $\bar{X}x^*$ and y^* solve the approximating problem. ∎

5.7.3 Theorem The solution x^*, y^*, θ^* and z^* of the system (5.81) - (5.85) is the following:

$$z^* = \theta^*(A\bar{X}^2 A^T)^{-1} B(y^* - \bar{y}) - (A\bar{X}^2 A^T)^{-1} A\bar{X}^2 c \quad (5.89)$$

$$y^* - \bar{y} = \frac{1}{\theta^*}(B^T(A\bar{X}^2 A^T)^{-1} B)^{-1}(B^T(A\bar{X}^2 A^T)^{-1} A\bar{X}^2 c - d) \quad (5.90)$$

$$x^* - e = -\frac{1}{\theta^*}\bar{X}(c + A^T z^*) \quad (5.91)$$

$$\theta^* = \|\bar{X}(c + A^T z^*)\|/\rho \quad (5.92)$$

Proof: Note that z^* is independent of θ^*, and thus θ^* is well defined by the Equation (5.92). The other values can be confirmed by direct substitution into the system (5.81) - (5.85). From our assumptions, matrices $A\bar{X}^2 A^T$ and $B^T(A\bar{X}^2 A^T)^{-1} B$ are invertible, and we are done. ∎

We can derive the primal affine scaling method by setting $B = 0$ and $d = 0$ in the Equations (5.89) - (5.92). This results in

$$z^* = -(A\bar{X}^2 A^T)^{-1} A\bar{X}^2 c$$

$$x^* - e = -\frac{1}{\theta^*}\bar{X}(c + A^T z^*)$$

$$\theta^* = \|\bar{X}(c + A^T z)\|/\rho$$

and from Theorem 5.7.2, the optimal solution to the approximating problem is $\bar{X}x^*$ and we obtain the primal affine scaling iterate by observing that $\bar{X}e = \bar{x}$.

We will discuss the derivation of the Dual Affine Scaling Method from the Equations (5.89) - (5.92) in the next section.

5.8 Dual Affine Scaling Method

The affine scaling strategy applied to the dual problem generates the dual affine scaling method. Thus, the strategy is applied to

$$\begin{aligned} \text{maximize} \quad & b^T y \\ & A^T y + s = c \\ & s \geq 0 \end{aligned}$$

5.8 DUAL AFFINE SCALING METHOD

the dual of the canonical linear program. The strategy here is to start with an interior point solution of the dual, create the approximating problem by replacing the non-negativity constraints by an ellipsoidal constraint, and solve to generate the dual affine scaling iterates. Since the variables y are unrestricted and the variables s are non-negative, we can use the results of the section 5.7, specifically, the Equations (5.89) - (5.92) to generate the solution to the approximating problem defined by the interior point (y^0, s^0) where $s^0 > 0$ and $A^T y^0 + s^0 = c$. This solution is:

$$z^* = \theta^* S_0^{-2} A^T (y^* - y^0)$$
$$y^* - y^0 = \frac{1}{\theta^*}(AS_0^{-2}A^T)^{-1} b$$
$$s^* - s^0 = -A^T(y^* - y^0)$$
$$\theta^* = \frac{1}{\rho}\|S_0^{-1} A^T (AS_0^{-2}A^T)^{-1} b\|$$

which generates the following dual affine scaling method:

5.8.1 Dual Affine Scaling Method

Step 0 Let (y^0, s^0) with $s^0 > 0$ be an interior point. k=0.

Step 1 Define

$$dy^k = (AS_k^{-2} A^T)^{-1} b$$
$$\theta_k = \|S_k^{-1} A^T dy^k\|$$

Step 2 Set

$$y^{k+1} = y^k + \frac{1}{\theta_k} dy^k$$
$$s^{k+1} = s^k - \frac{1}{\theta_k} A^T dy^k$$
$$x^{k+1} = S_k^{-2} A^T dy^k$$

Step 3 If $s_j^{k+1} = 0$ for some j, STOP. Otherwise, set $k = k + 1$, and go to Step 1.

We now show that if the algorithm stops in Step 3, then y^{k+1} and s^{k+1} solve the dual problem and x^{k+1} solves the primal problem.

5.8.2 Theorem
If $s_j^{k+1} = 0$ for some j then y^{k+1}, s^{k+1} solve the dual problem and x^{k+1} solves the primal problem.

Proof: Since $s_j^{k+1} = 0$,

$$s_j^k - \frac{(A^T dy^k)_j}{\|S_k^{-1} A^T dy^k\|} = 0$$

or

$$(s_j^k)^{-1}(A^T dy^k)_j = \|S_k^{-1} A^T dy^k\|.$$

Thus

$$(A^T dy^k)_j \begin{cases} = 0 & i \neq j \\ > 0 & i = j \end{cases}$$

and so $x^{k+1} = S_k^{-2} A^T dy^k \geq 0$. Also $Ax^{k+1} = b$ and $(s^{k+1})^T x^{k+1} = 0$ and the conditions of the complementary slackness theorem are satisfies, and we are done. ∎

We now present the large step dual affine scaling method where we will move beyond the boundary of the ellipsoid, but still stay interior to the non-negative orthant.

5.8.3 Large Step Dual Affine Scaling Method

Step 0 Let (y^0, s^0) with $s^0 > 0$ be an interior point of the dual, and let $0 < \alpha < 1$ be a constant. We will typically choose $\alpha = 0.95$. Set $k = 0$.

Step 1 Define

$$dy^k = (A S_k^{-2} A^T)^{-1} b$$

Step 2 Define

$$\begin{aligned} ds^k &= -A^T dy^k \\ x^k &= S_k^{-2} ds^k \end{aligned}$$

Step 3 Unbounded Solution: If

$$ds^k \geq 0$$

then STOP. The solution is unbounded.

5.8 DUAL AFFINE SCALING METHOD

Step 4 Min Ratio Test: Define

$$\theta_k = \min \left\{ \frac{s_j^k \|S_k^{-1} ds^k\|}{-ds_j^k} : ds_j^k < 0 \right\}$$

If $\theta_k = 1$ set $\alpha = 1$.

Step 5 Let

$$\bar{\theta}_k = \frac{\alpha \theta_k}{\|S_k^{-1} ds^k\|}$$

and set

$$\begin{aligned} y^{k+1} &= y^k + \bar{\theta}_k dy^k \\ s^{k+1} &= s^k + \bar{\theta}_k ds^k \end{aligned}$$

Step 6 If $s_j^{k+1} = 0$ for some j, then STOP. Otherwise set $k = k+1$ and go to Step 1.

5.8.4 Proof of Convergence of the Large Step Dual Affine Scaling Method

We will now prove the convergence of the dual affine scaling method with the following assumptions on the underlying dual problem:

1. The linear program has an interior point.

2. $b \neq 0$.

3. Every feasible solution of the dual is non-degenerate.

4. The primal has a feasible solution.

5. The matrix A is $m \times n$ and has rank m.

We will now assume that the infinite sequences $\{y^k\}$, $\{s^k\}$ and $\{x^k\}$ are generated by the large step dual affine scaling method. We now show that they converge to the solutions of their respective problems. Let $dy^k = y^{k+1} - y^k$ and $ds^k = s^{k+1} - s^k$.

5.8.5 Theorem $s^k > 0$ for all k, and the sequence $\{b^T y^k\}$ is monotone increasing, and converges. Also $S_k^{-1} ds^k = S_k x^k \to 0$ as $k \to \infty$, and $\sum_{i=1}^{\infty} b^T (y^{k+1} - y^k) < \infty$.

Proof: By the min ratio test at step 4, $s^k > 0$ for all k. Also, by a simple calculation $\|S_k^{-1} ds^k\| = \sqrt{b^T (AS_k^{-2} A^T)^{-1} b}$. Thus, from Step 5

$$b^T y^{k+1} = b^T y^k + \alpha \theta_k \sqrt{b^T (AS_k^{-2} A^T)^{-1} b}.$$

From Assumption (5), $AS_k A^T$ is positive definite, and Assumption (2) implies that the expression under the square root is positive, and we have the monotonicity of the sequence. Since the primal is feasible, this sequence is bounded, and thus must converge. Since $\alpha \theta_k \geq 0.95$ for all k, $b^T (AS_k^{-2} A^T)^{-1} b \to 0$ as $k \to \infty$. Thus $S_k^{-1} ds^k \to 0$ or $S_k . S_k^{-2} ds^k \to 0$. Also $\sum_{i=1}^{\infty} b^T (y^{k+1} - y^k) = b^* - b^T y^0$ where $b^T y^k \to b^*$, and we are done. ∎

We are now ready to prove the convergence of the dual affine scaling method.

5.8.6 Dual Affine Convergence Theorem Let the assumptions (1) - (5) hold. Then the sequences $\{x^k\}$, $\{y^k\}$ and $\{s^k\}$ converge to an optimum solution of the primal and the dual problems respectively.

Proof: Note that the vector dy^k computed at Step 1 can be obtained by solving

$$\text{maximize} \quad b^T u$$
$$u^T AS_k^{-2} A^T u \leq 1$$

This is seen by using the Lagrangian theory since the solution will lie on the boundary of the ellipsoid. The solution is

$$u^k = \frac{1}{\lambda_k} (AS_k^{-2} A^T)^{-1} b$$
$$\lambda_k = \sqrt{b^T (AS_k^{-2} A^T)^{-1} b}$$

and, thus for each k

$$dy^k = \alpha \theta_k u^k.$$

From Theorem 5.1.17, there is a constant $q(A,b) > 0$ such that $|b^T u^k| \geq q(A,b) \|u^k\|$. Thus, $|b^T dy^k| = \alpha \theta_k |b^T u^k| \geq \alpha \theta_k q(A,b) \|u^k\| = q(A,b) \|dy^k\|$. From Theorem 5.8.5,

$$\infty > \sum_{k=1}^{\infty} b^T dy^k \geq q(A,b) \sum_{k=1}^{\infty} \|dy^k\|.$$

Thus the sequence $\{y^k\}$ converges to, say, y^*. Since for each k, $s^k = c - A^T y^k$, the sequence $\{s^k\}$ also converges, say to $s^* \geq 0$.

Define $B = \{j : s_j^* = 0\}$. Thus $A_B^T y^* = c_B$. As all dual solutions are non-degenerate, A_B has full column rank m. Now $x^k = -S_k^{-2} A^T (A S_k^{-2} A^T)^{-1} b$. From Theorem 5.1.11, for each j, $|(A S_k^{-2} A^T)^{-1} A S_k^{-2} u_j|$, where u_j is the jth unit vector, is bounded by a constant $q(A, u_j) > 0$. Defining $q(A) = \max q(A, u_j) > 0$ we note that $\|S_k^{-2} A^T (A S_k^{-2} A^T)^{-1} b\| \leq \sqrt{n} q(A) \|b\|$ and thus $\{x^k\}$ is bounded. Let x^* be any cluster point of this sequence. Now consider $j \notin B$. From Theorem 5.8.5, $x_j^k s_j^k \to 0$. Hence $x_j^* = 0$ for each $j \notin B$. Thus
$$b = Ax^* = A_B x_B^*.$$

But this system has a unique solution (since A_B has full column rank) thus $\{x^k\}$ converges to x^*.

Now consider $j \in B$. If $x_j^* < 0$ then there is an $L \geq 1$ such that for all $k \geq L$, $x_j^k < 0$. But $0 > x_j^k = (s_j^k)^{-2} ds_j^k$ implies that $ds_j^k < 0$, and since
$$\begin{aligned} s_j^{k+1} &= s_j^k + \bar{\theta}_k ds_j^k \\ &> s_j^k \end{aligned}$$

contradicts $j \in B$. Since y^*, s^* are feasible for the dual and x^* is feasible for the primal and satisfy the complementary slackness conditions, we are done. ■

5.9 Primal-Dual Affine Scaling Method

The strategy of the primal-dual affine scaling method differs from that of the primal affine scaling method or the dual affine scaling method. Here, starting with an interior point of the primal and the dual, a step is taken in both the polyhedrons simultaneously. This step is determined by applying Newton's method (of section 2.7.1) to a system based on the complementary slackness theorem 3.3.3. The step taken in the primal and the dual polyhedrons differs, respectively, from the primal affine scaling step in the primal polyhedron (of section 5.1) and dual affine scaling step in the dual polyhedron (of section 5.8) in the scaling matrix. This justifies the name of this method. Thus, primal-dual affine scaling method deals with the following primal-dual pair of linear programs:
$$\begin{aligned} \text{minimize} \quad & c^T x \\ Ax &= b \\ x &\geq 0 \end{aligned}$$

and its dual

$$\begin{array}{rl} \text{maximize} & b^T y \\ A^T y + & s = c \\ & s \geq 0. \end{array}$$

We will investigate the primal-dual affine scaling method under the following assumptions:

1. The matrix A is $m \times n$ and has rank m.

2. Both the primal and the dual objective functions are not constant on their respective feasible regions.

3. Both the primal and the dual problems have non-empty interiors.

We will show that the primal-dual method generates sequences that converge either to an optimum solution for the pair of linear programs, or to a non-optimal solution where no variable satisfies the strict complementarity condition (of 3.3.4), and thus can get caught at some boundary region. Barrier, Homotopy or potential reduction methods avoid this by controlling the distance of the iterates to the boundary.

Given $x^0 > 0$ an interior point of the primal and (y^0, s^0), $s^0 > 0$ an interior point of the dual, the direction in which complementary slackness will decrease is chosen by applying Newton's method to the following system of non-linear equations, generated by the complementary slackness theorem 3.3.3:

$$\begin{array}{rl} Ax & = b \\ A^T y + s & = c \\ Xs & = 0 \\ (x \geq 0) \quad (s \geq 0) & \end{array} \quad (5.93)$$

which are the complementary slackness conditions satisfied by a pair of optimum solutions. The first two conditions represent primal and dual feasibility, and third, the complementary slackness condition. When Newton's method is applied to this system, starting at the interior points x^0 and (y^0, s^0), it generates the directions $(\Delta x, \Delta y, \Delta s)$ given by the following system of linear equations:

$$\begin{array}{rl} A \Delta x & = 0 \\ A^T \Delta y + \Delta s & = 0 \\ S \Delta x + X \Delta s & = -X_0 s^0. \end{array} \quad (5.94)$$

This then generates the following primal-dual affine scaling method:

5.9 PRIMAL-DUAL AFFINE METHOD

5.9.1 Primal-Dual Affine Scaling Method

Step 0 Let (x^0, y^0, s^0) be an interior solution, $0 < \alpha < 1$, and let $k = 0$.

Step 1 Compute direction $(\Delta x^k, \Delta y^k, \Delta s^k)$ by solving system (5.94). If $\Delta x^k = 0$, $\Delta y^k = 0$ and $\Delta s^k = 0$, then stop. The solution (x^k, y^k, s^k) is optimum.

Step 2 Minimum Ratio Test: Define:

$$\phi_k = \max\{\phi(-X_k^{-1}\Delta x^k), \phi(-S_k^{-1}\Delta s^k)\}$$

where $\phi(u) = \max_j u_j$ for any given vector u. If $\phi_k = 1$, set $\alpha = 1$.

Step 3 Next Interior Point:

$$x^{k+1} = x^k + \frac{\alpha}{\phi_k}\Delta x^k$$
$$y^{k+1} = y^k + \frac{\alpha}{\phi_k}\Delta y^k$$
$$s^{k+1} = s^k + \frac{\alpha}{\phi_k}\Delta s^k$$

Step 4 Iterative Step: Set $k = k + 1$. If $\phi_{k-1} = 1$ then stop. The solution (x^k, y^k, s^k) is an optimum solution. Otherwise, go to Step 1.

We now show that in case of finite termination, the last vectors found are optimum solutions of their respective problems. We first prove a simple lemma.

5.9.2 Lemma
For every $k = 1, 2, \cdots$ $(\Delta x^k)^T \Delta s^k = 0$.

Proof: Follows readily from the first two equations of system (5.94). ∎

We now show that the minimum ratio test in Step 2 is well defined, and also prove the optimality of the last solution found if the method terminates finitely at Steps 1 or 4.

5.9.3 Finite Convergence
Exactly one of the following happens:

(a) The method stops at Step 1. Solution x^k is optimum for primal and (y^k, s^k) is optimum for dual.

(b) The method stops at Step 4. Solution x^k is optimum for primal and (y^k, s^k) is optimum for dual.

(c) For each $k = 1, 2, \cdots$, $\phi_k > 1$.

Proof: The three parts are exhaustive. To see (a), we note that the Newton step is zero only at the solution of a non-linear system of equations. To see (b), note that from third equation of system (5.94), we obtain

$$x^k + \Delta x^k = -S_k^{-1} X_k \Delta s^k$$
$$s^k + \Delta s^k = -X_k^{-1} S_k \Delta x^k$$

and so if $\Delta x_j^k > 0$ or $\Delta s_j^k > 0$ for any j then $\phi_k > 1$. Thus $\Delta x^k \leq 0$ and $\Delta s^k \leq 0$. From Lemma 5.9.2, $\Delta x_j^k \Delta s_j^k = 0$ for each $j = 1, \cdots, n$. Thus, for each $j = 1, \cdots, n$ either $\Delta x_j^k = 0$ or $\Delta s_j^k = 0$ or both. Thus $\phi_k = 1$ and so $\alpha = 1$. Also $\Delta x_j^k = -x_j^k$ or $\Delta s_j^k = -s_j^k$ or both, and so either $x_j^{k+1} = 0$ or $s_j^{k+1} = 0$ or both and our result follows from the complementary slackness theorem. ∎

The previous result shows that if the sequences are finite, an optimum solution is found. Thus, we henceforth assume that the sequences generated by the method are infinite.

Properties of Sequences

We will now investigate the properties of sequences $\{x^k\}$ and $\{(y^k, s^k)\}$. We will show that these converge and that the objective functions are monotone along these sequences. These results are analogous to respective results obtained for primal affine scaling and the dual affine scaling methods.

We now consider the primal sequence $\{x^k\}$ and, for some k, define a diagonal matrix $D_k = S_k^{-1} X_k$. Then system (5.94) can be written as

$$A \Delta x = 0$$
$$A^T \Delta y + \Delta s = 0$$
$$D_k^{-1} \Delta x + \Delta s = -s$$

which, by substitution $c = A^T y^0 + s^0$ is readily seen equivalent to the following system:

$$-c - D_k^{-1} \Delta x + A^T \tilde{y} = 0$$
$$A \Delta x = 0.$$

where $\tilde{y} = y^0 + \Delta y$. The above system represents K.K.T. conditions of following convex quadratic programming problem:

$$\text{minimize} \quad c^T \Delta x + \tfrac{1}{2}(\Delta x)^T D_k^{-1} \Delta x$$
$$A \Delta x = 0 \tag{5.95}$$

5.9 PRIMAL-DUAL AFFINE METHOD

where $\tilde{y}$ is the Lagrange multiplier associated with constraint of quadratic program (5.95). Also, consider the following ellipsoidal approximating problem:

$$\begin{aligned} \text{maximize} \quad & -c^T v \\ Av &= 0 \\ \|D_k^{-\frac{1}{2}} v\| &\leq 1 \end{aligned}$$

The following lemma relates solution of the above system and (5.95):

5.9.4 Lemma The direction Δx^* generated by the quadratic program (5.95) is a positively scaled multiple of the direction v^* generated by the ellipsoidal approximating problem.

Proof: It is readily seen that the solution to the quadratic program (5.95) is

$$\begin{aligned} \tilde{y} &= (AD_0A^T)^{-1}AD_0c \\ \Delta x^* &= -D_0(c - A^T\tilde{y}) \end{aligned}$$

and the solution to the ellipsoidal approximating problem is

$$\begin{aligned} y' &= -(AD_0A^T)^{-1}AD_0c \\ v^* &= -\frac{D_0(c + A^T y')}{\|D_0^{\frac{1}{2}}(c + A^T y')\|} \end{aligned}$$

and we have our result. ∎

We are now ready to prove the main convergence result related to the primal sequence.

5.9.5 Primal Sequence Convergence Sequence $\{c^T x^k\}$ is strictly monotone decreasing and is bounded. Also

(a) $\{c^T x^k\}$ converges, say to c^*.

(b) $\{x^k\}$ converges, say to x^*.

(c) There is a constant $\delta > 0$ such that for every $k = 1, 2, \cdots$ and j with $x_j^* = 0$,

$$\frac{c^T x^k - c^*}{x_j^k} \geq \delta.$$

Proof: Note that
$$c^T x^{k+1} = c^T x^k + \frac{\alpha}{\phi_k} c^T \Delta x^k$$

and

$$\begin{aligned} c^T \Delta x^k &= -c^T D_k^{\frac{1}{2}} (I - D_k^{\frac{1}{2}} A^T (AD_k A^T)^{-1} AD_k^{\frac{1}{2}}) D_k^{\frac{1}{2}} c \\ &= -\|P_k D_k^{\frac{1}{2}} c\|^2 \\ &= -\|D_k^{\frac{1}{2}} (c - A^T \tilde{y}^k)\|^2, \end{aligned} \quad (5.96)$$

where P_k is the projection matrix into the null space $\mathcal{N}(AD_k^{\frac{1}{2}})$. From Assumption (2), c does not belong to the row space $\mathcal{R}(A^T)$ of A, and so $D_k^{\frac{1}{2}} c$ does not belong to the row space $\mathcal{R}(D_k^{\frac{1}{2}} A^T)$ of $AD_k^{\frac{1}{2}}$; thus $P_k D_k^{\frac{1}{2}} c$ is not zero, and the monotonicity follows. This sequence is bounded by Assumption (1), and since every bounded monotone sequence converges, we have (a).

To see (b), from Step 3, Theorem 5.1.17 (as done in Theorem 5.1.20) and Lemma 5.9.4 we note that for each $k = 1, 2, \cdots$, there is $\mu_k > 0$ such that
$$x^k - x^{k+1} = -\mu_k v^k.$$

Thus
$$\begin{aligned} \infty > c^T x^0 - c^* &= \sum_{k=0}^{\infty} c^T (x^k - x^{k+1}) = -\sum_{k=0}^{\infty} \mu_k c^T v^k \\ &\geq \frac{1}{\rho} \sum_{k=0}^{\infty} \mu_k \|v^k\| = \frac{1}{\rho} \sum_{k=0}^{\infty} \|x^{k+1} - x^k\| \end{aligned}$$

and thus the sequence $\{x^k\}$ is a Cauchy sequence and converges, say to x^*.

To see (c), let k be arbitrary. Then, from Step 3, Theorem 5.1.17 (as done in Theorem 5.1.20) and Lemma 5.9.4,

$$\begin{aligned} \infty > c^T x^k - c^* &= \sum_{l=k}^{\infty} c^T (x^l - x^{l+1}) \geq \frac{1}{\rho} \sum_{l=k}^{\infty} \|(x^l - x^{l+1})\| \\ &\geq \frac{1}{\rho} \|\sum_{l=k}^{\infty} (x^l - x^{l+1})\| = \frac{1}{\rho} \|x^k - x^*\| \end{aligned}$$

and our result follows. ∎

Consider now the dual sequence $\{(y^k, s^k)\}$. We show that as in sequence generated by dual affine scaling method, the dual objective function is monotone on this sequence and this sequence converges.

5.9 PRIMAL-DUAL AFFINE METHOD

From system (5.94), it is readily seen that

$$\Delta y = (ADA^T)^{-1}b.$$

Δy can also be obtained by solving the following quadratic program:

$$\begin{aligned}\text{maximize} \quad & b^T \Delta y \\ & \Delta y^T ADA^T \Delta y \leq 1.\end{aligned} \quad (5.97)$$

We can now prove that dual objective function is strictly monotone increasing on the sequence $\{y^k\}$, and the sequence converges.

5.9.6 Theorem The sequence $\{b^T y^k\}$ is strictly monotone increasing.

(a) $b^T y^k$ converges, say to b^*.

(b) The sequence $\{(y^k, s^k)\}$ converges, say to (y^*, s^*).

(c) There is a $\delta > 0$ such that for all $k = 1, 2 \cdots$ and j with $s_j^* = 0$

$$\frac{b^* - b^T y^k}{s_j^k} \geq \delta.$$

Proof: (a) follows from the fact that

$$b^T(y^{k+1} - y^k) = \frac{\alpha}{\phi_k} b^T \Delta y^k = \frac{\alpha}{\phi_k} b^T (ADA^T)^{-1} b \quad (5.98)$$

and that ADA^T is a positive definite matrix and from assumption (2), $b \neq 0$. (a) now follows since the sequence $\{b^T y^k\}$ is bounded (by the weak duality theorem) and every bounded monotone sequence converges.

From Step 3, (a) and Theorem 5.1.17, we see that

$$\infty > b^* - b^T y^k = \sum_{k=0}^{\infty} b^T(y^{k+1} - y^k) = \sum_{k=0}^{\infty} \frac{\alpha}{\phi_k} b^T \Delta y^k$$

$$\geq \sum_{k=0}^{\infty} \frac{\alpha}{\phi_k} \frac{1}{\rho} \|\Delta y^k\| = \frac{1}{\rho} \sum_{k=0}^{\infty} \|y^{k+1} - y^k\|$$

and thus $\{y^k\}$ is a Cauchy sequence and thus converges, say to y^*. But $s^k = c - A^T y^k$, and thus $\{s^k\}$ also converges, to $s^* = c - A^T y^*$, and (b) follows.

(c) follows exactly as the proof of 5.9.5 (c) and the fact that $\|s^k - s^*\| \leq \|A^T(y^* - y^k)\| \leq \mu \|y^k - y^*\|$. ∎

We now establish some important properties of other sequences. We have seen that there exist vectors $x^* \geq 0$, y^* and $s^* \geq 0$ such that they are respectively, the limit points of $\{x^k\}$, $\{y^k\}$ and $\{s^k\}$. Define

$$B = \{j : x_j^* > 0\}$$
$$N = \{j : s_j^* > 0\}. \tag{5.99}$$

We do not assume here that $B \cap N = \emptyset$ nor that $B \cup N = \{1, \cdots, n\}$. We will obtain some important results related to the primal-dual sequence $\{(x^k)^T s^k\}$.

We now prove two simple lemmas.

5.9.7 Lemma The sequence $\{(x^k)^T s^k\}$ is monotone decreasing.

Proof: We note that using system (5.94) and Step 3, we obtain

$$\begin{aligned}(x^{k+1})^T s^{k+1} &= (x^k + \frac{\alpha}{\phi_k}\Delta x^k)^T(s^k + \frac{\alpha}{\phi_k}\Delta s^k) \\ &= (1 - \frac{\alpha}{\phi_k})(x^k)^T s^k\end{aligned} \tag{5.100}$$

and our result follows from Theorem 5.9.3. ∎

5.9.8 Lemma For every $k = 1, 2, \cdots$

$$\|X_k^{-\frac{1}{2}} S_k^{\frac{1}{2}} \Delta x^k\| \leq \sqrt{(x^k)^T s^k}$$
$$\|X_k^{\frac{1}{2}} S_k^{-\frac{1}{2}} \Delta s^k\| \leq \sqrt{(x^k)^T s^k}.$$

Proof: This lemma readily follows from the following facts: for $\bar{D}_k = (X_k S_k^{-1})^{\frac{1}{2}}$, from system (5.94), we obtain

$$\bar{D}_k^{-1}\Delta x^k + \bar{D}_k \Delta s^k = -(X_k S_k)^{\frac{1}{2}} e,$$

and

$$(\bar{D}_k^{-1}\Delta x^k)^T(\bar{D}_k \Delta s^k) = (\Delta x^k)^T \Delta s^k = 0.$$

Thus $\|\bar{D}_k^{-1}\Delta x^k\| \leq \|(X_k S_k)^{\frac{1}{2}} e\| = \sqrt{(x^k)^T s^k}$. We are done, as the other result follows by the same argument. ∎

We now establish an important property of the sequences $\{\Delta x^k\}$, $\{\Delta s^k\}$ and $\{(x^k)^T s^k\}$.

5.9 PRIMAL-DUAL AFFINE METHOD

5.9.9 Theorem There exist constants $\beta_1 > 0$, $\beta_2 > 0$ and $\beta_3 > 0$ such that for all $k = 1, 2, \cdots$

$$\|\Delta x^k\| \leq \beta_1 (x^k)^T s^k$$
$$\|\Delta y^k\| \leq \beta_2 (x^k)^T s^k$$
$$\|\Delta s^k\| \leq \beta_3 (x^k)^T s^k.$$

Proof: From the relation (5.96) and Lemma 5.9.8

$$\begin{aligned} -c^T \Delta x^k &= \|D_k^{\frac{1}{2}}(c - A^T \tilde{y}^k)\|^2 \\ &= \|(X_k S_k)^{\frac{1}{2}} e + D_k^{-\frac{1}{2}} \Delta s^k\|^2 \\ &\leq 4(x^k)^T s^k \end{aligned}$$

and the first result of the theorem follows from Theorem 5.1.17.

From relation (5.98),

$$\begin{aligned} b^T \Delta y^k &= b^T (AD_k A^T)^{-1} b \\ &= e^T (X_k S_k)^{\frac{1}{2}} D_k^{\frac{1}{2}} A^T (AD_k A^T)^{-1} AD_k^{\frac{1}{2}} (X_k S_k)^{\frac{1}{2}} e \\ &= \|\bar{P}_k (X_k S_k)^{\frac{1}{2}} e\|^2 \\ &\leq \|(X_k S_k)^{\frac{1}{2}} e\|^2 \\ &= (x^k)^T s^k \end{aligned}$$

where $\bar{P}_k$ is the matrix that projects into the row space $\mathcal{R}(D_k^{\frac{1}{2}} A^T)$ of $AD_k^{\frac{1}{2}}$. The second part now follows from Theorem 5.1.17. Also, since $\Delta s^k = -A^T \Delta y^k$ the third part follows from the second. ∎

Next theorem characterizes convergence property of the sequences:

5.9.10 Theorem $(x^k)^T s^k \longrightarrow 0$ if and only if

$$X_{k,N}^{-1} \Delta x_N^k \longrightarrow -e$$
$$S_{k,B}^{-1} \Delta s_B^k \longrightarrow -e.$$

Proof: Let $(x^k)^T s^k \to 0$. Then $B \cap N = \emptyset$. From Theorem 5.9.9, $\Delta x^k \to 0$, and $\Delta s^k \to 0$. Also, from system (5.94), $X_{k,B}^{-1} \Delta x_B^k + S_{k,B}^{-1} \Delta s_B^k = -e$. Thus $x_B^* > 0$ implies the first result. Similarly, from (5.94), $X_{k,N}^{-1} \Delta x_N^k + S_{k,N}^{-1} \Delta s_N^k = -e$. Thus $s_N^* > 0$ implies the second result.

We complete the proof by noting that if the two conditions hold, then $\Delta s_N^k \to 0$ and $\Delta x_B^k \to 0$. Thus, if there is a $j \in B \cap N$ then

$$\frac{\Delta x_j^k}{x_j^k} + \frac{\Delta s_j^k}{s_j^k} \to 0$$

which is a contradiction. ∎

We now establish another important property:

5.9.11 Theorem For each $j = 1, \cdots, n$ and $k = 1, 2, \cdots$

$$-1 \leq \frac{1}{\phi_k^2} \frac{\Delta x_j^k}{x_j^k} \frac{\Delta s_j^k}{s_j^k} \leq 1.$$

Proof: Let k be arbitrary. From system (5.94), for each $j = 1, \cdots, n$

$$\frac{\Delta x_j^k}{x_j^k} + \frac{\Delta s_j^k}{s_j^k} = -1. \tag{5.101}$$

Thus, either $\Delta x_j^k < 0$ and $\Delta s_j^k < 0$; or, $\Delta x_j^k \Delta s_j^k < 0$. In the first case, by the definition of ϕ_k, we have

$$0 \geq \frac{\Delta x_j^k}{\phi_k x_j^k} \geq -1 \text{ and } 0 \geq \frac{\Delta s_j^k}{\phi_k s_j^k} \geq -1.$$

Thus their product satisfies the required inequality. To see the second case, without loss of generality assume that $\Delta x_j^k > 0$. Then $0 \geq \frac{\Delta s_j^k}{\phi_k s_j^k} \geq -1$, and so

$$0 \leq \frac{\Delta x_j^k}{\phi_k x_j^k} = -\frac{1}{\phi_k} - \frac{\Delta s_j^k}{\phi_k s_j^k} \leq 1$$

and we are done. ∎

On Convergence to Optimality

We establish two results here. We also show here that if the complementary slackness violation does not go to zero, then $B = N$.

5.9.12 Lemma Let $\alpha_k = \frac{\alpha}{\phi_k}$ for each $k = 1, 2, \cdots$. Then

(a) $(x^k)^T s^k \not\to 0$ if and only if $\sum_{j=1}^{\infty} \log(1 - \alpha_j) > -\infty$.

(b) $\sum_{j=1}^{\infty} \log(1 - \alpha_j) > -\infty$ if and only if $\sum_{j=1}^{\infty} \alpha_j < \infty$.

Proof: Note that

$$(x^{k+1})^T s^{k+1} = (1 - \alpha_k)(x^k)^T s^k = \prod_{j=0}^{k}(1 - \alpha_j)(x^0)^T s^0$$

5.9 PRIMAL-DUAL AFFINE METHOD

and (a) follows by taking logs of the above expression. To see the only if part of (b), from Lemma 5.2.2, $\log(1 - \alpha_j) \leq -\alpha_j$, note that

$$-\infty < \sum_{j=1}^{\infty} \log(1 - \alpha_j) < -\sum_{j=1}^{\infty} \alpha_j$$

To see the if part, we note that from Lemma 5.2.3,

$$\sum_{j=0}^{\infty} \log(1 - \alpha_j) \geq -\sum_{j=0}^{\infty} \alpha_j - \sum_{j=0}^{\infty} \frac{\alpha_j^2}{2(1 - \alpha_j)} > -\infty$$

and we are done. ∎

5.9.13 Non-convergence to Optimality Let $(x^k)^T s^k \not\to 0$ and $x_j^k s_j^k \to 0$ for some j. Then, for that j

1. $x_j^k \longrightarrow 0$

2. $s_j^k \longrightarrow 0$

3. $\frac{\Delta x_j^k \Delta s_j^k}{x_j^k s_j^k} \longrightarrow -\infty$

Proof: Let $a_j^k = \frac{\Delta x_j^k \Delta s_j^k}{x_j^k s_j^k}$. Then $a_j^k \not\to +\infty$ since, using Equation (5.101), we can conclude that $\frac{\Delta x_j^k}{x_j^k} < 0$ and $\frac{\Delta s_j^k}{s_j^k} < 0$ and thus $a_j^k \leq \frac{1}{4}$.

Now, assume the sequence $\{a_j^k\}$ is bounded, and for some $R > 0$, we have $|a_j^k| < R$. Under the hypothesis of the theorem and using Lemma 5.9.12, $\alpha_k \to 0$. Thus, there exists an $L \geq 1$ such that for all $k \geq L$, $\frac{\alpha_k^2}{1-\alpha_k}R < 1$. Using $x_j^{k+1} s_j^{k+1} = (1 - \alpha_k + \alpha_k^2 a_j^k)x_j^k s_j^k$ and defining $b_{j,k}^r = \frac{x_j^{k+r} s_j^{k+r}}{x_j^k s_j^k}$ we see that

$$b_{j,k}^{r+1} = \prod_{i=0}^{r}(1 - \alpha_{k+i} + \alpha_{k+i}^2 a_j^{k+i}).$$

Thus, from Lemma 5.9.12 (a), there exists $M > 0$ such that

$$\log(b_{j,k}^{r+1}) = \sum_{i=0}^{r} \log(1 - \alpha_{k+i} + \alpha_{k+i}^2 a_j^{k+i})$$

$$\geq \sum_{i=k}^{\infty} \log(1 - \alpha_i) + \sum_{i=k}^{r} \log(1 + \frac{\alpha_i^2}{1 - \alpha_i} a_j^i)$$

$$= -M + N$$

where $N = \sum_{i=k}^{r} \log(1 + \frac{\alpha_i^2}{1-\alpha_i} a_j^i)$. Using Lemma 5.2.3

$$\begin{aligned} N &\geq \sum_{i=k}^{r} \log(1 - \frac{\alpha_i^2}{1-\alpha_i} R) \\ &\geq -\sum_{i=k}^{\infty} (\frac{\alpha_i^2}{1-\alpha_i} R + \frac{R^2 \alpha_i^4}{(1-\alpha_i)^2} \frac{1}{2(1 - \frac{\alpha_i^2}{1-\alpha_i} R)}). \end{aligned}$$

Since $\alpha_i \to 0$, there is an $\bar{L} \geq 1$ such that for all $i \geq \bar{L}$ $\alpha_i < \frac{1}{2}$ and $1 - \alpha_i - \alpha_i^2 R > \frac{1}{2}$. Then

$$\frac{\alpha_i}{1-\alpha_i} < \frac{1}{2} \text{ and } \frac{\alpha_i^2}{1 - \alpha_i - \alpha_i^2 R} < \frac{1}{2}.$$

Thus, for some positive constants N_1, N_2 and N_3, from Lemma 5.9.12 we obtain

$$\begin{aligned} N &\geq -\sum_{i=k}^{\bar{L}-1} (\frac{\alpha_i^2}{1-\alpha_i} R + \frac{R^2 \alpha_i^4}{(1-\alpha_i)^2} \frac{1}{2(1 - \frac{\alpha_i^2}{1-\alpha_i} R)}) \\ &\geq -\sum_{i=\bar{L}}^{\infty} (\frac{\alpha_i^2}{1-\alpha_i} R + \frac{R^2 \alpha_i^4}{(1-\alpha_i)^2} \frac{1}{2(1 - \frac{\alpha_i^2}{1-\alpha_i} R)}) \\ &\geq -N_1 - RN_2 - R^2 N_3 \\ &> -\infty. \end{aligned}$$

Hence, for every $r \geq k$, $\log(b_{j,k}^{r+1}) > -\infty$. Thus, as $r \to \infty$, $b_{j,k}^{r+1} \not\to 0$, and this contradicts the hypothesis that $x_j^k s_j^k \to 0$. Thus (c) follows.

To see (a) and (b) note that from Equation (5.101), we can conclude that $\frac{|\Delta x_j^k|}{x_j^k} \to \infty$ if and only if $\frac{|\Delta s_j^k|}{s_j^k} \to \infty$. Since the numerator of these expressions is bounded we have parts (a) and (b) from part (c). ∎

We now state a simple corollary to the previous theorem.

5.9.14 Corollary $(x^k)^T s^k \not\to 0$ if and only if $B = N$.

Proof: Follows readily from Theorem 5.9.13 and definitions. ∎

5.10 Path Following or Homotopy Methods

These methods are based on the existence of a smooth paths, called central paths, in the interior of the primal and the dual polyhedrons, which converge to the analytic center of the respective optimal solution sets. These methods use a predictor-corrector strategy to follow this path.

5.10 PATH FOLLOWING OR HOMOTOPY METHOD

The Central Paths

We will investigate these paths for the following pair of dual linear programs:

$$\begin{aligned} \text{minimize} \quad & c^T x \\ Ax &= b \\ x &\geq 0. \end{aligned}$$

with the dual

$$\begin{aligned} \text{maximize} \quad & b^T y \\ A^T y + s &= c \\ s &\geq 0. \end{aligned}$$

The primal central path (trajectory) is defined by a parameter $t > 0$, and for each t, is the solution to following logarithmic barrier problem:

$$\begin{aligned} \text{minimize} \quad & c^T x - t \sum_{j=1}^{n} \log(x_j) \\ Ax &= b \qquad (5.102) \\ (x &> 0). \end{aligned}$$

We now prove a result establishing conditions for the existence of a solution to the above problem.

5.10.1 Theorem The primal and the dual have interior points if and only if for every $t > 0$, problem (5.102) has a solution.

Proof: Assume that both the primal and the dual have interior points. Then

$$\{x : Ax = b, x \geq 0\} \cap \{x : c^T x - t \sum_{j=1}^{n} \log(x_j) < \infty\} \neq \emptyset.$$

Also, let $(\bar{y}, \bar{s})$ with $\bar{s} > 0$ be an interior solution to the dual. Then

$$c^T x = \bar{s}^T x + b^T \bar{y}.$$

Since, for every i, $\bar{s}_i > 0$ and every u, $e^{\log(u)} = u$, it follows that for every $t \geq 0$, $\bar{s}_i x_i - t\log(x_i) \to \infty$ as $x_i \to \infty$. Thus $c^T x - t \sum_{j=1}^{n} \log(x_j)$ is bounded below. Since, for each $t > 0$, it is strictly convex problem (5.102) a unique solution.

Assume that the problem (5.102) has a solution for each $t > 0$. Then the following K.K.T. conditions

$$\begin{aligned} c - tX^{-1}e - A^T y &= 0 \\ Ax &= b \end{aligned} \qquad (5.103)$$

have the solution $\hat{x} > 0$, $\hat{y}$. Thus $\hat{x}$ is an interior solution of the primal and

$$\hat{s} = t\hat{X}^{-1}e > 0,\ \hat{y}$$

is an interior solution of the dual. ∎

Since the objective function of problem (5.102) is strictly convex, it has a unique solution obtained by solving the system (5.103). To see that this solution set is a path, define

$$F(x, y, t) = \begin{bmatrix} c - tX^{-1}e - A^T y \\ Ax - b \end{bmatrix}.$$

Also, let $\hat{t} > 0$ be arbitrary, and let $x(\hat{t})$ and $y(\hat{t})$ be the unique solution of system (5.103). Re-parameterizing with respect to the arc length μ, we note that this path is obtained by solving the initial value problem:

$$DF(x(\mu), y(\mu), t(\mu)) \begin{bmatrix} \dot{x}(\mu) \\ \dot{y}(\mu) \\ \dot{t}(\mu) \end{bmatrix} = 0$$

$$\|(\dot{x}(\mu), \dot{y}(\mu), \dot{t}(\mu))\| = 1$$

$$(x(0), y(0)) = (x(\hat{t}), y(\hat{t}))$$

where

$$DF(x(\mu), y(\mu), t(\mu)) = \begin{bmatrix} t(\mu)X(\mu)^{-2} & -A^T & -X(\mu)^{-1}e \\ A & 0 & 0 \end{bmatrix}.$$

We now show that $t(\mu)$ is monotonic in μ, and thus we can consider the path as parameterized by t.

5.10.2 Proposition $\dot{t}(\mu) \neq 0$ for each μ.

Proof: The proposition follows from the fact that if $\dot{t}(\mu) = 0$ for some μ, then for that μ, the matrix

$$\begin{bmatrix} t(\mu)X(\mu)^{-2} & -A^T \\ A & 0 \end{bmatrix}$$

is singular. This is not the case as $t(\mu) > 0$, $x(\mu) > 0$ and A has full row rank. ∎

A consequence of the above proposition is that the path can be parameterized with respect to the variable t. With this set up, strategy of the path following methods is to follow the path in decreasing t direction.

5.10 PATH FOLLOWING OR HOMOTOPY METHOD

By defining $s = c - A^T y$, system (5.103) is seen equivalent to

$$\begin{array}{rcl} Ax - b & = & 0 \\ A^T y + s - c & = & 0 \\ Xs - te & = & 0. \end{array} \tag{5.104}$$

The system (5.104) also represents the K.K.T. conditions of the following dual barrier problem:

$$\begin{array}{rl} \text{maximize} & b^T y + t \sum_{j=1}^n \log(s_j) \\ & A^T y + \phantom{t \sum_{j=1}^n \log(s_j)} s = c \\ & (s > 0). \end{array} \tag{5.105}$$

As a consequence of above facts, we see that for each $t > 0$, $\{x(t)\}$ is a path in interior of the primal polyhedron and $\{(y(t), s(t))\}$ is a path in interior of the dual polyhedron. We now investigate convergence properties of these paths when t approaches 0 or ∞.

5.10.3 Convergence to Analytic Center

Let both the primal and dual have interior points, and $\{x(t)\}$, $\{(y(t), s(t))\}$ be the respective paths generated by the barrier method. Then

(a) $c^T x(t)$ strictly increases and $b^T y(t)$ strictly decreases in t.

(b) For each $\hat{t} > 0$, central paths for $t < \hat{t}$ are bounded.

(c) When the respective polyhedron is bounded, and $t \longrightarrow \infty$, central path converges to the analytic center of the polytope.

(d) As $t \longrightarrow 0$, central paths converge to the analytic centers of optimal faces of their respective polyhedrons.

Proof: We will prove results for the central path of primal. The proof for the central path of dual follows by similar arguments.

To see (a), consider $t > \bar{t} > 0$. For each $t > 0$, since $x(t)$ uniquely solves (5.102), we obtain

$$c^T x(t) - t \sum_{j=1}^n \log(x_j(t)) < c^T x(\bar{t}) - t \sum_{j=1}^n \log(x_j(\bar{t}))$$

and

$$c^T x(\bar{t}) - \bar{t} \sum_{j=1}^n \log(x_j(\bar{t})) < c^T x(t) - \bar{t} \sum_{j=1}^n \log(x_j(t)).$$

The second inequality can be re-arranged as

$$c^T x(\bar{t}) - c^T x(t) < \bar{t} \sum_{j=1}^{n} \log(\frac{x_j(\bar{t})}{x_j(t)}).$$

Adding the two relations gives

$$t \sum_{j=1}^{n} \log(\frac{x_j(\bar{t})}{x_j(t)}) < \bar{t} \sum_{j=1}^{n} \log(\frac{x_j(\bar{t})}{x_j(t)})$$

Since $t > \bar{t}$, $\sum_{j=1}^{n} \log(\frac{x_j(\bar{t})}{x_j(t)}) < 0$, and the strict monotonicity of $c^T x(t)$ in t follows.

To see (b), from Corollary 3.4.2, the optimal solution set of primal problem is bounded. Thus, from Theorem 2.6.16, for any $\hat{t} > 0$, the set $\Gamma = \{x : Ax = b, c^T x \leq c^T x(\hat{t})\}$ is bounded. To see (b), for $t < \hat{t}$, consider the set $\Gamma_t = \{x : Ax = b, c^T x = c^T x(t)\}$. Note that solution $x(t)$ to (5.102) also solves

$$\begin{aligned}\text{maximize} \quad & \sum_{j=1}^{n} \log(x_j) \\ Ax &= b \\ c^T x &= c^T x(t) \\ (x > 0).& \end{aligned}$$

Thus, for every $t < \hat{t}$, from Part (a), $x(t)$ belongs to Γ, and (b) follows.

To see (c), consider the following problem equivalent to (5.102):

$$\begin{aligned}\text{minimize} \quad & \frac{c^T x}{t} - \sum_{j=1}^{n} \log(x_j) \\ Ax &= b \\ (x > 0).& \end{aligned}$$

For every $t > 0$, this problem has a bounded solution when the set $\{x : Ax = b, x \geq 0\}$ is bounded. In that case, any cluster point of the bounded sequence $\{x(t_k)\}$ as $t_k \longrightarrow \infty$ solves

$$\begin{aligned}\text{maximize} \quad & \sum_{j=1}^{n} \log(x_j) \\ Ax &= b \\ (x > 0)& \end{aligned}$$

which has the analytic center of the polyhedron as the unique solution. Thus the sequence $\{x(t)\}$ converges to this center and (c) follows.

To see (d), we first note that from Theorem 3.3.5 we conclude that there is a set N and a set B such that for every optimal solution of the primal, x, $x_j = 0$ for each $j \in N$ and for every optimal solution of the dual, (y, s),

5.10 PATH FOLLOWING OR HOMOTOPY METHOD

$s_j = 0$ for each $j \in B$, and $B \cup N = \{1, 2, \cdots, n\}$. Also, there is a pair of optimal solutions $\bar{x}$ and $(\bar{y}, \bar{s})$ which satisfy the strict complementarity condition, i.e., $\bar{x}_B > 0$ and $\bar{s}_N > 0$. For each $t > 0$, using the fact that $c^T x = \bar{s}_N^T x_N$ for every x, we see that, by substituting $\bar{s}_N^T x_N$ for $c^T x$ and setting $x_N = x_N(t)$, $x_B(t)$ solves the following problem:

$$\text{maximize } \sum_{j \in B} \log(x_j)$$
$$A_B x_B = b - A_N x_N(t)$$
$$(x_B > 0).$$

For $b(t) = b - A_N x_N(t)$, K.K.T. conditions for the above problem are

$$X_B^{-1} e - A_B^T u = 0$$
$$A_B x_B - b(t) = 0. \tag{5.106}$$

Now, let $t_k \longrightarrow 0$, and on a common subsequence, let x^* and (y^*, s^*) be limits of $\{x(t_k)\}$ and $\{(y(t_k), s(t_k))\}$ respectively. From the third relation of system (5.104), we conclude that $X_* s^* = 0$. And from Complementary Slackness Theorem 3.3.3, we conclude that these limits are optimal solutions of their respective problems. Thus $x_N^* = 0$, and thus $x(t)_N \longrightarrow 0$ as $t \longrightarrow 0$.

Let $\{u(t)\}$ be bounded for every $t < \hat{t}$. On a common subsequence, let x^* and u^* be limits of $\{x(t)\}$ and $\{u(t)\}$ respectively as $t \longrightarrow 0$. Taking limits in the system (5.106), we obtain

$$X_{*,B}^{-1} e - A_B^T u^* = 0$$
$$A_B x_B^* - b = 0.$$

which are the K.K.T conditions defining the analytic center of the optimal primal face. Since this center is unique (d) follows.

To see the boundedness of $\{u(t)\}$, we note that the polytope $F = \{x : A_B x_B = b, x_N = 0\}$ is bounded, and has the analytic center $x_B^* > 0$. Thus, for some $\delta > 0$, $x_j^* \geq \delta$ for each $j \in B$. Using Corollary 2.6.27, there is an $\epsilon > 0$ such that for each $\bar{b}$ with $\|\bar{b} - b\| < \epsilon$ there exists $\hat{x}_B \in \{x_B : A_B x_B = \bar{b}\}$ such that $\hat{x}_j \geq \frac{\delta}{2}$ for each $j \in B$. Since $b(t) \longrightarrow b$ as $t \longrightarrow 0$, for all $t > 0$ sufficiently small, $u(t) = (A_B X_B^2 A_B^T)^{-1} b(t)$ is seen as bounded. ∎

Primal-Dual Path Following Method

We will now present a path following method that tracks the central trajectory in space of primal and dual variables. The smooth path is followed by a predictor - corrector strategy, in which the corrector step is designed to reach a neighborhood of the central path. This neighborhood approaches the solution as the parameter t approaches 0. When the data is integral, we will show that this method finds the optimal solution of the primal and dual linear programs in time proportional to $O(L\sqrt{n})$, where L is the size of the binary string generated when coding the integer data A, b and c; and n is the number of variables in the problem. This neighborhood consists of all solutions to the system

$$\begin{aligned} Ax - b &= 0 \\ A^T y + s - c &= 0 \\ Xs - te &= E \end{aligned}$$

where $e^T E = 0$, $\|E\| \leq \gamma t$ for some $\gamma > 0$. We will use the value $\gamma = 0.091136652$ in the analysis that follows. Let $\mathcal{N}$ be the neighborhood defined by set of all solutions to the above system. From system (5.104), we see that if $E = 0$, the resulting solution is on the central path. For a given $(x, y, s, t) \in \mathcal{N}$, we take a predictor step of length $\alpha > 0$ in a direction approximately tangent to the central path. This changes the value of the parameter t to $(1-\alpha)t$. Then a corrector step is taken to assure that the next iterate is in $\mathcal{N}$. During this step, value of t is kept fixed and the direction is determined by Newton's method solving for the zero of the resulting system (5.104).

Assume that a vector (x, y, s, t) in $\mathcal{N}$ is available. We will describe later a Big M procedure to obtain such a vector. Also, assume that

$$\|Xs - te\| = \|E\| \leq \frac{\delta}{\sqrt{8}} \frac{\alpha^2}{1-\alpha} nt. \tag{5.107}$$

where α is size of step taken in predictor direction, and $\delta > 0$ is a constant. Values of these constants will be specified later. We are now ready to present the predictor corrector path following method.

5.10.4 Primal-Dual Path Following Method

Step 0 Let (x^0, y^0, s^0, t_0) be in $\mathcal{N}$, $0 < \alpha < 1$, $\epsilon > 0$ and let $k = 0$.

Step 1 Predictor Step: Compute $(\Delta x^k, \Delta y^k, \Delta s^k)$ by solving the following linear system:

$$\begin{aligned} A \Delta x^k &= 0 \\ A^T \Delta y^k + \Delta s^k &= 0 \\ S_k \Delta x^k + X_k \Delta s^k &= -X_k s^k \end{aligned} \tag{5.108}$$

5.10 PATH FOLLOWING OR HOMOTOPY METHOD

and define
$$\begin{aligned}
\bar{x}^k &= x^k + \alpha \Delta x^k \\
\bar{y}^k &= y^k + \alpha \Delta y^k \\
\bar{s}^k &= s^k + \alpha \Delta s^k \\
\bar{t}_k &= (1-\alpha)t_k
\end{aligned}$$

Step 2 Corrector Step: Compute $(\Delta \bar{x}^k, \Delta \bar{y}^k, \Delta \bar{s}^k)$ by solving the following linear system:

$$\begin{aligned}
A\Delta \bar{x}^k &&&= 0 \\
& A^T \Delta \bar{y}^k &+ \Delta \bar{s}^k &= 0 \\
\bar{S}_k \Delta \bar{x}^k && + \bar{X}_k \Delta \bar{s}^k &= \bar{t}_k e - \bar{X}_k \bar{s}^k
\end{aligned} \quad (5.109)$$

and define
$$\begin{aligned}
x^{k+1} &= \bar{x}^k + \Delta \bar{x}^k \\
y^{k+1} &= \bar{y}^k + \Delta \bar{y}^k \\
s^{k+1} &= \bar{s}^k + \Delta \bar{s}^k \\
t_{k+1} &= \bar{t}_k
\end{aligned}$$

Step 3 Iterative Step: Set $k = k+1$. If $(s^k)^T x^k < \epsilon$, stop. Otherwise, go to Step 1.

We note here that the predictor step is the same as the primal-dual affine scaling step defined by (5.94). Here, we are taking a fixed step size $\alpha > 0$ in this direction. The next proposition justifies this as a predictor direction.

5.10.5 The Predictor Step Direction computed during the predictor step is a scaled version of the classical predictor direction as described in 2.7.14 such that $\Delta t = -t$.

Proof: Letting
$$F(x, y, s, t) = \begin{bmatrix} Ax \\ A^T y + s \\ Xs - te \end{bmatrix}.$$

The classical predictor direction is given by solving $DF(x, y, s, t) \Delta w = -F(x, y, s, t)$ where $\Delta w = (\Delta x, \Delta y, \Delta s \Delta t)^T$, and
$$DF(x, y, s, t) = \begin{bmatrix} A & 0 & 0 & 0 \\ 0 & A^T & I & 0 \\ S & 0 & X & -e \end{bmatrix}.$$

The result follows by setting $\Delta t = -t$. ∎

Convergence in Polynomial Time

We now show that there is an $0 < \alpha < 1$ such that for each k, the iterates $\{x^k\}$, $\{(y^k, s^k)\}$ and $\{t_k\}$ are in $\mathcal{N}$ with $t_k \longrightarrow 0$. Thus the cluster point of these sequences converge to the optimal faces of their respective problems. And when the data A, b and c is integral, this convergence occurs in time proportional to $O(\sqrt{n}L)$, i.e., in time that is a polynomial function of the data. We now establish three straight forward results:

5.10.6 Lemma Let a, b, c be three vectors in $\mathbf{R}^n$ such that $a + b = c$ and $a^T b \geq 0$. Then $\|Ab\| \leq \frac{1}{\sqrt{8}}\|c\|^2$. Also $\|a\| \leq \|c\|$ and $\|b\| \leq \|c\|$.

Proof: Consider the following inequalities:

$$\begin{aligned}
\|Ab\|^2 &= \sum_{j=1}^n a_j^2 b_j^2 \\
&\leq (\sum_{a_j b_j > 0} a_j b_j)^2 + (\sum_{a_j b_j < 0} a_j b_j)^2 \\
&\leq 2(\sum_{a_j b_j > 0} a_j b_j)^2 \qquad \text{(since } a^T b \geq 0\text{)} \\
&\leq 2(0.25 \sum_{a_j b_j > 0} (a_j + b_j)^2)^2 \qquad \text{(use } (a_j + b_j)^2 \geq 4 a_j b_j\text{)}
\end{aligned}$$

and the first result follows. The remaining part follows from the observation that the largest side of a triangle is opposite the largest angle. ∎

5.10.7 Lemma $(\Delta x^k)^T \Delta s^k = 0$, $(\Delta \bar{x}^k)^T \Delta \bar{s}^k = 0$, $(\Delta \bar{x}^k)^T \Delta s^k = 0$ and $(\Delta x^k)^T \Delta \bar{s}^k = 0$.

Proof: The proof is identical to that of Lemma 5.9.2. ∎

5.10.8 Lemma For each $k = 1, 2, \cdots$ let $e^T E^k = 0$ and $D_k = X_k^{\frac{1}{2}} S_k^{-\frac{1}{2}}$. Then

(a) $\|D_k^{-1} \Delta x^k\| \leq \sqrt{n t_k}$.

(b) $\|D_k \Delta s^k\| \leq \sqrt{n t_k}$.

(c) $\|\Delta X_k \Delta s^k\| \leq \frac{1}{\sqrt{8}} n t_k$, where ΔX_k is the diagonal matrix whose jth diagonal entry is Δx_j^k.

Proof: Multiplying the third equation of system (5.108) by $(XS)^{-\frac{1}{2}}$ we obtain
$$D_k^{-1} \Delta x^k + D_k \Delta s^k = (X_k S_k)^{\frac{1}{2}} e. \tag{5.110}$$
The lemma now follows from Lemmas 5.10.7 and 5.10.6 since $\|(X_k S_k)^{\frac{1}{2}} e\|^2 \leq (s^k)^T x^k = n t_k + e^T E^k = n t_k$. ∎

The Predictor Step

We investigate the predictor step, and resulting iterate $(\bar{x}^k, \bar{y}^k, \bar{s}^k, \bar{t}_k)$.

5.10.9 Theorem Let $(x^k, y^k, s^k, t_k) \in \mathcal{N}$ with $e^T E = 0$, and satisfy equation (5.107). Let $\bar{E} = \bar{X}_k \bar{s}_k - \bar{t}_k e$. Then $e^T \bar{E}^k = 0$ and $\|\bar{E}^k\| \leq \frac{\delta+1}{\sqrt{8}} \alpha^2 n t_k$.

Proof: Note that

$$\begin{aligned}
\bar{X}_k \bar{s}^k &= (X_k + \alpha \Delta X_k)(s^k + \alpha \Delta s^k) \\
&= X_k s_k + \alpha(S_k \Delta x^k + X_k \Delta s^k) + \alpha^2 \Delta X_k \Delta s^k \\
&= (1-\alpha)(t_k e + E^k) + \alpha^2 \Delta X_k \Delta s^k
\end{aligned}$$

Thus $\bar{E}^k = (1-\alpha)E^k + \alpha^2 \Delta X_k \Delta s^k$. The first part follows from Lemma 5.10.7 and second part (c) of Lemma 5.10.8. ∎

We now obtain a condition on the parameters α and δ such that the iterate after the predictor step remains feasible.

5.10.10 Proposition Let $(1 - \frac{\delta}{\sqrt{8}} \frac{\alpha^2}{1-\alpha} n)^{\frac{1}{2}} - \alpha\sqrt{n} > 0$. Then $\bar{x}^k > 0$ and $\bar{s}^k > 0$.

Proof: We will show this for $\bar{x}_j^k$, and the proof is similar for $\bar{s}_j^k$. Note that, from (5.107),

$$x_j^k s_j^k = t_k + E_j^k \geq t_k - \|E^k\| \geq (1 - \frac{\delta}{\sqrt{8}} \frac{\alpha^2}{1-\alpha} n) t_k.$$

Now, from (5.110), and Step 1, $\bar{x}_j^k > 0$ if $\sqrt{\frac{s_j^k}{x_j^k}} \bar{x}_j^k = (x_j^k s_j^k)^{\frac{1}{2}} + \alpha \sqrt{\frac{s_j^k}{x_j^k}} \Delta x_j^k \geq ((1 - \frac{\delta}{\sqrt{8}} \frac{\alpha^2}{1-\alpha} n)^{\frac{1}{2}} - \alpha\sqrt{n})\sqrt{t_k} > 0$. The inequality follows from part (a) of Lemma 5.10.8. ∎

The Corrector Step

We now investigate the corrector step and then obtain values of parameters α and δ such that all iterates are feasible interior points and and, the iterate after the corrector step is in the neighborhood $\mathcal{N}$.

5.10.11 Theorem Let $E^{k+1} = X_{k+1} s^{k+1} - t^{k+1} e$. Then $e^T E^{k+1} = 0$ and $\|E^{k+1}\| \leq \frac{1}{\sqrt{8}} \frac{\theta^2}{1-\theta} \bar{t}_k$ where $\theta = \frac{1+\delta}{\sqrt{8}} \frac{\alpha^2}{1-\alpha} n$.

Proof: We note that

$$\begin{aligned}X_{k+1}s^{k+1} &= (\bar{X}_k + \Delta\bar{X}_k)(\bar{s}^k + \Delta\bar{s}^k)\\ &= \bar{t}_k e + \Delta\bar{X}_k\Delta\bar{s}^k\\ &= t_{k+1}e + \Delta\bar{X}_k\Delta\bar{s}^k\end{aligned}$$

Thus, $e^T E^{k+1} = (\Delta\bar{x}^k)^T\Delta\bar{s}^k = 0$. Also, from proof of part (c) of Lemma 5.10.8, we obtain

$$\|E^{k+1}\| = \|\Delta\bar{X}_k\Delta\bar{s}^k\| \le \frac{1}{\sqrt{8}}\|(\bar{X}_k\bar{S}_k)^{-\frac{1}{2}}\|^2\|\bar{E}^k\|^2.$$

But $\bar{x}_j^k\bar{s}_j^k \ge \bar{t}_k - \|\bar{E}^k\| \ge (1 - \frac{1+\delta}{\sqrt{8}}\frac{\alpha^2}{1-\alpha}n)\bar{t}_k$. Thus this theorem follows from Theorem 5.10.9. ∎

We now establish a condition such that the iterate after the corrector step is feasible.

5.10.12 Proposition Let $\theta = \frac{1+\delta}{\sqrt{8}}\frac{\alpha^2}{1-\alpha}n < \frac{1}{2}$. Then $x^{k+1} > 0$ and $s^{k+1} > 0$.

Proof: We will show this for x^{k+1}. The argument is the same for s^{k+1}. Now $x_j^{k+1} > 0$ if and only if $\sqrt{\frac{\bar{s}_j^k}{\bar{x}_j^k}}x_j^{k+1} = (\bar{x}_j^k\bar{s}_j^k)^{\frac{1}{2}} + \sqrt{\frac{\bar{s}_j^k}{\bar{x}_j^k}}\Delta\bar{x}_j^k > 0$. But a lower bound on this term is $(1-\theta)^{\frac{1}{2}}\bar{t}_k^{\frac{1}{2}} - \|(\bar{X}_k\bar{S}_k)^{-\frac{1}{2}}\|\|\bar{E}^k\|$. The result now follows from Theorem 5.10.9. ∎

The Big M Strategy

We now describe a Big M strategy for obtaining a starting vector on the central trajectory. Let $M > 0$ be large, and consider the following modification of the primal problem:

$$\begin{array}{llll}\text{minimize} & c^T x & + & Mx_{n+1}\\ & Ax & + & (\frac{1}{M}b - Ae)x_{n+1} & = b\\ & -(e - \frac{1}{M}c)^T x & & & - x_{n+2} = -\bar{M}\\ & x \ge 0 & & x_{n+1} \ge 0 & x_{n+2} \ge 0\end{array}$$

and its dual

$$\begin{array}{llll}\text{maximize} & b^T y & & - \bar{M}s_{n+2}\\ & A^T y & +s & - (e - \frac{1}{M}c)s_{n+2} = c\\ & (\frac{1}{M}b - Ae)^T y & & +s_{n+1} & = M\\ & & s \ge 0 \quad s_{n+1} \ge 0 & s_{n+2} \ge 0.\end{array}$$

5.10 PATH FOLLOWING OR HOMOTOPY METHOD

where $\bar{M} = (n+1)M + e^T c$. It is easily seen that $x^0 = Me$, $x^0_{n+1} = M$ and $x^0_{n+2} = M$ is an interior solution of the primal, and $y^0 = 0$, $s^0 = Me$, $s^0_{n+1} = M$ and $s^0_{n+2} = M$ is an interior solution of the dual. Also, setting $t_0 = M^2$, we see that $E^0 = 0$, and thus this pair of solutions lies on the central path. We can thus initiate the path following method with this pair of interior solutions.

Convergence to Optimality

We now show that there exist values for constants δ and α such that the sequences converge.

5.10.13 Convergence Theorem There exist values for constants $\delta > 0$ and $\alpha > 0$ such that for every $k = 1, 2, \cdots$ $x^k > 0$, $s^k > 0$, belong to the neighborhood $\mathcal{N}$, and $t_{k+1} = (1-\alpha)t_k$. Also, any cluster points x^* and (y^*, s^*) of the sequences $\{x^k\}$ and (y^*, s^*) solve the primal and the dual linear programs respectively.

Proof: Choose $\delta = 0.30$, $\theta = \frac{\sqrt{8}\delta}{1+(1+\sqrt{8})\delta}$, $a = \frac{\sqrt{8}\theta}{1+\delta}$, $\gamma = \frac{\delta}{1+\delta}a$ and α as the positive root of the quadratic equation $n\alpha^2 + a\alpha - a = 0$. By straight forward calculation it is seen that

$$\sqrt{\frac{a}{n}} - \frac{2a}{n} \leq \alpha \leq \sqrt{\frac{a}{n}}. \tag{5.111}$$

A simple calculation shows that hypothesis of Propositions 5.10.10 and 5.10.12 are satisfied, and the first part follows. Also, from Theorem 5.10.11, this value of θ ensures that 5.107 holds for E^{k+1}. Since $\alpha > 0$, $\{t_k\}$ converges to zero linearly. Thus, as $t_k \longrightarrow 0$, from 5.107, $X_k s^k \longrightarrow 0$, and our result follows. ∎

We are now ready to prove the finite convergence of this method.

5.10.14 Finite and Polynomial Convergence Let the data A, b and c be integral and let L be the binary string needed to code this data. Then the optimal solution x^* and (y^*, s^*) is identified after at most $O(\sqrt{n}L)$ predictor and $O(\sqrt{n}L)$ corrector steps.

Proof: Letting $L = \lceil(\sum_{i=1}^m \sum_{j=1}^m \log_2(1+|A_{i,j}|) + \sum_{i=1}^m \log_2(1+|b_i|) + \sum_{j=1}^n \log_2(1+|c_j|) + \log(n+2))\rceil$ and $M = 2^L$. We see that $\bar{M} = M + e^T c \leq 2^L + n2^L \leq 2^{2L}$ and $|Ae| < n^2 2^L < 2^{3L}$. Thus the data of the Big M problem can be code by a binary array of length no more than $8L$. Now, consider

$$x_j^k s_j^k \leq (x^k)^T s^k = (n+2)t_k = (n+2)(1-\alpha)^k t_0 \leq (1-\alpha)^k 2^{3L}$$

Now, let k be so large that $(1-\alpha)^k 2^{3L} = 2^{-16L}$. Then

$$-\alpha k \geq k\log_2(1-\alpha) = -19L$$

and so $k \leq 19\frac{L}{\alpha}$. Also, $x_j^k > 2^{-8L}$ implies that $s_j^k < 2^{-8L}$, and from Corollary 2.6.30, $x_j^k = 0$. Thus, for any k of the order $O(\frac{L}{\alpha})$, $x_j^k s_j^k = 0$ for each j. But for all sufficiently large n, from (5.111), we note that $\frac{1}{\alpha}$ is of the order $O(\sqrt{n})$ and result thus follows. ∎

Modified Corrector Step

Step 2 of path following method requires an inversion of the matrix associated with the system (5.109). A variant that uses the same matrix as the system (5.108) has the advantage of saving this inversion. We will now show that such a variant also solves the linear program in $O(\sqrt{n}L)$ iterations. For this purpose, we modify Step 2 of the path following algorithm to:

Step 2 Corrector Step: Compute $(\Delta \bar{x}^k, \Delta \bar{y}^k, \Delta \bar{s}^k)$ by solving the following linear system:

$$\begin{aligned} A\Delta\bar{x}^k &= 0 \\ A^T\Delta\bar{y}^k + \Delta\bar{s}^k &= 0 \\ S_k\Delta\bar{x}^k + X_k\Delta\bar{s}^k &= -\bar{t}_k e - \bar{X}_k\bar{s}^k \end{aligned} \quad (5.112)$$

and define

$$\begin{aligned} x^{k+1} &= \bar{x}^k + \Delta\bar{x}^k \\ y^{k+1} &= \bar{y}^k + \Delta\bar{y}^k \\ s^{k+1} &= \bar{s}^k + \Delta\bar{s}^k \\ t_{k+1} &= \bar{t}_k \end{aligned}$$

We are ready to prove the first result regarding this modified corrector step.

5.10.15 Theorem Let $E^{k+1} = X_{k+1}s^{k+1} - t_{k+1}e$. Then $e^T E^{k+1} = 0$ and $\|E^{k+1}\| \leq \frac{1}{\sqrt{2}}(\delta+1)\theta\alpha^3 n^{\frac{3}{2}}t_k + \frac{1}{8\sqrt{8}}(\delta+1)^2\theta^2\alpha^4 n^2 t_k$ where $\theta = (1 - \frac{\delta}{\sqrt{8}}\frac{\alpha^2}{1-\alpha}n)^{-\frac{1}{2}}$.

Proof: Now

$$X_{k+1}s^{k+1} = (\bar{X}_k + \Delta\bar{X}_k)(\bar{s}^k + \Delta\bar{s}^k)$$

5.10 PATH FOLLOWING OR HOMOTOPY METHOD

$$\begin{aligned}&= \bar{X}_k \bar{s}^k + \bar{X}_k \Delta \bar{s}^k + \bar{S}_k \Delta \bar{x}^k + \Delta \bar{X}_k \Delta \bar{s}^k \\ &= \bar{X}_k \bar{s}^k + X_k \Delta \bar{s}^k + S_k \Delta \bar{x}^k + (\bar{X}_k - X_k)\Delta \bar{s}^k + (\bar{S}_k - S_k)\Delta \bar{x}^k \\ &\quad + \Delta \bar{X}_k \Delta \bar{s}^k \end{aligned}$$

and thus $E^{k+1} = \alpha(\Delta X_k \Delta \bar{s}^k + \Delta S_k \Delta \bar{x}^k) + \Delta \bar{X}_k \Delta \bar{s}^k$. Now, from Lemma 5.10.8 and its proof, we see that

$$\|\Delta X_k \Delta \bar{s}^k\| = \|\Delta X_k D_k^{-1} D_k \Delta \bar{s}^k\| \leq \|(X_k S_k)^{\frac{1}{2}} e\| \|(X_k S_k)^{-\frac{1}{2}} \bar{E}^k\|$$

and

$$\|\Delta \bar{X}_k \Delta \bar{s}^k\| \leq \frac{1}{\sqrt{8}} \|(X_k S_k)^{-\frac{1}{2}} \bar{E}^k\|^2.$$

The result follows from Theorem 5.10.9. ∎

5.10.16 Proposition If $1 - \frac{\alpha\sqrt{n}}{\sqrt{1-\alpha}} - \frac{2\delta+1}{\sqrt{8}} \frac{\alpha^2 n}{1-\alpha} > 0$ then $\bar{x}^k > 0$, $\bar{s}^k > 0$, $x^{k+1} > 0$ and $s^{k+1} > 0$.

Proof: We will prove this proposition for $\bar{x}^k$ and x^{k+1}. The proof of the others is the same. Now, $x_j^{k+1} > 0$ if and only if $\sqrt{\frac{s_j^k}{x_j^k}} x_j^{k+1} > 0$. But

$$\begin{aligned}\sqrt{\frac{s_j^k}{x_j^k}} x_j^{k+1} &= \sqrt{\frac{s_j^k}{x_j^k}} \bar{x}_j^k + \sqrt{\frac{s_j^k}{x_j^k}} \Delta \bar{x}_j^k \\ &= \sqrt{\frac{s_j^k}{x_j^k}} x_j^k + \alpha \sqrt{\frac{s_j^k}{x_j^k}} \Delta x_j^k + \sqrt{\frac{s_j^k}{x_j^k}} \Delta \bar{x}_j^k \\ &\geq \sqrt{x_j^k s_j^k} - \alpha \|(X_k S_k)^{\frac{1}{2}} e\| - \|(X_k S_k)^{-\frac{1}{2}} \bar{E}^k\| \\ &\geq (1 - \frac{\delta}{\sqrt{8}} \frac{\alpha^2}{1-\alpha} n)^{\frac{1}{2}} \sqrt{t_k} - \alpha \sqrt{n t_k} - \\ &\quad \frac{\delta+1}{\sqrt{8}} (1 - \frac{\delta}{\sqrt{8}} \frac{\alpha^2}{1-\alpha} n)^{-\frac{1}{2}} \frac{\alpha^2}{1-\alpha} n \sqrt{t_k}.\end{aligned}$$

Thus, both $x_j^{k+1} > 0$ and $\bar{x}_j^k > 0$ are positive if

$$1 - \frac{2\delta+1}{\sqrt{8}} \frac{\alpha^2}{1-\alpha} n - \frac{\alpha\sqrt{n}}{\sqrt{1-\alpha}} (1 - \frac{\delta}{\sqrt{8}} \frac{\alpha^2}{1-\alpha} n)^{\frac{1}{2}} > 0.$$

From the hypothesis we can conclude that $(1 - \frac{\delta}{\sqrt{8}} \frac{\alpha^2}{1-\alpha} n) > 0$. As $(1 - \frac{\delta}{\sqrt{8}} \frac{\alpha^2}{1-\alpha} n) < 1$, our result follows from the hypothesis. ∎

Convergence of Modified Method

We now show that the modified method also converges in a finite number of iterations, which is of the order $O(\sqrt{n}L)$.

5.10.17 Convergence Theorem There exist values for constants $\delta > 0$ and $\alpha > 0$ such that for every $k = 1, 2, \cdots$ $x^k > 0$, $s^k > 0$, belong to the neighborhood $\mathcal{N}$, and $t_{k+1} = (1-\alpha)t_k$. Also, if x^* and (y^*, s^*) cluster points of the sequences $\{x^k\}$ and $\{(y^k, s^k)\}$, they respectively solve primal and dual linear programs.

Proof: Choose $\delta = 4.0$, and $a = \frac{\alpha\sqrt{n}}{\sqrt{1-\alpha}} = \frac{8(\sqrt{(\delta+1)(1+\sqrt{8}\delta)+\delta^3}-(\delta+1))}{(\delta+1)^2+\sqrt{8}\delta^2} = 0.313622...$. It is then seen, by simple calculation that the hypothesis of Proposition 5.10.16 is satisfied, and that the bounds on E^{k+1} in Theorem 5.10.15 is less than $\frac{\delta}{\sqrt{8}}\frac{\alpha^2}{1-\alpha}nt_{k+1}$. The rest of the theorem follows by the same argument as of the proof of Theorem 5.10.13. ∎

Infeasible Start Path Following Method

We have seen that obtaining a vector on central path requires a Big M strategy to initiate the path following method. Computationally this poses problems. We now discuss a predictor - corrector method that can be initiated at any positive vector, which is not necessarily a solution to primal or dual linear program. A convenient choice is $x^0 = \rho e$, $y^0 = 0$ and $s^0 = \rho e$ for some $\rho > 0$, generally large and where e is the vector of all ones. We present a predictor - corrector strategy in which the predictor step is taken both to decrease infeasibility and improve optimality, while the corrector step is taken towards the central path. Note that the central path is relevant only when infeasibility is small. We will show that, in $O(nL)$ iterations, this method finds an optimal solution or shows that none exists in the region $\{(x, s) : x \leq \rho e, s \leq \rho e\}$. Thus infeasibility is discovered when ρ is large. The method follows:

5.10.18 Infeasible Start Path Following Method

Step 0 Let $\rho > 0$ be arbitrary, $x^0 = \rho e$, $y^0 = 0$ and $s^0 = \rho e$. Also let $1 > \alpha > 0$, $1 > \beta_1 > \beta > 0$, $\lambda \geq 1$ and $\theta_0 = 1$. Set $k = 0$.

5.10 PATH FOLLOWING OR HOMOTOPY METHOD

Step 1 Predictor Step. Let Δx^k, Δy^k and Δs^k solve the system:

$$\begin{aligned} A\Delta x^k &= -\lambda(Ax^k - b) \\ A^T \Delta y^k + \Delta s^k &= -\lambda(A^T y^k + s^k - c) \\ S_k \Delta x^k + X_k \Delta s^k &= -X_k s^k + \beta \frac{(x^k)^T s^k}{n} e \end{aligned} \quad (5.113)$$

and define

$$\begin{aligned} \bar{x}^k &= x^k + \alpha \Delta x^k \\ \bar{y}^k &= y^k + \alpha \Delta y^k \\ \bar{s}^k &= s^k + \alpha \Delta s^k \\ \theta_{k+1} &= (1 - \alpha\lambda)\theta_k \end{aligned}$$

Step 2 Corrector Step: Compute $(\Delta \bar{x}^k, \Delta \bar{y}^k, \Delta \bar{s}^k)$ by solving the following linear system:

$$\begin{aligned} A\Delta \bar{x}^k &= 0 \\ A^T \Delta \bar{y}^k + \Delta \bar{s}^k &= 0 \\ S_k \Delta \bar{x}^k + X_k \Delta \bar{s}^k &= -\bar{X}_k \bar{s}^k + \frac{(\bar{x}^k)^T \bar{s}^k}{n} e \end{aligned} \quad (5.114)$$

and define

$$\begin{aligned} x^{k+1} &= \bar{x}^k + \Delta \bar{x}^k \\ y^{k+1} &= \bar{y}^k + \Delta \bar{y}^k \\ s^{k+1} &= \bar{s}^k + \Delta \bar{s}^k \end{aligned}$$

Step 3 Iterative Step: Set $k = k + 1$. If $\|(x^k, s^k)\|_1 > \frac{3(x^k)^T s^k}{\rho \theta_k}$ then stop. There is no optimal solution in the set $\{(x, s) : 0 \leq x \leq \rho e, 0 \leq x \leq \rho e\}$. Otherwise, go to Step 1.

Let $0 < \gamma < 1$ be a constant. Define the neighborhood

$$\mathcal{N} = \{(x, s) : x > 0, s > 0 \text{ and } \|Xs - \frac{x^T s}{n} e\| \leq \gamma \frac{x^T s}{n}\}.$$

We note that $(x^0, s^0) \in \mathcal{N}$. We will show that there is a choice of the parameters $\alpha = O(\frac{1}{n})$, $\beta > 0$, $\beta_1 > 0$, $\gamma > 0$ and $\lambda \geq 1$ such that the iterates $\{(x^k, s^k)\}$ are in $\mathcal{N}$. Also, after at most $O(nL)$ iterations of the above method, we will either discover that the linear program has no optimal solution within the region $\{(x, s) : x \leq \rho e, s \leq \rho e\}$, or the iterates converge to an optimum solution. For this purpose, we will assume henceforth that α is to be determined and a value has not been specified for it. Also, we define

$$\bar{x}(\alpha) = x^k + \alpha \Delta x^k$$
$$\bar{y}(\alpha) = y^k + \alpha \Delta y^k$$
$$\bar{s}(\alpha) = s^k + \alpha \Delta s^k$$

and note that x^{k+1}, y^{k+1} and s^{k+1} are also functions of α, but the dependence is not explicit in the defining relations of Step 2. Thus, whenever there is no confusion, we will write x^{k+1} in place of $x^{k+1}(\alpha)$. We are now ready to investigate the predictor step.

Infeasible Start Predictor Step

We now prove three simple lemmas and then investigate the predictor step.

5.10.19 Lemma Let u and (v,w) be such that $Au = b$ and $A^T v + w = c$. Then for each $k = 1, 2, \cdots$,

(a) $Ax^k - b = \theta_k A(x^0 - u)$.

(b) $A^T y^k + s^k - c = \theta_k(A^T(y^0 - v) + (s^0 - w))$.

Proof: We now show (a). (b) follows by a similar argument. The result holds for $k = 0$. Now assume it holds for $k \geq 0$, and we now show it for $k+1$, and our result follows by an induction argument. Thus consider

$$\begin{aligned} Ax^{k+1} - b &= A(x^k + \alpha \Delta x^k) - b \\ &= (1 - \alpha\lambda)(Ax^k - b) \\ &= (1 - \alpha\lambda)\theta_k A(x^0 - u) \\ &= \theta_{k+1} A(x^0 - u) \end{aligned}$$

and (a) follows. ∎

5.10.20 Lemma Let $\mu = \beta \frac{(x^k)^T s^k}{n}$, u and (v, w) be such that $Au = b$ and $A^T v + w = c$, $D_k^2 = S_k^{-1} X_k$ and $P_k = D_k A^T (A D_k^2 A^T)^{-1} A D_k$ the projection matrix into the row space $\mathcal{R}(D_k A^T)$ of $A D_k$. Then for each $k = 1, 2, \cdots$

$$D_k^{-1} \Delta x^k = -\lambda \theta_k P_k D_k^{-1}(x^0 - u) + \lambda \theta_k (I - P_k) D_k (s^0 - w)$$

5.10 PATH FOLLOWING OR HOMOTOPY METHOD

$$\Delta y^k = \begin{aligned}&+ (I - P_k)(X_k S_k)^{-\frac{1}{2}}(X_k s^k - \mu e)\\&-\lambda\theta_k(y^0 - v) - \lambda\theta_k(AD_k^2 A^T)^{-1}AD_k(D_k^{-1}(x^0 - u))\\&+ \lambda\theta_k(AD_k^2 A^T)^{-1}AD_k^2(s^0 - w)\\&- (AD_k^2 A^T)^{-1}AD_k(X_k S_k)^{-\frac{1}{2}}(X_k s^k - \mu e)\end{aligned}$$

$$D_k \Delta s^k = \lambda\theta_k P_k D_k^{-1}(x^0 - u) - \lambda\theta_k(I - P_k)D_k(s^0 - w)$$
$$+ P_k(X_k S_k)^{-\frac{1}{2}}(X_k s^k - \mu e)$$

Proof: Using the result of Lemma 5.10.19, we can rewrite the system (5.113) defining the predictor step as

$$\begin{bmatrix} A & 0 & 0 \\ 0 & A^T & I \\ S_k & 0 & X_k \end{bmatrix} \begin{bmatrix} \Delta x^k + \lambda\theta_k(x^0 - u) \\ \Delta y^k + \lambda\theta_k(y^0 - v) \\ \Delta s^k + \lambda\theta_k(s^0 - w) \end{bmatrix} = \begin{bmatrix} 0 \\ 0 \\ p^k \end{bmatrix}$$

where $p^k = -X_k s^k + \beta \frac{(x^k)^T s^k}{n} e + \lambda\theta_k S_k(x^0 - u) + \lambda\theta_k X_k(s^0 - w)$. The lemma follows by solving this new system. ∎

We now prove an important lemma relating to the predictor step.

5.10.21 Lemma Let $-\rho e \le u \le \rho e$ and (v, w) with $w \le \rho e$ be such that $Au = b$ and $A^T v + w = c$, $(x^k)^T s^k \in \mathcal{N}$. Also, let $\theta_k \rho \|(x^k, s^k)\|_1 \le 3(x^k)^T s^k$. Then $\|D_k^{-1} \Delta x^k\| \le (6\lambda + 1)\sqrt{\frac{n(x^k)^T s^k}{1-\gamma}}$ and $\|D_k \Delta s^k\| \le (6\lambda + 1)\sqrt{\frac{n(x^k)^T s^k}{1-\gamma}}$.

Proof: We will only show this for $D_k \Delta s^k$. Since $\|P_k\| \le 1$, $\|I - P_k\| \le 1$, $-2\rho e \le x^0 - u \le 2\rho e$ and $-2\rho e \le s^0 - w \le 2\rho e$, from Lemma 5.10.20, we see that

$$\|D_k \Delta s^k\| \le \lambda\theta_k(\|D_k^{-1}(x^0 - u)\| + \|D_k(s^0 - w)\|) +$$
$$\|(X_k S_k)^{-\frac{1}{2}}(X_k s^k - \beta\frac{(x^k)^T s^k}{n} e)\|$$

But

$$\|(XS)^{-\frac{1}{2}}(Xs - \mu e)\|^2 = \sum_{j=1}^{n}((x_j s_j)^{\frac{1}{2}} - \mu(x_j s_j)^{-\frac{1}{2}})^2.$$
$$= x^T s - 2\mu n + \sum_{j=1}^{n} \mu^2(x_j s_j)^{-1}. \quad (5.115)$$

Thus

$$\|D_k \Delta s^k\| \le 2\lambda\rho\theta_k(\|D_k e\| + \|D_k^{-1} e\|) + ((1 - 2\beta + \frac{\beta^2}{1-\gamma})(x^k)^T s^k)^{\frac{1}{2}}$$

Since $(x^k, s^k) \in \mathcal{N}$, from the definition of E^k, we see that

$$x_j^k s_j^k = \frac{(x^k)^T s^k}{n} + E_i^k \geq (1-\gamma)\frac{(x^k)^T s^k}{n}. \tag{5.116}$$

Now $\|D_k e\| + \|D_k^{-1} e\| = \|(X_k S_k)^{-\frac{1}{2}} x^k\| + \|(X_k S_k)^{-\frac{1}{2}} s^k\| \leq \|(X_k S_k)^{-\frac{1}{2}}\| (\|x^k\| + \|s^k\|) \leq \sqrt{\frac{n}{(1-\gamma)(x^k)^T s^k}} \|(x^k, s^k)\|_1$. Using $\theta_k \rho \|(x^k, s^k)\|_1 \leq 3(x^k)^T s^k$, we obtain

$$\|D_k \Delta s^k\| \leq 6\lambda \left(\frac{n(x^k)^T s^k}{1-\gamma}\right)^{\frac{1}{2}} + \left((1 - 2\beta + \frac{\beta^2}{1-\gamma})(x^k)^T s^k\right)^{\frac{1}{2}}$$

and our result follows. ∎

We are now ready to prove the error bound after the first predictor step. This bound will indicate that for any $\alpha > 0$, $(\bar{x}(\alpha), \bar{s}(\alpha))$ may not belong to the neighborhood $\mathcal{N}$, thus necessitating a corrector step.

5.10.22 Theorem Let $(x^k, s^k) \in \mathcal{N}$, $\theta_k \rho \|(x^k, s^k)\|_1 \leq 3(x^k)^T s^k$ $\bar{E}(\alpha) = \bar{X}(\alpha)\bar{s}(\alpha) - \frac{\bar{x}(\alpha)^T \bar{s}(\alpha)}{n} e$ and $\alpha_1^* = \frac{\sqrt{2}\gamma(1-\gamma)}{(6\lambda+1)n}\left(1 - \frac{1}{\sqrt{8n(6\lambda+1)}}\right)$. Then, for every $\alpha \leq \alpha_1^*$, $\|\bar{E}(\alpha)\| \leq 2(1-\alpha)\gamma \frac{(x^k)^T s^k}{n}$.

Proof: Consider

$$\begin{aligned}
\bar{x}(\alpha)^T \bar{s}(\alpha) &= (x^k)^T s^k + \alpha((s^k)^T \Delta x^k + (x^k)^T \Delta s^k) + \alpha^2 (\Delta x^k)^T \Delta s^k \\
&= (1 - \alpha + \beta\alpha)(x^k)^T s^k + \alpha^2 (\Delta x^k)^T \Delta s^k. \tag{5.117}
\end{aligned}$$

A simple calculation shows that

$$\bar{E}(\alpha) = (1-\alpha)E^k + \alpha^2 \left(\Delta X_k \Delta s^k - \frac{(\Delta x^k)^T \Delta s^k}{n} e\right).$$

From the definition of $\mathcal{N}$ and Lemma 5.10.21 we get

$$\begin{aligned}
\|\bar{E}(\alpha)\| &\leq (1-\alpha)\gamma \frac{(x^k)^T s^k}{n} + \alpha^2 (6\lambda + 1)^2 \left(\frac{n(x^k)^T s^k}{1-\gamma} + \frac{\sqrt{n}(x^k)^T s^k}{1-\gamma}\right) \\
&\leq \left((1-\alpha)\gamma + \frac{2(6\lambda+1)^2 \alpha^2 n^2}{1-\gamma}\right)\frac{(x^k)^T s^k}{n}.
\end{aligned}$$

α_1^* is computed such that $2(6\lambda + 1)^2 \alpha^2 n^2 - (1-\alpha)\gamma(1-\gamma) \leq 0$ for every $\alpha \leq \alpha_1^*$, and our result follows. ∎

5.10 PATH FOLLOWING OR HOMOTOPY METHOD

5.10.23 Proposition Let $\theta_k \rho \|(x^k, s^k)\|_1 \leq 3(x^k)^T s^k$ and $1 - \gamma - (6\lambda + 1)n\alpha > 0$. Then $\bar{x}(\alpha) > 0$ and $\bar{s}(\alpha) > 0$.

Proof: We will show this for $\bar{x}(\alpha)_j$. But $\bar{x}(\alpha) > 0$ if and only if $\sqrt{\frac{s_j^k}{x_j^k}}\bar{x}(\alpha) > 0$, or $\sqrt{x_j^k s_j^k} + \alpha\sqrt{\frac{s_j^k}{x_j^k}}\Delta x_j^k > 0$. The result now follows from Lemma 5.10.21 and the fact that $s_j^k x_j^k \geq (1 - \gamma)\frac{(x^k)^T s^k}{n}$. ∎

We are now ready to investigate the corrector step and prove the main convergence theorem.

Infeasible Start Corrector Step

We are now ready to investigate the corrector step and prove the main convergence theorem.

5.10.24 Lemma Let $(x^k)^T s^k \in \mathcal{N}$ and $\theta_k \rho \|(x^k, s^k)\|_1 \leq 3(x^k)^T s^k$. There is an $\alpha_2^* > 0$ such that for each $k = 1, 2, \cdots$ and all $0 < \alpha < \alpha_2^*$,

$$(1 - \alpha)(x^k)^T s^k \leq (x^{k+1})^T s^{k+1} \leq (1 - \alpha(1 - \beta_1))(x^k)^T s^k.$$

Proof: Using equation (5.117), we see that

$$\begin{aligned}(x^{k+1})^T s^{k+1} &= (\bar{x}(\alpha) + \Delta\bar{x}(\alpha))^T(\bar{s}(\alpha) + \Delta\bar{s}(\alpha)) \\ &= \bar{x}(\alpha)^T \bar{s}(\alpha) + \alpha(\Delta\bar{x}(\alpha)^T \Delta x^k + \Delta\bar{s}(\alpha)^T \Delta x^k) \quad (5.118) \\ &= (1 - \alpha + \beta\alpha)(x^k)^T s^k + \alpha^2(\Delta x^k)^T \Delta s^k + \\ &\quad \alpha(\Delta\bar{x}(\alpha)^T \Delta s^k + \Delta\bar{s}(\alpha)^T \Delta x^k)\end{aligned}$$

Let

$$e(\alpha) = \Delta\bar{x}(\alpha)^T \Delta s^k + \Delta\bar{s}(\alpha)^T \Delta x^k. \quad (5.119)$$

From Lemmas 5.10.6 and 5.10.21 we note that

$$\|e(\alpha)\| \leq 2(6\lambda + 1)\sqrt{\frac{n(x^k)^T s^k}{1 - \gamma}}\|(X_k S_k)^{-\frac{1}{2}}\bar{E}(\alpha)\|. \quad (5.120)$$

From Theorem 5.10.22 and Equation (5.116) there is an $\alpha_1^* > 0$ such that for all $0 < \alpha < \alpha_1^*$

$$\|e(\alpha)\| \leq \frac{4(6\lambda + 1)(1 - \alpha)\gamma}{1 - \gamma}(x^k)^T s^k \leq \frac{4(6\lambda + 1)\gamma}{1 - \gamma}(x^k)^T s^k.$$

Also, from Lemma 5.10.21, $\|(\Delta x^k)^T \Delta s^k\| \leq \frac{(6\lambda+1)^2 n (x^k)^T s^k}{1-\gamma}$. Thus, to see the lower bound note that

$$(x^{k+1})^T s^{k+1} \geq (1 - \alpha + \beta\alpha - \alpha \|e(\alpha)\|) - \alpha^2 \|\|(\Delta x^k)^T \Delta s^k\|)(x^k)^T s^k.$$

Thus for $\bar{\alpha}_2 = \min\{\alpha_1^*, \frac{(1-\gamma)\beta - 4(6\lambda+1)\gamma}{(6\lambda+1)^2 n}\}$, we have the lower bound. The second term in above definition is obtained by setting $(1-\gamma)\beta = 16\alpha n - 16\gamma$.

To see the upper bound note that $(x^{k+1})^T s^{k+1} \leq (1 - \alpha + \beta\alpha + \alpha \|e(\alpha)\| + \alpha^2 \|\|\Delta X_k \Delta s^k\|)(x^k)^T s^k$. By setting $\beta_1 - \beta = \frac{(6\lambda+1)^2 \alpha n}{1-\gamma} + \frac{4(6\lambda+1)\gamma}{1-\gamma}$, defining $\alpha_2^* = \min\{\bar{\alpha}_2, \frac{(1-\gamma)(\beta_1 - \beta) - 4(6\lambda+1)\gamma}{(6\lambda+1)^2 n}\}$, the upper bound follows. ∎

5.10.25 Theorem Let $(x^k)^T s^k \in \mathcal{N}$ and $\theta_k \rho \|(x^k, s^k)\|_1 \leq 3(x^k)^T s^k$ and $E^{k+1} = X_{k+1} s^{k+1} - \frac{(x^{k+1})^T s^{k+1}}{n} e$ and $\alpha_3^* = \min\{\alpha_2^*, \frac{1 - \sqrt{2}\gamma - \gamma}{4(6\lambda+1)(n+\gamma)}\}$ where α_2^* is as in Lemma 5.10.24. For each $\alpha \leq \alpha_3^*$, $\|E^{k+1}\| \leq \gamma \frac{(x^{k+1})^T s^{k+1}}{n}$.

Proof: By a simple calculation we see that $X_{k+1} s^{k+1} = \frac{\bar{x}(\alpha)^T \bar{s}(\alpha)}{n} e + E(\alpha)$ where

$$E(\alpha) = \alpha(\Delta X_k \Delta \bar{s}(\alpha) + \Delta S_k \Delta \bar{x}(\alpha)) + \Delta \bar{X}(\alpha) \Delta \bar{s}(\alpha).$$

Using Equations (5.118), and (5.119), we see that

$$E^{k+1} = E(\alpha) - \frac{\alpha}{n} e(\alpha).$$

From Lemmas 5.10.6 and 5.10.21 and Equation (5.120) we see that

$$\|E^{k+1}\| \leq \alpha \|(X_k S_k)^{-\frac{1}{2}} \bar{E}(\alpha)\|(\|D_k^{-1} \Delta x^k\| + \|D_k^{-1} \Delta x^k\|) + \frac{1}{\sqrt{8}} \|(X_k S_k)^{-\frac{1}{2}} \bar{E}(\alpha)\|^2 + \frac{\alpha}{n} \|e(\alpha)\|$$

The right hand side of above expression is bounded by

$$(\frac{4(6\lambda+1)\alpha(1-\alpha)\gamma n}{1-\gamma} + \frac{\sqrt{2}(1-\alpha)^2 \gamma^2}{1-\gamma} + \frac{4(6\lambda+1)\alpha(1-\alpha)\gamma^2}{1-\gamma}) \frac{(x^k)^T s^k}{n}$$

Using the lower bound of Lemma 5.10.24, we obtain our result by setting the above upper bound to $\gamma \frac{(x^{k+1})^T s^{k+1}}{n}$. ∎

We now obtain a condition to make every iterate positive.

5.10.26 Proposition Let $\alpha_4^* = \min\{\alpha_3^*, \frac{1 - 3\gamma}{(6\lambda+1)n - 2\gamma}\}$, where α_3^* is as in Theorem 5.10.25. Then, for each $\alpha \leq \alpha_4^*$, $\bar{x}(\alpha) > 0$, $\bar{s}(\alpha) > 0$, $x^{k+1} > 0$ and $s^{k+1} > 0$.

5.10 PATH FOLLOWING OR HOMOTOPY METHOD

Proof: Note that $x_j^{k+1} > 0$ if and only if $\sqrt{\frac{s_j^k}{x_j^k}} x_j^{k+1} > 0$. But the second term is equal to $\sqrt{x_j^k s_j^k} + \alpha \sqrt{\frac{s_j^k}{x_j^k}} \Delta x_j^k + \sqrt{\frac{s_j^k}{x_j^k}} \Delta \bar{x}(\alpha)_j$, which is bounded below by

$$\sqrt{\frac{(1-\gamma)(x^k)^T s^k}{n}} - (6\lambda+1)\alpha \sqrt{\frac{n(x^k)^T s^k}{1-\gamma}} - 2(1-\alpha)\gamma \sqrt{\frac{(x^k)^T s^k}{n(1-\gamma)}}.$$

The expression above is positive if and only if $(1-\gamma)-(6\lambda+1)n\alpha-2(1-\alpha)\gamma > 0$. The second value in the definition of α_4^* makes this expression zero, and we have the result from Proposition 5.10.23. ∎

5.10.27 Infeasible Problem Detection For k, let $\|(x^k, s^k)\|_1 > \frac{3(x^k)^T x^k}{\rho \theta_k}$ but for $l < k$, $\|(x_l, s_l)\|_1 \leq \frac{3(x^l)^T s^l}{\rho \theta_l}$. Then there is no optimal solution in the set $\{(x, s) : 0 \leq x \leq \rho e, 0 \leq s \leq \rho e\}$.

Proof: We first show that for each $l = 1, 2, \cdots, k$, $\theta_l \rho^2 n \leq (x^l)^T s^l$. This is true for $l = 0$. Now assume it is true for some $l \geq 0$ and consider $(x^{l+1})^T s^{l+1}$. The result follows from lower bound of Lemma 5.10.24 and induction.

Now, assume the contrary, and let optimal solution $x^* \leq \rho e$, y^* and $s^* \leq \rho e$ exist. Then $(x^*)^T s^* = 0$. Also, for $\hat{x} = \theta_k x^0 + (1-\theta_k)x^*$ and $\hat{y} = \theta_k y^0 + (1-\theta_k)y^*$ and $\hat{s} = \theta_k s^0 + (1-\theta_k)s^*$ we note from part (a) of Lemma 5.10.19 that $A(\hat{x} - x^k) = 0$ and, from part (b), $A^T(\hat{y} - y^k) + (\hat{s} - s^k) = 0$. Thus $(\hat{x} - x^k)^T(\hat{s} - s^k) = 0$. Now:

$$\begin{aligned}
\theta_k \rho \|(x^k, s^k)\|_1 &= \theta_k \rho (e^T x^k + e^T s^k) \\
&\leq \hat{x}^T s^k + \hat{s}^T x^k \\
&= \hat{x}^T \hat{s} + (x^k)^T s^k \\
&= \theta_k^2 (x^0)^T s^0 + \theta_k(1-\theta_k)((x^0)^T s^* + (s^0)^T x^*) + (x^k)^T s^k \\
&\leq \theta_k^2 n \rho^2 + 2\theta_k(1-\theta_k)n\rho^2 + (x^k)^T s^k \\
&\leq 2\theta_k n \rho^2 + (x^k)^T s^k \\
&\leq 3(x^k)^T s^k
\end{aligned}$$

We thus have a contradiction. ∎

We are now ready to prove the main convergence theorem, and also establish its polynomial convergence.

5.10.28 Polynomial Convergence Let the data A, b and c of the linear program be integer. There exist values for constants α, β, β_1 and γ

such that in $O(nL)$ iterations the method will either discover that there is no solution in $\{(x,s) : 0 < s \le \rho e, 0 < s \le \rho e\}$; or find one.

Proof: Set values to the constants $\beta_1 > \beta$, $\gamma > 0$ and $\lambda \ge 1$ such that α_1^*, α_2^*, α_3^* and α_4^* are positive and defined. Then set $\alpha = \alpha_4^*$. From 5.10.27, the first time the method stops at Step 3, we can conclude the linear program has no feasible solution. Thus assume that the method does not stop at Step 3. It follows that $\alpha = O(\frac{1}{n}) > 0$ for each $k = 1, 2 \cdots$. Thus, from the upper bound of Lemma 5.10.24, and Lemma 5.10.19, we note that as $\theta_k \longrightarrow 0$ and $\alpha(1 - \beta_1) > 0$. Thus

$$(x^k)^T s^k \longrightarrow 0$$
$$\|Ax^k - b\| \longrightarrow 0$$
$$\|A^T y^k + s^k - c\| \longrightarrow 0.$$

Since the data is integer, we finish the proof of this theorem using the argument of Theorem 5.10.14. ∎

5.11 Projective Transformation Method

The algorithm proposed by Karmarkar in 1984 involved the use of a projective transformation of a simplex instead of an affine transformation of the non-negative orthant. We have studied algorithms based on the later, in the earlier sections of this chapter. There are certain fundamental ideas of the projective transformation method that are essential to the understanding of interior point methods, and we will now present them in this and the next few sections.

The projective transformation method assumes the following canonical form for the linear program:

$$\begin{aligned} \text{minimize} \quad & c^T x \\ Ax &= 0 \\ e^T x &= 1 \\ x &\ge 0 \end{aligned}$$

where A is an $m \times n$ matrix and e is a n-vector of all 1's. In addition we assume that

1. The matrix A has full row rank m.

2. The linear program has an interior point $x^0 > 0$ with $Ax^0 = 0$ and $e^T x^0 = 1$.

5.11 PROJECTIVE TRANSFORMATION METHOD

3. The optimal value of the objective function is 0.

4. The objective function is not constant on the feasible region.

This special form of the linear program and the assumption that the optimal value of the objective function be 0 are major disadvantages of this method when considered from the implementation prospective. Thus this is not a preferred method for this purpose. These assumptions however give an algorithm which takes time growing as a provably low order polynomial of the size of the data of the problem. The affine scaling methods do not have any such known bound. It must be noted, however, that the methods based on affine scaling perform well when implemented.

The dual of the above linear program is:

$$\text{maximize } v$$
$$A^T u + ev \leq c.$$

We now prove a simple lemma:

5.11.1 Lemma $A^T u < c$ has no solution.

Proof: If this system has a solution, then the dual has a solution with $v > 0$, and thus the minimum value of the objective function of the primal cannot be zero. ∎

5.11.2 Projective Transformation We now introduce the projective transformation. Let D be a diagonal matrix with positive diagonal entries, and let

$$S = \{x : e^T x = 1, x \geq 0\}$$

be the standard $(n-1)$-dimensional simplex. Then the projective transformation of S we will deal with is the mapping

$$T : S \longrightarrow S$$

given by

$$T(x) = \frac{D^{-1} x}{e^T D^{-1} x}.$$

It is readily seen that T carries a face of S into the corresponding face of S, and that T has the inverse

$$T^{-1}(x') = \frac{D x'}{e^T D x'}.$$

Given an interior point $x^0 > 0$, say, through the projective transformation defined by a diagonal matrix D such that its jth diagonal entry $D_{jj} = x_j^0$ for each $j = 1, \cdots, n$; the feasible region

$$P = \{x : Ax = 0, e^T x = 1, x \geq 0\}$$

of the linear program is transformed into the region

$$\begin{aligned} P' &= TP \\ &= \{x' : ADx' = 0, e^T x' = 1, x' \geq 0\}. \end{aligned}$$

In this transformed polyhedron, x^0 is mapped to the center $\frac{1}{n} e$ of the simplex S. The idea now is to take a step towards the boundary of P' from the center of S in the transformed space. When x^0 in P is close to the boundary of S, a very small step may be allowed in the direction of decreasing objective function value. But, in the projected polyhedron P' a considerably larger step may be possible. Unfortunately, the objective function is not transformed into a linear function. In any case, we define the following transformed linear program which is not equivalent to the original:

5.11.3 Transformed Linear Program Given the projective transformation T we define the following transformed linear program and caution the reader that this is not equivalent to the problem being solved:

$$\begin{aligned} \text{minimize} \quad & c^T D x' \\ A D x' &= 0 \\ e^T x' &= 1 \\ x' &\geq 0. \end{aligned}$$

The center of the simplex S is feasible for this transformed linear program, and thus a sizable step may be taken at the expense of possibly increasing the value of the original objective function. As was the case for the affine scaling method, this step is computed by defining an approximating problem where the non-negativity constraints are replaced by an ellipsoidal constraint. In this method it is an inscribing hypersphere of the simplex. We are thus lead to the following approximating problem:

5.11.4 Approximating Problem Projective transormation method takes a step not defined by an equivalent problem but by a transformed

5.11 PROJECTIVE TRANSFORMATION METHOD

approximating problem. The approximating problem is the following:

$$\begin{aligned}
\text{minimize} \quad & c^T D x' \\
A D x' &= 0 \\
e^T x' &= 1 \\
\|x' - \tfrac{1}{n} e\|^2 &\leq \alpha^2 r^2
\end{aligned}$$

where r is chosen so that it is the radius of the largest hypersphere that can be inscribed in S, and $0 < \alpha < 1$ is the step size. It is readily verified that

$$r = \frac{1}{\sqrt{n(n-1)}}.$$

The objective function value at the solution to the approximating problem is clearly seen as less than its value at the starting point $\frac{1}{n}e$. Because this objective function is not equivalent to the original, solution to this problem, when transformed back to the original polyhedron, may not decrease its objective function. Thus, unlike the affine scaling methods, the projective transformation methods do not necessarily generate a sequence of feasible solutions for which the objective function value is monotonically decreasing. We will, later, introduce the concept of a "potential" to analyze this sequence.

5.11.5 Solution of the Approximating Problem

The geometry of the solution of the approximating problem is very similar to that developed for the affine scaling method, so we will adopt a different approach for solving this problem here. Noting that the quadratic constraint must be satisfied with an equality, the solution to the approximating problem must satisfy the following optimality conditions generated by setting up a Langragean problem:

$$Dc - DA^T u - ve + \theta(x' - \tfrac{1}{n}e) = 0 \qquad (5.121)$$

$$AD(x' - \tfrac{1}{n}e) = 0 \qquad (5.122)$$

$$e^T(x' - \tfrac{1}{n}e) = 0 \qquad (5.123)$$

$$\|x' - \tfrac{1}{n}e\| = \alpha r \qquad (5.124)$$

where u, v and θ are the Lagrange multipliers on the equality constraints. The solution to this system is given in the next lemma:

5.11.6 Lemma The solution to the above system of equations is:

$$u = (AD^2A^T)^{-1}AD^2c$$

$$v = \frac{c^T x^k}{n} = \bar{c}^T a$$

$$\theta = \frac{1}{\alpha r}\|D(c - A^T u) - ve\|$$

and

$$\bar{x} = a - \frac{1}{\theta}(D(c - A^T u) - ve)$$

where $\bar{c} = Dc$ and $a = \frac{1}{n}e$. Also, $\bar{x} > 0$.

Proof: Multiply Equation 5.121 by e^T and generate

$$e^T Dc - e^T DA^T u - nv + \theta e^T(x' - \frac{1}{n}e) = 0$$

and using Equation 5.123 and the fact that $Ax^k = 0$ we simplify the above equation to

$$c^T x^k - nv = 0$$

and the second result follows.

To obtain the first result, substitute

$$x' - \frac{1}{n}e = \frac{-1}{\theta}(D(c - A^T u) - ve)$$

obtained from Equation 5.121 into equation 5.122 to obtain

$$AD^2(c - A^T u) - vADe = 0.$$

Since $ADe = Ax^k = 0$, we have the first result. The third result readily follows from Equations 5.121 and 5.124, and the last result from equation 5.121. Since the solution $\bar{x}$ lies strictly inside the largest inscribing hypersphere of S, it is strictly positive. Thus we are done. ∎

We are now ready to develop a projective transformation method.

5.11 PROJECTIVE TRANSFORMATION METHOD

5.11.7 Primal Projective Transformation Method

The solution found in the Lemma 5.11.6 generates the primal projective transformation method, and the steps of this method follow:

Step 0 $x^0 > 0$ an arbitrary interior point. $0 < \alpha < \sqrt{\frac{n-1}{n}}$, $k = 0$ and $r = \frac{1}{\sqrt{n(n-1)}}$.

Step 1 Define the diagonal matrix X_k with x_j^k as its jth diagonal element for each $j = 1, \cdots, n$.

Step 2 Tentative Solution to the Dual: Compute

$$u^k = (AX_k^2 A^T)^{-1} AX_k^2 c$$
$$v_k = \frac{c^T x^k}{n}$$

Step 3 Solution in the transformed space:

$$\bar{x}^k = a - \alpha r \frac{X_k(c - A^T u^k) - v_k e}{\|X_k(c - A^T u^k) - v_k e\|}$$

Step 4 The Next Iterate:

$$x^{k+1} = T^{-1}(\bar{x}^k)$$
$$= \frac{X_k \bar{x}^k}{e^T X_k \bar{x}^k}$$

Step 5 $k = k + 1$. Go to Step 1.

Convergence of Sequences

The algorithm generates the sequences $\{c^T x^k\}$, $\{\bar{x}^k\}$, $\{x^k\}$, and $\{u'\}$ that satisfy all the conditions 5.121 through 5.124. We now investigate the convergence properties of the first sequence.

Consider the sequence $\{c^T x^k\}$. This sequence may not be monotone, and to prove that this sequence converges to the value 0 (thus any cluster point of $\{x^k\}$ is an optimal solution) we will show that a certain potential associated with the primal solution x^k decreases by at least a constant. This will enable us to show that the sequence $\{c^T x^k\}$ converges to 0. We now introduce this potential:

5.11.8 Potential of an Interior Point

We associate with each interior feasible solution, $x > 0$ in P, of the linear program, the potential $F(x)$ given by

$$F(x) = \sum_{j=1}^{n} \log(\frac{c^T x}{x_j})$$

$$= n\log(c^T x) - \sum_{j=1}^{n} \log(x_j).$$

Compare this to one defined by (5.32). The projective transformation does not preserve the linearity of the objective function, and thus the transformed problem is not equivalent to the original. However, it modifies the potential by a constant. To see this, let

$$T(x) = \frac{D^{-1} x}{e^T D^{-1} x}$$

be the projective transformation defined by a diagonal matrix D with positive diagonal elements. Now note that the potential $F'(x')$ for each feasible solution $x' > 0$ of the transformed problem is given by

$$F'(x') = n\log(\bar{c}^T x') - \sum_{j=1}^{n} \log(x'_j)$$

where we recall $\bar{c} = Dc$. The following lemma establishes the required result:

5.11.9 Lemma

$$F'(x') = F(x) + \log(\det(D))$$

Proof: Note that

$$F'(x') = n\log(\frac{c^T D D^{-1} x}{e^T D^{-1} x}) - \sum_{j=1}^{n} \log(\frac{x_j/D_{jj}}{e^T D^{-1} x})$$

$$= n\log(c^T x) - \sum_{j=1}^{n} \log(x_j) + \log(\det(D))$$

$$= F(x) + \log(\det(D))$$

and we are done. ∎

5.11.10 Convergence of $\{c^T x^k\}$

In this section we prove the convergence of the sequences $\{c^T x^k\}$ and $\{\bar{c}^T \bar{x}^k\}$.

We now establish an important lemma.

5.11.11 Lemma

$$\frac{\bar{c}^T \bar{x}^k}{\bar{c}^T a} \leq 1 - \alpha r.$$

Proof: Multiplying Equation 5.121 by $(\bar{x}^k - a)^T$, and using the result of Equations 5.122 and 5.123 we see that

$$\begin{aligned}c^T(\bar{x}^k - a) &= -\theta_k\|\bar{x}^k - a\|^2 + (u^k)^T A X_k(\bar{x}^k - a) - v_k e^T(\bar{x}^k - a) \\ &= -\theta_k\|\bar{x}^k - a\|^2\end{aligned}$$

Thus from lemma 5.11.6 we see that

$$\begin{aligned}c^T(\bar{x}^k - a) &= -\frac{1}{\theta_k}\|X_k(c - A^T u^k) - v_k e\|^2 \\ &= -\alpha r \|X_k(c - A^T u^k) - v_k e\|\end{aligned}$$

Thus we have

$$\bar{c}^T \bar{x}^k = \bar{c}^T a - \alpha r \|X_k(c - A^T u^k) - v_k e\|$$

Now, from lemma 5.11.1, there is a j such that $(c - A^T u^k)_j \leq 0$. Thus

$$\begin{aligned}\|X_k(c - A^T u^k) - v_k e\| &\geq |x_j^k(c - A^T u^k)_j - v_k| \\ &\geq v_k.\end{aligned}$$

Our result now follows from lemma 5.11.6. ∎

We are now ready to prove our main convergence theorem:

5.11.12 Theorem
There exist constants $M > 0$ and $\delta > 0$ such that

$$c^T x^k \leq M(c^T x^0) e^{-k\delta/n}.$$

Hence $c^T x^k \longrightarrow 0$ as $k \longrightarrow \infty$.

Proof: It is readily seen that the potential difference

$$F'(\bar{x}^k) - F'(a) = n\log(\frac{\bar{c}^T \bar{x}^k}{\bar{c}^T a}) - \sum_{j=1}^{n} \log(\frac{\bar{x}_j^k}{a_j}).$$

Also,

$$\frac{\bar{x}_j^k}{a_j} = 1 + \frac{\bar{x}_j^k - a_j}{a_j}$$

and note that if w is a vector such that $w_j = \frac{\bar{x}_j^k - a_j}{a_j}$, then $e^T w = 0$; and, using $a_j = \frac{1}{n}$ and $r = \frac{1}{\sqrt{n(n-1)}}$ we see that

$$\sqrt{w^T w} = n\alpha r = \sqrt{\frac{n}{n-1}}\alpha = \beta < 1.$$

for $\alpha < \sqrt{\frac{n-1}{n}}$. Using the results of the lemmas 5.2.3, 5.2.2 and 5.11.11, we see that

$$\begin{aligned} F'(\bar{x}^k) - F'(a) &\leq n\log(1 - \alpha r) + \frac{\beta^2}{2(1-\beta)^2} \\ &\leq -\beta + \frac{\beta^2}{2(1-\beta)^2} \\ &= -\delta \end{aligned}$$

It is readily verified that for $\beta = 0.40$, $\delta = 0.2667$.

Thus, from Lemma 5.11.9, we see that $F(x^{k+1}) - F(x^k) \leq -\delta$ and thus

$$F(x^k) \leq F(x^0) - k\delta.$$

So $\log(c^T x^k) = \log(c^T x^0) + \frac{1}{n}\sum_{j=1}^n \log(\frac{x_j^k}{x_j^0}) - \frac{k\delta}{n}$. Since $0 \leq x_j^k \leq 1$, we have

$$c^T x^k \leq M(c^T x^0) e^{-k\delta/n}$$

where $M = e^\theta$ with $\theta = \frac{-1}{n}\sum_{j=1}^n \log(x_j^0) \geq \log(n)$, and we are done. ∎

We are now ready to prove the finite and polynomial convergence of this method.

5.11.13 Polynomial Convergence Let the data A, b and c of the linear program be integer. There exists a constant $\delta > 0$ such that in at most $O(nL)$ iterations the optimum solution of the linear program is identified.

Proof: The result follows from Theorem 5.11.12 where it can be seen that $\log(M)$ is linear in L, and the argument of Theorem 5.10.14. ∎

5.12 Method and Unrestricted Variables

The projective transformation method as presented in the previous section, has several major disadvantages. Two of these are the requirements that the optimum value of the objective function be zero and that the system of equations be homogeneous. When a lower bound on the optimum value

5.12 WITH UNRESTRICTED VARIABLES

of the objective function is available, a simpler methodology, not requiring these two assumptions, can be designed. This method is based on the *sliding objective function value*. We will develop this idea for the following form for the linear program:

$$\begin{aligned} \text{minimize} \quad & c^T x + d^T y \\ & Ax + By = b \\ & x \geq 0 \end{aligned}$$

where the variables y are not required to be non-negative. As was the case for the affine scaling methods, this will enable us to derive a dual version of the projective transformation method. We assume, for this case, the following:

1. The matrix A is $m \times n$ and has full row rank m.

2. The matrix B is $m \times k$ and has full column rank k.

3. The linear program has an interior point (x^0, y^0) with $x^0 > 0$ and $Ax^0 + By^0 = b$.

4. The objective function is not constant on the feasible region.

5. A lower bound z_0 on the optimal value of the objective function is known or it is a value such that if $c^T x + d^T y - z_0 < \epsilon$ for some $\epsilon > 0$ for some feasible solution (x, y), then we can declare the problem unbounded, i.e., that its dual has no feasible solution.

The dual of this linear program is

$$\begin{aligned} \text{maximize} \quad & b^T u \\ & A^T u \leq c \\ & B^T u = d. \end{aligned}$$

Using the lower bound, we will transform, through a projective transformation, the linear program into the following:

$$\begin{aligned} \text{minimize} \quad & c^T D x' + d^T y' - t'z \\ & ADx' + By' - t'b = 0 \\ & e^T x' + t' = 1 \\ & x' \geq 0 t' \geq 0 \end{aligned}$$

which is in a form with homogeneous equations. And, if z is the optimum value of the objective function, the other requirement is also met. Otherwise, we use the sliding objective function methodology where the value of z is updated as new information about the problem is obtained. We now introduce the projective transformation:

5.12.1 The Projective Transformation Given an interior point (x^0, y^0) with $x^0 > 0$, let D be a diagonal matrix with $D_{jj} = x_j^0$ for each $j = 1, \cdots, n$. Then, define the transformation:

$$T : \boldsymbol{R}_{++}^n \times \boldsymbol{R}^k \longrightarrow S^n \times \boldsymbol{R}^k$$

where $S^n = \{x \in \boldsymbol{R}^{n+1} : e^T x = 1, x \geq 0\}$ is a n dimensional simplex, and $\boldsymbol{R}_{++}^n = \{x \in \boldsymbol{R}^n : x > 0\}$ is the positive orthant in $\boldsymbol{R}^n$; by

$$T(x, y) = (x', t', y')$$

with

$$x' = \frac{D^{-1} x}{1 + e^T D^{-1} x}$$
$$y' = \frac{y}{1 + e^T D^{-1} x}$$
$$t' = \frac{1}{1 + e^T D^{-1} x}.$$

It is readily confirmed that this transformation has an inverse, and

$$T^{-1}(x', t', y') = (x, y)$$

with

$$x = \frac{D x'}{t'}$$
$$y = \frac{y'}{t'}$$

Also, with this transformation, the set of feasible solutions

$$P = \{(x, y) : Ax + By = b, x \geq 0\}$$

of the linear program is transformed into an equivalent set of feasible solutions

$$\begin{aligned} P' &= TP \\ &= \{(x', y', t') : ADx' + By' - t'b = 0, e^T x' + t' = 1, x' \geq 0, t' \geq 0\} \end{aligned}$$

of the transformed problem. We now present the approximating problem.

5.12 WITH UNRESTRICTED VARIABLES

5.12.2 The Approximating Problem
Using the projective transformation, generate the following approximating problem:

$$
\begin{aligned}
\text{minimize} \quad & c^T D x' + d^T y' - t'z \\
& A D x' + B y' - t'b = 0 \\
& e^T x' + t' = 1 \\
& \|x' - a\| + \|t' - \beta\| \le \alpha^2 r^2
\end{aligned}
$$

where $a = \frac{1}{n+1}e$, $0 < \alpha < 1$, $\beta = \frac{1}{n+1}$, $r = \frac{1}{\sqrt{n(n+1)}}$ the radius of the largest inscribing hypersphere in S^n and $(a, /beta, \bar{y}) = T(x^0, y^0)$. This problem is very similar to the one considered in section 5.7. As the solution to this problem lies on the boundary of the non-linear constraint, using Lagrangian theory, we can derive the following necessary conditions for the solution:

$$
\begin{aligned}
Dc - DA^T u - ve + \theta(x' - a) &= 0 & (5.125) \\
d - B^T u &= 0 & (5.126) \\
-z + b^T u - v + \theta(t' - \beta) &= 0 & (5.127) \\
AD(x' - a) + B(y' - \bar{y}) - b(t' - \beta) &= 0 & (5.128) \\
e^T(x' - a) + (t' - \beta) &= 0 & (5.129) \\
\|x' - a\|^2 + (t' - \beta)^2 &= \alpha^2 r^2 & (5.130)
\end{aligned}
$$

We are now ready to prove an important lemma:

5.12.3 Lemma
Value of v in the above system is

$$v^* = \frac{c^T x^0 + d^T y^0 - z}{n+1}.$$

Proof: Multiply the Equation 5.125 by e^T and add Equation 5.127 to obtain

$$(e^T Dc - z) - (e^T DA^T - b^T)u - (n+1)v + \theta(e^T x' + t' - e^T a - \beta) = 0$$

Note that equation 5.129 eliminates the term involving θ, and $e^T Dc = c^T x^0$. From 5.126

$$u^T(ADe - b) = -u^T B y^0 = -d^T y^0$$

and thus the result follows. ■

5.12.4 Solution of the Approximating Problem Rewriting the system (5.125) - (5.130) in a form similar to the system (5.81) - (5.85), we obtain

$$\bar{c} - \bar{A}^T \bar{u} + \theta(\bar{x}' - a) = 0$$
$$d - \bar{B}^T \bar{u} = 0$$
$$\bar{A}(\bar{x}' - a) + \bar{B}(y' - \bar{y}) = 0$$
$$\|\bar{x}' - a\| = \alpha r$$

where

$$\bar{A} = \begin{pmatrix} AD & -b \\ e^T & 1 \end{pmatrix}, \bar{B} = \begin{pmatrix} B \\ 0 \end{pmatrix}, \bar{c} = \begin{pmatrix} Dc \\ -z \end{pmatrix}, \bar{u} = \begin{pmatrix} u \\ v \end{pmatrix}, \bar{x}' = \begin{pmatrix} x' \\ t' \end{pmatrix}$$

5.12.5 Theorem The solution of the above equations is

$$\bar{u}^* = -\theta^*(\bar{A}\bar{A}^T)^{-1}\bar{B}(\bar{y}^* - \bar{y}) + (\bar{A}\bar{A}^T)^{-1}\bar{A}\bar{c} \tag{5.131}$$

$$\bar{y}^* - \bar{y} = \frac{1}{\theta^*}(\bar{B}^T(\bar{A}\bar{A}^T)^{-1}\bar{B})^{-1}(\bar{B}^T(\bar{A}\bar{A}^T)^{-1}\bar{A}\bar{c} - d) \tag{5.132}$$

$$\bar{x}^* - a = -\frac{1}{\theta^*}(\bar{c} - \bar{A}^T\bar{u}^*) \tag{5.133}$$

$$\theta^* = \frac{1}{\alpha r}\|\bar{c} - \bar{A}^T\bar{u}^*\| \tag{5.134}$$

Proof: The proof is similar to that of theorem 5.7.3. ∎

5.12.6 On the lower bound z The solution of the approximating problem allows an updating of the lower bound z. We now show how this can be done. We first establish a simple lemma:

5.12.7 Lemma There exist vectors $\bar{u}^1$ and $\bar{u}^2$, independent of z, such that

$$\bar{u}^* = \bar{u}^1 + z\bar{u}^2$$

and $\bar{B}^T\bar{u}^2 = B^T u^2 = 0$.

Proof: From equations 5.131 and 5.132, we obtain

$$\bar{u}^* = -(\bar{A}\bar{A}^T)^{-1}\bar{B}(\bar{B}^T(\bar{A}\bar{A}^T)^{-1}\bar{B})^{-1}(\bar{B}^T(\bar{A}\bar{A}^T)^{-1}\bar{A}\bar{c} - d)$$
$$+ (\bar{A}\bar{A}^T)^{-1}\bar{A}\bar{c}. \tag{5.135}$$

Since

$$\bar{c} = \begin{pmatrix} Dc \\ 0 \end{pmatrix} - z\begin{pmatrix} 0 \\ 1 \end{pmatrix},$$

5.12 WITH UNRESTRICTED VARIABLES

which is linear in z; and $\bar{u}^*$ is linear in $\bar{c}$, we readily see that

$$\bar{u}^* = \bar{u}^1 + z\bar{u}^2$$

for some vectors $\bar{u}^1$ and $\bar{u}^2$. Since $\bar{B}^T \bar{u}^* = d$ for all $\bar{c}$ and, thus for all z, we are done. ∎

Now consider equation 5.133. We can write

$$\bar{c} - \bar{A}^T \bar{u}^* = \begin{pmatrix} Dc \\ -z \end{pmatrix} - \begin{pmatrix} DA^T & e \\ -b^T & 1 \end{pmatrix} \begin{pmatrix} u^* \\ v^* \end{pmatrix} \quad (5.136)$$

$$= \begin{pmatrix} Dc - DA^T u^* \\ b^T u^* - z \end{pmatrix} - v^* e. \quad (5.137)$$

5.12.8 Lemma Let $c - A^T u^* > 0$ and $b^T u^* > z$. Then there exists a vector u' and a number z' such that $c - A^T u' \geq 0$, $B^T u' = d$ and $z' = b^T u' > z$ with $(\bar{c} - \bar{A}^T u')_j = 0$ for some j.

Proof: Using the result of the lemma 5.12.7, and with

$$w^1 = \begin{pmatrix} c - A^T u^1 \\ b^T u^1 \end{pmatrix}, \quad w^2 = -\begin{pmatrix} A^T u^2 \\ b^T u^2 - 1 \end{pmatrix} \quad (5.138)$$

we have $w^1 + zw^2 > 0$. Now, $w^2 \not> 0$, since the contrary implies that $w^1 + z'w^2 > 0$ for all $z' > 0$ and thus the dual is unbounded. This violates, by the Strong Duality Theorem 3.2.3, the assumption that the primal is feasible. Thus, let

$$z' = \max\{\frac{w^1_j}{-w^2_j} : w^2_j < 0\}, \quad (5.139)$$

and $\bar{u}' = \bar{u}^1 + z'\bar{u}^2$. It is readily seen that it, along with z', satisfies the conditions of the lemma. ∎

We are now ready to state the projective transformation method with unrestricted variables.

Method with Unrestricted Variables

We now present the projective transformation method with unrestricted variables. This summarizes the discussion and solution of the approximating problem for this linear program, and involves updating of the lower bounds on the objective function:

Step 0 Let (x^0, y^0), $x^0 > 0$ with $Ax^0 + By^0 = b$ be an arbitrary interior point. z_0 be a given lower bound on the optimum value of the objective function or the value determining unboundedness of the linear program; and, $u^0 = 0$ be a tentative dual solution. $0 < \alpha < 1$, $r = \frac{1}{\sqrt{n(n+1)}}$ and $k = 0$.

Step 1 Define a diagonal matrix D with $D_{jj} = x_j^k$ the jth diagonal element of D, for each $j = 1, \cdots, n$.

Step 2 Tentative Solution to the Dual: Compute $\bar{u}^k$ by the formula 5.135, and set
$$u^k = u_1^k + z_k u_2^k$$
Using the result of the lemma 5.12.3, set
$$v_k = \frac{c^T x^k + d^T y^k - z_k}{n+1}$$

Step 3 Updating lower bound and associated dual solution:
If $c - A^T u^k \not> 0$, let
$$\begin{aligned} z_{k+1} &= z_k \\ u^{k+1} &= u^k \end{aligned}$$
Otherwise define z' by equation 5.139 and set
$$\begin{aligned} z_{k+1} &= z' \\ u^{k+1} &= u_1^k + z' u_2^k \end{aligned}$$

Step 4 Update transformed primal solution:
$$(\bar{x}')^k = \begin{pmatrix} (x')^k \\ t'_k \end{pmatrix} = a - \alpha r \frac{\bar{c} - \bar{A}^T \bar{u}^k}{\|\bar{c} - \bar{A}^T \bar{u}^k\|}$$
$$(\bar{y}')^k = \bar{y} - \alpha r \frac{(\bar{B}^T(\bar{A}\bar{A}^T)^{-1}\bar{B})^{-1}(\bar{B}^T(\bar{A}\bar{A}^T)^{-1}\bar{A}\bar{c} - d)}{\|\bar{c} - \bar{A}^T \bar{u}^k\|}$$

Step 5 Update primal solution:
$$\begin{aligned} x^{k+1} &= \frac{D(x')^k}{t'_k} \\ y^{k+1} &= \frac{(y')^k}{t'_k} \end{aligned}$$

5.12 WITH UNRESTRICTED VARIABLES

Step 6 Unboundedness of the objective function:

If $c^T x^{k+1} + d^T y^{k+1} - z_{k+1} \leq 0$, STOP.

Otherwise, go to Step 1

Convergence of Sequences

We now study the convergence properties of the sequences $\{x^k\}$, $\{y^k\}$, $\{u^k\}$, $\{z_k\}$. We will first show the convergence of the sequence $\{c^T x^k + d^T y^k - z_k\}$ to zero. Since $\{z_k\}$ is a bounded monotone sequence, it converges, and thus, when the dual has a feasible solution, any cluster point of $\{(x^k, y^k)\}$ converges to the optimum solution of the primal, and any cluster point of $\{u^k\}$ converges to the optimum solution of the dual. Otherwise, the optimum solution of the primal is unbounded below. To prove the convergence of the objective function value, we introduce the concept of the potential of an interior point.

5.12.9 Potential of an Interior Point Given a speculated lower bound, z, on the optimum value of the objective function or the number, z, determining unboundedness of the linear program, we associate with each interior feasible solution (x, y), $x > 0$, in P, with $c^T x + d^T y - z > 0$, the potential $F(x, y; z)$ given by

$$F(x, y; z) = (n+1)\log(c^T x + d^T y - z) - \sum_{j=1}^{n} \log(x_j)$$

When P is transformed into an equivalent polyhedron P' by the projective transformation of section 5.12.1 the corresponding potential of the interior points of the transformed polyhedron is

$$F'(x', y', t'; z) = (n+1)\log(\bar{c}^T x' + d^T y' - zt') - \sum_{j=1}^{n} \log(x'_j) - \log(t')$$

We now establish some important properties of this potential function.

5.12.10 Lemma $F(x, y; z)$ is a decreasing function in z.

Proof: Follows form the fact that $\log(\eta)$ is a monotone increasing function of η. ∎

5.12.11 Lemma $F'(x', y', t'; z) = F(x, y; z) - \log(\det(D))$.

Proof: By definition

$$\begin{aligned}
F'(x', y', t'; z) &= \log\left(\frac{c^T Dx'/t' + d^T y'/t' - z}{x'_j/t'}\right) + \log\left(\frac{c^T Dx' + d^T y' - t'z}{t'}\right) \\
&= \log\left(\frac{c^T x + d^T y - z}{x_j/d_j}\right) + \log(c^T x + d^T y - z) \\
&= F(x, y; z) + \log(\det(D)),
\end{aligned}$$

and we are done. ∎

We now prove an important lemma:

5.12.12 Lemma At step k, let $((x')^k, (y')^k, t'_k, z_{k+1})$ solve equations 5.125 - 5.130. Then, for $T(x, y) = (a, \beta, \bar{y})$

$$\frac{\bar{c}^T (x')^k + d^T (y')^k - z_{k+1} t'}{\bar{c}^T a + d^T \bar{y} - \beta z_{k+1}} \leq 1 - \alpha r.$$

Proof: Multiplying equation 5.128 by u^T, equation 5.125 by $(x' - a)^T$, equation 5.127 with $(t' - \beta)$; adding, and substituting equations 5.127 and 5.129 we obtain the identity

$$\bar{c}^T (x' - a) + d^T (y' - \bar{y}) - z(t' - \beta) + \theta \|\bar{x}' - a\|^2 = 0$$

We note that $((x')^k, (y')^k, t'_k, z_{k+1})$ satisfies the above identity. Also, from equations 5.133 and 5.134,

$$\theta \|\bar{x}' - a\|^2 = \alpha r \|\bar{c} - \bar{A}^T \bar{u}\|.$$

Thus

$$\bar{c}^T x' + d^T y' - z_{k+1} t' = \bar{c}^T a + d^T \bar{y} - \beta z_{k+1} - \alpha r \|\bar{c} - \bar{A}^T \bar{u}\|.$$

Now, $\bar{c} - \bar{A}^T \bar{u} = \begin{pmatrix} D(c - A^T u) \\ -z_{k+1} + b^T u \end{pmatrix}$ – ve. Also, from lemma 5.12.3,

$$\begin{aligned}
v &= \frac{c^T x^k + d^T y^k - z_{k+1}}{n + 1} \\
&= \bar{c}^T a + d^T \bar{y} - \beta z_{k+1}.
\end{aligned}$$

5.12 WITH UNRESTRICTED VARIABLES

By the choice of z_{k+1}, we note that $(c - A^T u)_j \leq 0$ for some j. Hence

$$\|\bar{c} - \bar{A}^T \bar{u}\| \geq |x_j^k (c - A^T u)_j - v|$$
$$\geq v,$$

and we have the result. ∎

We are now ready to prove the main convergence result.

5.12.13 Theorem Either the sequence $\{c^T x^k + d^T y^k\}$ converges to the optimum value of the objective function, z^*, or to z_0. In the later case, the dual has no feasible solution, and thus the problem has an unbounded solution.

Proof: Consider the sequence $\{z_k\}$. Since this is a bounded monotone sequence, it converges, say to z^*. In case the dual has no feasible solution, it is clear from Step 3 that $z_k = z_0$ for all k, and thus $z^* = z_0$. We now show that $c^T x^k + d^T y^k \longrightarrow z^*$, and thus our theorem follows.

After some simple algebra, it can be readily established that

$$F'((x')^k, (y')^k, t'_k; z_{k+1}) - F'(a, \bar{y}, \beta; z_{k+1}) =$$
$$(n+1)\log\left(\frac{\bar{c}^T (x')^k + d^T (y')^k - z_{k+1} t'}{\bar{c}^T a + d^T \bar{y} - \beta z_{k+1}}\right) - \sum_{j=1}^{n} \log\left(\frac{x'_j}{a_j}\right) - \log\left(\frac{t'}{\beta}\right)$$

Also, $a_j = \beta = \frac{1}{n+1}$ for each $j = 1, \cdots, n$; and if

$$w_j = \begin{cases} \frac{x'_j - a_j}{a_j} & j \leq n \\ \frac{t' - \beta}{\beta} & j = n+1 \end{cases}$$

we note that $e^T w = 0$ and $w^T w = \frac{n+1}{n} \alpha^2$. Thus for $\beta = \sqrt{\frac{n+1}{n}} \alpha$, from lemma 5.2.3 we obtain

$$\sum_{j=1}^{n} \log\left(\frac{x'_j}{a_j}\right) + \log\left(\frac{t'}{\beta}\right) \geq -\frac{\beta^2}{2(1-\beta)}.$$

But from lemmas 5.12.10 and 5.12.11,

$$F(x^{k+1}, y^{k+1}; z_{k+1}) - F(x^k, y^k; z_k) \leq F(x^{k+1}, y^{k+1}; z_{k+1}) - F(x^k, y^k; z_{k+1})$$
$$= F'((x')^k, (y')^k, t'_k; z_{k+1}) -$$
$$F'(a, \bar{y}, \beta; z_{k+1})$$
$$\leq (n+1)\log(1 - \alpha r) + \frac{\beta^2}{2(1-\beta)}$$

$$\leq -\beta + \frac{\beta^2}{2(1-\beta)}$$
$$< -0.2667$$

for all $\beta = 0.40$. Thus

$$F(x^k, y^k; z_k) \leq F(x^0, y^0; z_0) - 0.2667k$$

And, so for some $M > 0$,

$$c^T x^k + d^T y^k - z_k \leq M(c^T x^0 + d^T y^0 - z_0)e^{-\frac{0.2667k}{n+1}}$$

Thus $c^T x^k + d^T y^k - z_k \longrightarrow 0$ as $k \longrightarrow \infty$. Since $z_k \longrightarrow z^*$, we have our result. ∎

5.13 Notes

Since the work of Narinder K. Karmarkar in 1984, considerable attention has been paid to interior point methods. This chapter covers the more important ones of these, namely, affine scaling methods, pathfollowing methods, and projective transformation methods. Successful methods start 'inside' the polyhedron, and approach the boundary from the 'inside', i.e., in a direction perpendicular rather than parallel to the boundary. When the data A, b and c is integer, many of these also solve the linear program in polynomial time.

Section 2.1 presents the primal affine scaling method. This is the earliest of these methods and was discovered by I. I. Dikin in 1967, Dikin [53]. In [54], Dikin proved that under primal non-degeneracy, the sequence converges to an optimum solution. His proof can also be found in Vanderbei and Lagarias [241]. Since then, there have been many developments in the convergence theory of this method. Mascarenhas [141] has produced an example for which the long step version of this method fails to converge to the optimal solution when the step size α to the boundary is 0.999, i.e., when the algorithm stays too close to the boundary. Hall and Vanderbei [97] give an example for which the dual sequence does not converge when the step size α is strictly greater than two-thirds. This example appears in section 5.1.33. Tsuchiya and Muramatsu [237] show that when α is less than and equal to two thirds, the dual sequence converges to the analytic center of the optimal dual face, and the primal sequence to a point in the relative interior of the optimal primal face with convergence of a derived primal sequence to the analytic center of a certain polytope. This result uses a 'local' version of the potential function used by Karmarkar. This function was introduced by Tsuchiya

5.13 NOTES

[233], [235], and the proof is based on the earlier works of Dikin [56], [55]. Simpler proofs have been developed by Monteiro, Tsuchiya, and Wang [163] and Saigal [201]. In Saigal [201], convergence to optimality for the larger step size of $\frac{2q}{3q-1}$ is shown. It is conjectured that this is the largest step size for which this method must converge to an optimal solution. Section 5.2 is based on the work of Saigal [201]. To date, it is not known if this method will solve any linear program in polynomial time.

Primal affine scaling can be shown to converge superlinearly. This was done by Tsuchiya and Monteiro [237] and Saigal [203]. The development in section 5.3 follows the work of Saigal [203]. By establishing a connection between the affine scaling step and the step of Newton's method applied to finding an analytic center of certain polyhedron, a predictor corrector strategy is developed. This strategy is implemented by manipulating the step size α, during the asymptotic phase of the algorithm. A predictor step is generated by choosing a large value (close to 1) of the step size and the corrector step by choosing a value close to half.

Section 5.4 presents the power variant of the affine scaling method. This variant is generated by choosing a real $r > \frac{1}{2}$. The affine scaling method is recovered when $r = 1$. For this method, when the step size is strictly less than $\frac{2}{2r+1}$, the dual sequence converges to the power center of the optimal dual face, and the primal sequence to the relative interior of the optimal primal face, with a modified primal sequence converging to the power center of a certain polytope. This section is based on the work of Saigal [205], and follows the development there.

Section 5.5 shows how the affine scaling method can be initiated with both the Big M strategy and a phase 1-phase 2 strategy. In a recent work of Muramatsu and Tsuchiya [167], an infeasible start for the affine scaling method has been developed. This can have a significant impact on the effectiveness of the implementations of affine scaling.

Section 5.6 presents the upper-bounded variable version of the primal affine scaling method, while section 5.7 lays the foundation of the dual affine scaling method, which is discussed in section 5.8. The dual method traces its roots to the work of Adler, Resende, Veiga, and Karmarkar [1] and Chandru and Kochar [26]. Section 5.9 presents the primal-dual affine scaling method and is based on the work of Saigal [206]. The proof given here is for every $\alpha < 1$, while in Saigal [206], a specialization of the potential function considered by Tanabe [219], Todd and Ye [228] and Mizuno and Nagasawa [152] is used to prove the result for $\alpha \leq \frac{\sqrt{5}-1}{2}$.

The path following or homotopy methods are studied in section 5.10. These methods are based on the classical globalization of Newton's method by embedding or continuation as discussed in section 2.7. Its application to linear programming was initiated by the work of Megiddo [143]. A polynomial time path following method based on this was developed by Kojima, Mizuno, and Yoshise [119]. The central trajectory is based on the logarithmic barrier function, which was introduced by Fiacco and McCormick [66]. The predictor-corrector method presented in this section is based on the talk Saigal [200]. The infeasible start method was developed by Kojima, Megiddo, and Mizuno [121] to show the convergence of the methods using this idea, as developed by Lustig [133] and Lustig, Marsten, and Shanno [135]. We follow here the work of Mizuno [151].

The classical method of Karmarkar [113] is presented in section 5.11. In section 5.12, we present a method that incorporates unrestricted variables, and thus generate dual versions of the projective transformation method. This section is based on the work of Todd and Burrell [226].

Chapter 6

Implementation

Efficient and stable implementations of linear programming methods, both boundary and interior point, present many challenges, including sparsity preservation, stability and adaptations to special structures. Interior point methods appear to have an advantage over boundary methods for the first two, since the sparsity patterns are invariant during the iterations and the linear systems to be solved are symmetric and positive definite. Boundary methods have an advantage for special structures, since these methods can take advantage of the unimodularity of the underlying matrix A, and thus deal with integral and triangular systems, while interior point methods must necessarily work with a real positive definite and symmetric system of equations. It is not clear if this advantage will carry over when the underlying matrix is very large. Variants of the simplex method that are specialized for special structures are well developed, almost no effort has been given to interior point methods for such implementations.

6.1 Implementation of Boundary Methods

At each iteration of the simplex method and its variants, two linear systems

$$Bx_B = b \tag{6.1}$$

called the primal system; and

$$B^T y = c_B \tag{6.2}$$

called the dual system, are solved. Here B is an $m \times m$ nonsingular submatrix of A. In addition, during the next iteration, the systems to be solved differ from (6.1) and (6.2) by exactly one column. Exploiting this property is the main theme of implementing these methods. We now discuss two popular implementations.

LU Factorization Method

The LU factorization of a non-singular matrix was introduced in Section 2.3.12 and uses Gaussian elimination discussed in Section 2.3 as the basic tool to obtain the relation

$$M_{m-1}P_{m-1}\cdots M_1 P_1 B = U \qquad (6.3)$$

where, for each $i = 1, \cdots, m-1$, P_i is an $m \times m$ permutation matrix, M_i is a lower triangular elementary matrix (discussed in Section 2.3.9) whose ith column is different from the identity. Also, U is an upper triangular matrix and the product $M_{m-1}P_{m-1}\cdots M_1 P_1$ is a lower triangular matrix. The basic idea behind this method is to eliminate entries below the diagonal by Gaussian elimination, and if for some column, the encountered diagonal entry is zero, permute a below diagonal row with the diagonal row to guarantee that after the permutation the diagonal entry is non-zero. It has been shown that permuting to make diagonal entry the largest of all below diagonal entries makes this method more stable.

Solution to the primal system (6.1) is obtained by forward substitution in triangular system

$$Ux_B = M_{m-1}P_{m-1}\cdots M_1 P_1 b.$$

Solution to the dual system (6.2) is obtained by backward substitution in

$$U^T \bar{y} = c_B$$

and the required vector y is obtained by observing that

$$y = P_1^T M_1^T \cdots P_{m-1}^T M_{m-1}^T \bar{y}.$$

The advantage of maintaining a decomposition of the basis matrix B in form (6.3) becomes evident when this basis changes by exactly one column. In that case the form of the decomposition remains the same, only the number of elementary and permutation matrices in the string are increased. We now show how this is done and then discuss its implications. Assume that at some iteration k of the simplex method, we have the decomposition

$$L^{(k)}B = U$$

where $L^{(k)} = M_{i_1} P_{i_1} M_{i_2} P_{i_2} \cdots M_{i_l} P_{i_l}$ is the string of elementary and permutation matrices, with M_{i_j} a lower triangular elementary matrix with column i_j different from the identity matrix. Also, let simplex method determine that the column A_r of matrix A must enter the basis in place of column B_s. Let $\hat{B} = (B_1, B_2, \cdots, B_{s-1}, B_{s+1}, \cdots, B_m, A_r)$ and

$$\bar{U} = L^{(k)} \hat{B}.$$

Matrix $\bar{U}$ is upper Hessenberg, and has at most one non-zero entry just below the diagonal in the columns $s, \cdots, m$. Also, $\hat{B}$ is the new basis matrix. Eliminating these entries by Gaussian elimination, the next upper triangular matrix $\hat{U}$ can be generated. This matrix has the decomposition

$$\hat{U} = \bar{M}_{m-1} \bar{P}_{m-1} \cdots \bar{M}_s \bar{P}_s L^{(k)} \hat{B}.$$

It is seen that for each $i = s, \cdots, m-1$ one of the matrices $\bar{M}_i$ or $\bar{P}_i$ is the identity.

This representation of the basis inverse has the advantage that changes of basis during simplex method are readily accommodated. A major disadvantage is that the needed memory for storing this string grows. Also, as the string $L^{(k)}$ become large more computations are needed to solve systems (6.1) and (6.2). Thus, frequent re-inversions may be needed when there is not enough memory available. A tradeoff between re-inversion time and extra computations for large strings may also require re-inversions. Since the simplex method is expected to take many iterations, this can lead to a major overhead on the overall performance of an implementation using this strategy.

QR Factorization method

We see from Theorem 2.3.17 part 2 that there is an upper triangular matrix R and an orthogonal matrix Q such that $B = QR$ and $Q^T Q = I$. Systems (6.1) and (6.2) can be solved by the following sequence of steps: For the primal system note that

$$\begin{aligned} x_B &= B^{-1} b \\ &= (B^T B)^{-1} B^T b \\ &= (R^T R)^{-1} B^T b \end{aligned}$$

From Section 2.3.13, we see that the inversion of $R^T R$ is not necessary and x_B can be obtained by a backward solve, a forward solve and a multiplication by B^T. Solution y to the dual system can be obtained as

$$y = B(B^T B)^{-1} c_B$$
$$= B(R^T R)^{-1} c_B.$$

QR factors of a matrix are generated by the use of Householder matrix which is unitary, symmetric and of the form $I - \frac{2uu^T}{u^T u}$. Like Gaussian elimination, u is chosen in a manner that the below diagonal entries of a column of B are made zero. We now show how such a vector u can be generated.

Let a be a given vector, and assume that a matrix of the form

$$H = I - \frac{2uu^T}{u^T u}$$

is to be defined so that $Ha = \gamma u^1$ where $u^1 = (1, 0, \cdots, 0)^T$, and γ is a constant. By substitution, it can be confirmed that for $u = a + \delta \|a\| u^1$ where

$$\delta = \begin{cases} 1 & \text{if } a_1 \geq 0 \\ -1 & \text{if } a_1 < 0 \end{cases}$$

satisfies the requirement for $\gamma = -\delta \|a\|$. The above choice of δ makes the magnitude of the first element of v as large as possible, so v will not become zero.

By defining the matrix

$$H_r = \begin{bmatrix} I & 0 \\ 0 & I - \frac{2uu^T}{u^T u} \end{bmatrix}$$

appropriate vector u can be found so that $H_r a = b$, where $b_j = 0$ for each $j > r$, $b_j = a_j$ for each $j < r$ and $b_r = \gamma a_r$. Thus H_r makes all entries below a_r equal to zero but changes no entries above.

Thus by the use of appropriate Householder matrices, we obtain the following decomposition:

$$H_{m-1} \cdots H_1 B = R$$

where $Q^{-1} = H_{m-1} \cdots H_1$. Also $Q^T = Q^{-1}$. Here H_1 converts the first column of B into the first column of R and thus has a non-zero entry only in the first position. H_2 makes all below diagonal entries of the second column of B zero, and so on. Since Q is not needed in the solution to systems (6.1) and (6.2), it need not be stored.

The factor R can be updated when one column of B changes. To see this, let the sth column B_s of B be replaced by A_r. Given R, we will find the

6.1 IMPLEMENTATION OF BOUNDARY METHODS

factor of the matrix $\bar{B} = (B_1, \cdots, B_{s-1}, B_{s+1}, \cdots, B_m, A_r)$. For the simplex method, this is the new basis matrix, and the method can be continued with this factor.

Define the permutation matrix P_s such that

$$BP_s = (B_1, \cdots, B_{s-1}, B_{s+1}, \cdots, B_m, B_s)$$

and note that $\bar{B} = BP_s + (A_r - B_s)(u^m)^T$, where $u^m = (0, \cdots, 0, 1)^T$. Thus, for $v = A_r - B_s$

$$\begin{aligned} Q^T \bar{B} &= Q^T BP_s + Q^T v(u^m)^T \\ &= RP_s - \bar{v}(u^m)^T \end{aligned}$$

is an upper Hessenberg matrix having non-zero entries just below the diagonal in columns $s+1, \cdots, m-1$. Also

$$\begin{aligned} Q^T v &= Q^T B(B^T B)^{-1} B^T v \\ &= R(R^T R)^{-1} B^T v \\ &= (R^T)^{-1} B^T v \end{aligned}$$

These below diagonal entries can be eliminated by Givens matrices of the type

$$G_r = \begin{bmatrix} I & 0 & 0 & 0 \\ 0 & \alpha & \beta & 0 \\ 0 & \beta & -\alpha & 0 \\ 0 & 0 & 0 & I \end{bmatrix}.$$

To see this consider the vector $w = (w_1, \cdots, w_r, \cdots, w_m)^T$. By simple calculation it can be seen that for

$$\alpha = \frac{w_r}{\sqrt{w_r^2 + w_{r+1}^2}}$$

$$\beta = \frac{w_{r+1}}{\sqrt{w_r^2 + w_{r+1}^2}}$$

$G_r w = (w_1, \cdots, \sqrt{w_r^2 + w_{r+1}^2}, 0, \cdots, w_m)^T$. Thus, by defining the appropriate Givens matrices, we can obtain

$$G_{m-1} G_{m-2} \cdots G_s Q^T \bar{B} = \bar{R}$$

where $\bar{R}$ is an upper triangular matrix.

Since QR factorization method uses orthogonal transformations, it is expected to be more stable than the LU factorization method. In addition, since factor Q is not used in computations, re-inversions of the basis are not required, as was seen in case of LU factorization. On the other hand the factors are expected to be more dense than those in LU factorization. Also, an additional multiplication by the matrix B and B^T is required during each iteration.

6.2 Implementation of Interior Point Methods

The main challenge in all interior point methods discussed in Chapter 5 is to effectively solve a system of the form

$$Cz = u \qquad (6.4)$$

where $C = ADA^T$, A is an $m \times n$ matrix of rank m, D is a diagonal matrix with positive diagonal entries and u is a given vector. A major advantage of this is that the underlying matrix involved in system (6.4) is symmetric and positive definite and, is thus expected to be more stable; and effective methods exist for its solution. In addition, during iterations of interior point methods, only the diagonal matrix D of the system changes. This preserves the sparsity pattern of the system to be solved during each iteration. Thus, effective data structures can be set up to exploit the resulting sparsity pattern. This one-time overhead generates the subroutines exploiting sparsity pattern and these are used in all iterations.

We will discuss several methods for solving system (6.4). These include sparsity preserving Cholesky factorizations, conjugate gradient method and infinitely summable series methods based on LQ factorizations and partitioning. We now present the use of Cholesky factorization.

Cholesky Factorization

One of the popular methods used in implementation is Cholesky factorization of the matrix C. This method is specially useful for implementation of interior point methods, as the sparsity pattern of the matrix C does not change from iteration to iteration, and effective methods exist for generating data structures to exploit this property.

6.2.1 Factorization Methods We now present the popular methods for obtaining the Cholesky factor of a symmetric positive definite matrix.

6.2 IMPLEMENTATION OF INTERIOR POINT METHODS

Understanding these methods also gives insights into the problem of finding a permutation matrix P such that the matrix L is as sparse as possible.

The first method, called the *bordering* method was presented in Chapter 2, Theorem 2.3.15. We will now discuss the second method, called the *outer product* method. To understand this method, we partition

$$C = \begin{bmatrix} \beta^1 & a^1 \\ (a^1)^T & \bar{C}_1 \end{bmatrix}$$

and note that

$$C = \begin{bmatrix} \sqrt{\beta_1} & 0 \\ \frac{(a^1)^T}{\sqrt{\beta_1}} & I \end{bmatrix} \begin{bmatrix} 1 & 0 \\ 0 & C_1 \end{bmatrix} \begin{bmatrix} \sqrt{\beta_1} & \frac{a^1}{\sqrt{\beta_1}} \\ 0 & I \end{bmatrix} = L_1 \hat{C}_1 L_1^T$$

where $C_1 = \bar{C}_1 - \frac{1}{\beta_1} a^1 (a^1)^T$. C_1 is clearly symmetric. To see that it is positive definite, for any x, note that

$$(-\frac{x^T a^1}{\beta_1}, x^T) \begin{bmatrix} \beta_1 & a^1 \\ (a^1)^T & \bar{C}_1 \end{bmatrix} \begin{bmatrix} -\frac{x^T a^1}{\beta_1} \\ x \end{bmatrix} = x^T (\bar{C}_1 - \frac{1}{\beta_1} a^1 (a^1)^T) x > 0.$$

Continuing this process one more step, we will see how this method obtains the Cholesky factor. To do this, let

$$\hat{C}_1 = \begin{bmatrix} 1 & 0 \\ 0 & C_1 \end{bmatrix} = \begin{bmatrix} 1 & 0 & 0 \\ 0 & \beta_2 & (a^2)^T \\ 0 & a^2 & \bar{C}_2 \end{bmatrix}$$

then

$$\hat{C}_1 = \begin{bmatrix} 1 & 0 & 0 \\ 0 & \sqrt{\beta_2} & 0 \\ 0 & \frac{(a^2)^T}{\sqrt{\beta_2}} & I \end{bmatrix} \begin{bmatrix} 1 & 0 & 0 \\ 0 & 1 & 0 \\ 0 & 0 & C_2 \end{bmatrix} \begin{bmatrix} 1 & 0 & 0 \\ 0 & \sqrt{\beta_2} & \frac{a^2}{\sqrt{\beta_2}} \\ 0 & 0 & I \end{bmatrix} = L_2 \hat{C}_2 L_2^T$$

Thus, by induction

$$\hat{C}_{k-1} = L_k \hat{C}_k L_k^T$$

with $\hat{C}_m = I$. Thus, we have generated lower triangular matrices L_i such that

$$C = L_1 L_2 \cdots L_m L_m^T L_{m-1}^T \cdots L_1^T$$

and in addition, using the structure of the lower triangular matrices L_i, the Cholesky factor of C is

$$L = L_1 + L_2 + \cdots L_m - (m-1)I$$

Third method for finding Cholesky factors is the *inner product* method. Here we obtain the unknown matrix L by solving the triangular system

$$LL^T = C$$

and equating the unknown terms in L with the corresponding known entries in C. This generates L as follows: For $j = 1, 2, \cdots, m$

$$L_{jj} = (C_{jj} - \sum_{l=1}^{j-1} L_{jl}^2)^{\frac{1}{2}}$$

and $i = j+1, j+2, \cdots, m$

$$L_{ij} = \frac{(C_{i,j} - \sum_{l=1}^{j-1} L_{il} L_{jl})}{L_{jj}}.$$

6.2.2 Sparse Cholesky Factorization Given a full rank matrix A, in Theorem 2.3.15 we show that there is a unique lower triangular matrix L such that

$$C = LL^T. \tag{6.5}$$

L is called the Cholesky factor of C. Even though L is unique, the sparsity pattern of L is effected by permutations of the type $PADA^T P^T$ where P is a permutation matrix. A classical example to demonstrate this is the following matrix

$$C = \begin{bmatrix} 16 & 4 & 4 & 4 \\ 4 & 2 & & \\ 4 & & 3 & \\ 4 & & & 7 \end{bmatrix} \text{ with } L = \begin{bmatrix} 4 & & & \\ 1 & 1 & & \\ 1 & -1 & 1 & \\ 1 & -1 & -2 & 1 \end{bmatrix}$$

but choosing P such that

$$PCP^T = \begin{bmatrix} 2 & & & 4 \\ & 3 & & 4 \\ & & 7 & 4 \\ 4 & 4 & 4 & 16 \end{bmatrix} ; \text{ then } L = \begin{bmatrix} \sqrt{2} & & & \\ & \sqrt{3} & & \\ & & \sqrt{7} & \\ \frac{4}{\sqrt{2}} & \frac{4}{\sqrt{3}} & \frac{4}{\sqrt{7}} & \frac{4}{\sqrt{42}} \end{bmatrix}.$$

It is clear from above illustration that the unique Cholesky factor of PCP^T is sparser than the factor of C. This is the foundation of sparse Cholesky factorization. In these methods, careful attention is paid in choosing the next column/row of C to be factored, and is chosen in a manner that the fill-in is minimized. The basic technique employed is heuristic in nature.

6.2 IMPLEMENTATION OF INTERIOR POINT METHODS

Given a permutation matrix P and that L is a cholesky factor of PCP^T, the system 6.4 can be solved using the identity:

$$PCP^T(Pz) = LL^T(Pz) = Pu.$$

We will not discuss these methods in any detail here, and will give only a brief introduction. These methods are covered in great detail by George and Liu [78], and we refer the reader to that book for further study.

As an overview, there are two basic heuristics that generate a permutation matrix P to reduce the fill-in when the Cholesky factor L of PCP^T is generated. First, the *band or envelope* methods are based on finding a permutation matrix P such that the non-zero entries of PCP^T are 'near' the diagonal. This property is retained in the Cholesky factor L, and such orderings result in reduced fill-in. The second, *minimum degree* method is more extensive. It sets up a graph representing the connection between non-zero entries of C, and then generates a permutation matrix based on the 'degree' of the nodes of this graph. Computationally, these heuristics have been found to be very successful in controlling the fill in during the factorization.

Since the system to be solved during various iterations has the same sparsity pattern in the factor L, successful implementations make a one-time factorization, called *symbolic factorization*, to generate the data structures for storing the sparsity pattern of L. They then use this structure and compute, at each iteration, only the values of non-zero coefficients of L.

6.2.3 Factors in Vector Parallel Environments

In such environments, the effectiveness of an algorithm is determined by how many computations are done asynchronously, i.e., how many computations are done at each processor without regard to or communication with other processors. Some communication is inevitable, but more each processor can do independently of other processors, the greater the utilization of the parallel system. As an example consider the problem of computing the solution after factorization, $C = LL^T$. From Section 2.3.13, this involves the following steps:

(Forward Solve) $Lz' = u$

(Backward Solve) $L^T z = z'$

Since most of the computations in these systems are done synchronously, i.e., $z'_k(z_k)$ can only be computed after $z'_1, \cdots, z'_{k-1}(z_{k+1}, \cdots, z_m)$ has been computed, it is difficult to develop an asynchronous algorithm. The computations are essentially done serially. On the other hand, if we rewrite

$$z = \bar{L}^T \bar{L} u$$

where $\bar{L}$ is the inverse of L, given m processors, solution z can be computed by the following algorithm:

1. To each processor j, $j = 1, \cdots, m$, assign the jth row $\bar{L}_j$ of the inverse $\bar{L}$ of L and the vector u.

2. Asynchronously, at processor $j = 1, \cdots, m$, compute

$$\begin{aligned} \bar{v}_j &= <\bar{L}_j, u> \\ \hat{L}_j &= \bar{v}_j \bar{L}_j \end{aligned}$$

3. From each processor $j = 2, \cdots, m$, communicate $\hat{L}_j$ to Processor 1, and, at processor 1, compute

$$z = \sum_{j=1}^{m} \hat{L}_j.$$

At step 2, u is multiplied by the appropriate row of $\bar{L}$, and then the corresponding column of $\bar{L}^T$ is multiplied by the result. After receiving the result of the asynchronous computation at each processor $j = 2, \cdots, m$, Processor 1 at step 3 then generates the solution by adding the vectors received from these processors. We note that there is only one communication between each processor $j = 2, \cdots, m$ and processor 1, and most work is done asynchronously. Also, both operations of step 2 can be done effectively in a vector environment.

In general, $\bar{L}$ will be considerably more dense than L. Thus, sparsity has been sacrificed in the above algorithm. This may not be an important concern in vector environments.

Using the bordering method used in the proof of Theorem 2.3.15, we can develop a method for finding inverse of the factor in parallel and vector environments. Thus, given inverse $\bar{L}$ of L, the factor of matrix F, the inverse of the factor of

$$C = \begin{bmatrix} F & f \\ f^T & \alpha \end{bmatrix}$$

is the matrix

$$\begin{bmatrix} \bar{L} & 0 \\ -\beta b^T \bar{L} & \beta \end{bmatrix}$$

6.2 IMPLEMENTATION OF INTERIOR POINT METHODS

where $b = \bar{L}f$ and $\beta^{-2} = \alpha - b^T b$. We now present an algorithm, based on the above, that computes the inverse of the factor of ADA^T. This algorithm also computes the elements of $C = ADA^T$.

1. To each processor, assign matrix $\bar{A} = AD^{\frac{1}{2}}$.

2. At processor k

   (a) For each $j = k+1, \cdots, m$, compute asynchronously the k vectors:
   $$a_k^{(j)} = \begin{bmatrix} <\bar{A}_1, \bar{A}_j> \\ <\bar{A}_2, \bar{A}_j> \\ \vdots \\ <\bar{A}_k, \bar{A}_j> \end{bmatrix}$$

   (b) After receiving from each processor $1 \leq j \leq k-1$ the scalars $b_k^{(j)}$ and $k-1$ vectors $\hat{L}_k^{(j)}$ compute the k vector
   $$\bar{L}_k = \beta(\sum_{j=1}^{k-1} \hat{L}_k^{(j)}, 1)$$
   where $\beta^{-2} = <\bar{A}_k, \bar{A}_k> - \sum_{j=1}^{k-1}(b_k^{(j)})^2$.

   (c) Compute asynchronously, for each $j = k+1, \cdots, m$ the scalars $b_j^{(k)} = <\bar{L}_k, a_k^{(j)}>$ and the j vector $\hat{L}_j^{(k)} = -b_j^{(k)}(\bar{L}_k, 0)$.

   (d) For each $j = k+1, \cdots, m$, send $b_j^{(k)}$ and $\hat{L}_j^{(k)}$ to the processor j.

3. Stop when all processor are done. Row $\bar{L}_j$ of the inverse of L is obtained at processor j.

Step 2(a) and (c) are asynchronous, but the processor k waits to receive information from each processor $j = 1, \cdots, k-1$. After receiving this information, it computes $b_j^{(k)}$ and $\hat{L}_j^{(k)}$, which need the completion of step 2(a) as well. These vectors are then communicated to processors $j = k+1, \cdots, m$. The above method does not assume any specific architecture, and must be specialized to one given. Processors generating intermediate rows of $\bar{L}$ do more computations than ones generating the top or the bottom rows of $\bar{L}$.

Conjugate Gradient Method

This method finds the solution to system (6.4) by solving the following convex quadratic optimization problem:

$$\text{minimize} \quad \frac{1}{2}x^T C x - u^T x \qquad (6.6)$$

whose solution is the same as the solution to (6.4). The basic idea in the conjugate gradient method is that if the linearly independent set of vectors $v^1, \cdots, v^m$ are C - conjugate, i.e.,

$$(v^i)^T C v^j = 0 \quad \text{for all} \quad j \neq i$$

then, by a suitable change of variables, the convex quadratic program can be made separable, and thus single variable problems can be solved. This is done as follows: given the linear independence of the m vectors, we can express

$$x = \sum_{j=1}^{m} \lambda_j v^j$$

and substituting in (6.6), we see that an equivalent problem is

$$\text{minimize} \quad \frac{1}{2} \sum_{j=1}^{m} (\lambda_j^2 (v^j)^T C v^j - \lambda_j u^T v^j).$$

It is seen that this problem is separable in the variables λ_j, and thus can be solved, for each $j = 1, \cdots, m$, by finding the solution to

$$\text{minimize} \quad \frac{1}{2} \lambda_j^2 (v^j)^T C v^j - \lambda_j u^T v^j.$$

The solution to the above problem is

$$\lambda_j^* = \frac{u^T v^j}{(v^j)^T C v^j}$$

and thus the solution z to problem (6.6) and thus to (6.4) is

$$z = \sum_{j=1}^{m} \lambda_j^* v^j.$$

Implementing this method requires a generation of the set of linearly independent vectors that are C - conjugate. This can be done by modifying the Gram-Schmidt orthogonalization process, as shown below:

Step 0 Let $a^1, \cdots, a^m$ be a set of linearly independent vectors in $\boldsymbol{R}^m$. Set $v^1 = a^1$, $k = 1$ and go to Step 1.

Step 1 Compute

$$v^k = a^k - \frac{(v^1)^T C a^k}{(v^1)^T C v^1} v^1 - \cdots - \frac{(v^{k-1})^T C a^k}{(v^{k-1})^T C v^{k-1}} v^{k-1}.$$

6.2 IMPLEMENTATION OF INTERIOR POINT METHODS

Step 2 $k = k+1$. If $k \leq m$ go to step 1. Otherwise, stop.

C - conjugacy of the generated vectors is confirmed by multiplying the formula at step 1 by $(v^i)^T C$ for $i < k$, and using induction hypothesis.

Conjugate gradient method is used in implementations to reduce the number of Cholesky factorizations during the later iterations. To see how this can be done, assume that for some large k, the Cholesky factor L_k of $C_k = A D_k A^T (= C)$ has been computed. Instead of computing the factor of C_{k+1} during the next iteration, we could use

$$\begin{aligned} C_{k+1} &= A D_{k+1} A^T \\ &= L_k(L_k^{-1} A D_k^{\frac{1}{2}}) D_k^{-1} D_{k+1} (D_k^{\frac{1}{2}} A^T (L_k^T)^{-1}) L_k^T \end{aligned} \quad (6.7)$$

for the form of the factorization. Note here that L_k is the "L" part of the LQ factorization of $A D_k^{\frac{1}{2}}$. (LQ factorization is introduced in Section 2.3.16). Also, if $P_k = L_k^{-1} A D_k^{\frac{1}{2}}$, then by straightforward calculation it is seen that $P_k P_k^T = I$, i.e., the rows of $L_k^{-1} A D_k^{\frac{1}{2}}$ are mutually othogonal. Now, the system $C_{k+1} z = u$ can be solved as follows: First find u' such that

$$L_k u' = u.$$

Then solve for z' such that

$$P_k \bar{D}_{k+1} P_k^T z' = u',$$

and, finally obtain the solution z by solving

$$L_k^T z = z'.$$

where $\bar{D}_{k+1} = D_k^{-1} D_{k+1}$. The first and third systems above are triangular. The second system is solved by conjugate gradient method where the initial set of linearly independent vectors are chosen such that the *steepest descent* method of optimization is implemented. This can be done, by using $C = P_k \bar{D}_{k+1} P_k^T$, as follows:

6.2.4 Conjugate Gradient Method

Step 0 Let z^0 be an arbitrary guess of the solution, $\epsilon > 0$, Set $k = 1$.

Step 1 Set $a^k = u - C z^{k-1}$. If $\|a^k\| \leq \epsilon$, then stop. z^{k-1} is an acceptable solution.

Step 2 Determine direction v^k which is C - conjugate to the directions $v^1, \cdots, v^{k-1}$, i.e.,

$$v^k = a^k - \frac{(v^1)^T C a^k}{(v^1)^T C v^1} v^1 - \cdots - \frac{(v^{k-1})^T C a^k}{(v^{k-1})^T C v^{k-1}} v^{k-1}.$$

Step 3 Compute

$$\lambda_k^* = \frac{u^T v^k}{(v^k)^T C v^k}.$$

Step 4 Generate

$$z^k = z^{k-1} + \lambda_k^* v^k.$$

Step 5 Set $k = k+1$. If $k \leq m$, got to step 1; otherwise stop.

Computation of a^k at step 1 shows that it is negative of the gradient of objective function of (6.6). This guarantees that if C is the identity matrix, using direction $a^1 = v^1$ will yield the solution z in one iteration of conjugate gradient method. C is generally not an identity matrix, but in case it is close to one, savings in computational work may result by using this method over factorization. We will show that, asymptotically, this happens in the case of the large step primal affine scaling method, Section 5.1.9, when $\alpha \leq \frac{2}{3}$. For this purpose we consider the situation where the Cholesky factor L_k of $AX_k^2 A^T$ is used in the iterations $k+1, \cdots, r$.

Assume that the sequence $\{x^k\}$ is generated by the primal affine scaling method. From Theorem 5.1.20, we see that there exists x^* such that $x^k \longrightarrow x^*$. Define $B = \{j : x_j^* > 0\}$ and $N = \{j : x_j^* = 0\}$. The following lemma can be proved:

6.2.5   Lemma Let $\alpha \leq \frac{2}{3}$ in the large step primal affine scaling method, and let the method generate sequence $\{x^k\}$ with x^* its limit. Let B and N be defined as above. Then

$$\lim_{k \to \infty} \frac{x_j^{k+1}}{x_j^k} = \begin{cases} 1 & j \in B \\ 1 - \alpha & j \in N \end{cases}$$

Proof: Note that for $j \in N$

$$\frac{x_j^{k+1}}{x_j^k} = \frac{v_j^{k+1}}{v_j^k} \frac{c^T x^{k+1} - c^*}{c^T x^k - c^*},$$

6.2 IMPLEMENTATION OF INTERIOR POINT METHODS

where $v_j^k = \frac{x_j^k}{c^T x^k - c^*}$. As has been seen in Section 5.3, the sequence $\{v_N^k\}$ converges to the analytic center, v_N^*, of the polyhedron $\mathcal{P}_N = \{v_N : A_B v_B + A_N v_N = 0, (\bar{s}_N)^T v_N = 1, v_N \geq 0\}$. Thus, as its analytic center v_N^* has strictly positive components ($v_j^* > 0$),

$$\frac{v_j^{k+1}}{v_j^k} \longrightarrow 1.$$

Also, from Theorem 5.2.11,

$$\frac{c^T x^{k+1} - c^*}{c^T x^k - c^*} \longrightarrow 1 - \alpha$$

and thus the conclusion for $j \in N$ follows. The conclusion for $j \in B$ also follows as $x_j^k \longrightarrow x_j^* > 0$. ■

Using the identity $\frac{x_j^{k+r}}{x_j^k} = \frac{x_j^{k+1}}{x_j^k} \cdot \frac{x_j^{k+2}}{x_j^{k+1}} \cdots \frac{x_j^{k+r}}{x_j^{k+r-1}}$, it is seen that a consequence of Lemma 6.2.5 is that for each $r \geq 1$,

$$\frac{x_j^{k+r}}{x_j^k} \longrightarrow \begin{cases} 1 & j \in B \\ (1-\alpha)^r & j \in N \end{cases}.$$

Defining a diagonal matrix Q whose jth diagonal element is $\lim_{k \to \infty} \frac{x_j^{k+r}}{x_j^k}$, we see that $\bar{D}_{k+r} = Q + \Xi_{k+r}$, where $\Xi_{k+r} \longrightarrow 0$, and, is a diagonal matrix. We thus see that

$$\begin{aligned} P_k \bar{D}_{k+r} P_k^T &= P_k (Q + \Xi_{k+r}) P_k^T \\ &= I + ((1-\alpha)^{2r} - 1) P_{k,N} P_{k,N}^T + P_k \Xi_{k+r} P_k^T. \end{aligned} \quad (6.8)$$

We can then prove the following result:

6.2.6 Theorem Let $\alpha \leq \frac{2}{3}$ and $x^* = 0$, i.e., $B = \emptyset$. Then, for every $r \geq 0$

$$P_k \bar{D}_{k+r} P_k^T \longrightarrow (1-\alpha)^{2r} I$$

Proof: Since $B = \emptyset$, $P_{k,N} = P_k$ for every k. This result follows from identity (6.8), since $\Xi_{k+r} \longrightarrow 0$ for every $r \geq 1$. ■

The previous theorem holds only for homogeneous linear programs which have the vector 0 as the only solution. It does show, however, that if the sequences are converging nicely, i.e., to analytic centers or some other well defined vectors in the interior of the optimal faces, then the numerical computations also exhibit nice properties.

Partial Cholesky Factorization

The technique of partial Cholesky factorization is generated when only a part of the factor is required to be triangular. For example, the outer product method can be terminated before the complete factor L has been generated. We will now show how this strategy can be effectively used to enhance sparsity or to exploit a given natural partition of A.

Assume that A has the *natural* partition

$$A = \begin{bmatrix} G \\ H \end{bmatrix}. \qquad (6.9)$$

where G is an $m_1 \times n$ and H is an $m_2 \times n$ matrices with $m_1 + m_2 = m$. Thus,

$$\begin{aligned} C &= ADA^T \\ &= \begin{bmatrix} GDG^T & GDH^T \\ HDG^T & HDH^T \end{bmatrix}. \end{aligned}$$

Sparsity of the Cholesky factor can be enhanced by factoring GDG^T and HDH^T independently of each other, i.e., find factors L_1 and L_2 such that

$$\begin{aligned} GDG^T &= L_1 L_1^T \\ HDH^T &= L_2 L_2^T \end{aligned}$$

and, using partial Cholesky factorization technique, we connect these two factors to solve the resulting system of equations (6.4). This is done in the lemma below:

6.2.7 Lemma Let L_1 and L_2 be respective Cholesky factors of GDG^T and HDH^T. Then, there are matrices $\bar{C}$, $\hat{C}$, L and Q, where L is lower triangular, such that

$$C = ADA^T = L\hat{C}L^T$$

where
$$\begin{aligned} L_1 Q^T &= GDH^T, \\ \bar{C} &= L_2(I - L_2^{-1} Q Q^T L_2^{-T}) L_2^T, \\ \hat{C} &= \begin{bmatrix} I & 0 \\ 0 & \bar{C} \end{bmatrix}, \\ L &= \begin{bmatrix} L_1 & 0 \\ Q & L_2 \end{bmatrix}. \end{aligned}$$

Proof: Can be verified by a direct multiplication of L, $\hat{C}$ and L^T. ∎

6.2 IMPLEMENTATION OF INTERIOR POINT METHODS

Define
$$z = \begin{bmatrix} z^1 \\ z^2 \end{bmatrix} \quad \text{and} \quad u = \begin{bmatrix} u^1 \\ u^2 \end{bmatrix}$$
such that (6.4) corresponds to the partition of A. Its solution can be generated by the following steps:

Step 1 Find $(z^1)'$ such that
$$L_1(z^1)' = u^1$$

Step 2 Define
$$(z^2)' = u^2 - Q(z^1)'$$

Step 3 Find $(z^2)''$ by solving
$$L_2(z^2)'' = (z^2)'$$

Step 4 Solve
$$(I - EE^T)(z^2)''' = (z^2)'' \tag{6.10}$$
where $L_2 E = Q$.

Step 5 Find z^2 by solving
$$L_2^T z^2 = (z^2)'''$$

Step 6 Find z^1 by solving
$$L_1^T z^1 = (z^1)' - Q^T z^2$$

The above steps are generated by, first, using the factored form of C in Lemma 6.2.7; then, the structure of L and $\hat{C}$. Other than Step 4, all steps use the factors L_1 and L_2, which are expected to be sparser than the factor for C. The price paid is in solving the generally denser system of Step 4. We can show the following result about this system:

6.2.8 Theorem Let E be defined by Step 4. It is an $m_2 \times m_1$ matrix and the spectral radius of EE^T is less than 1.

Proof: The size follows from definitions. Since ADA^T is positive definite, so is $L\hat{C}L^T$, and, thus $\hat{C}$ is positive definite. From this we derive that $\bar{C}$ is also positive definite. From equation (6.10), $x^T(I - EE^T)x > 0$ for all x. Thus $x^T EE^T x < x^T x$ for all x, and the result follows. ∎

We now show how the system at Step 4 can be solved by a recursive use of the Sherman-Morrison-Woodbury formula. The resulting method is well suited for implementation in a distributed/vector computing environment. Let, for each $j = 1, \cdots, m_1$, E_j be the jth column of E, and let

$$E_{m_1+1} = (z_2)''$$

Also, let $B_1 = I$, and for each $k = 1, \cdots, m_1$,

$$B_{k+1} = B_k - E_k E_k^T.$$

Note that $B_{m_1+1} = I - EE^T$. Now, for each $k = 1, \cdots, m_1 + 1$ and $j = 1, \cdots, m_1 + 1$ define

$$B_k E_j^{(k)} = E_j.$$

Thus, for each $j = 1, \cdots, m_1$ and $j = 1, \cdots, m_1 + 1$

$$B_{k+1} E_j^{(k+1)} = E_j$$

or,

$$(B_k - E_k E_k^T) E_j^{(k+1)} = E_j$$

or,

$$(I - E_k^{(k)} E_k^T) E_j^{(k+1)} = E_j^{(k)} \tag{6.11}$$

Using the Sherman-Woodbury-Morrison formula, we can write the solution to equation (6.11) as

$$\begin{aligned}
E_j^{(k+1)} &= E_j^{(k)} + \frac{<E_j, E_k^{(k)}>}{1 - <E_k, E_k^{(k)}>} E_k^{(k)} \\
&= E_j^{(k)} + \frac{<E_k, E_j^{(k)}>}{1 - <E_k, E_k^{(k)}>} E_k^{(k)}.
\end{aligned}$$

We can conclude

6.2.9 Theorem The solution $(z^2)'''$ to be computed at step 4 is $E_{m_1+1}^{(m_1+1)}$.

Proof: By definitions

$$B_{m_1+1} E_{m_1+1}^{(m_1+1)} = E_{m_1+1}$$

and the theorem follows by substituting the identities $B_{m_1+1} = I - EE^T$ and $E_{m_1+1} = (z^2)''$. ∎

The above inductive procedure (on k) suggests the following algorithm in a vector/parallel environment:

6.2 IMPLEMENTATION OF INTERIOR POINT METHODS

Step 0 To each processor $j = 1, \cdots, m_1 + 1$, assign E and set $E_j^{(1)} = E_j$, $k = 1$.

Step 1 From processor k, communicate $E_k^{(k)}$ and $<E_k, E_k^{(k)}>$ to each of the processors $j = k + 1, \cdots, m_1 + 1$.

Step 2 At processor j, $j = k + 1, \cdots, m_1 + 1$, compute

$$E_j^{(k+1)} = E_j^{(k)} + \frac{<K_k, E_j^{(k)}>}{1 - <K_k, E_k^{(k)}>} E_k^{(k)}$$

Step 3 Set $k = k + 1$. If $k \leq m_1 + 1$ go to step 1, otherwise declare $E_{k+1}^{(k+1)}$ as the solution $(z^2)'''$ of equation (6.10).

LQ Factorization

There is another case where we can use an alternative factorization of C to solve system (6.4). This is the case when the rows of A are mutually orthogonal, i.e,

$$A_i A_j^T = 0 \quad \text{for all} \quad i \neq j.$$

In this case, for $\alpha > 0$, we have the factorization

$$\begin{aligned} C &= ADA^T \\ &= \alpha(I - A(I - \frac{1}{\alpha}D)A^T) \end{aligned}$$

and the system (6.4) can be equivalently written as

$$(I - N)z = \frac{1}{\alpha}u \tag{6.12}$$

where $N = A(I - \frac{1}{\alpha}D)A^T = EE^T$, $E = A(I - \frac{1}{\alpha}D)^{\frac{1}{2}}$, provided $\alpha > \max_j D_{jj}$. We can prove the following result:

6.2.10 Theorem Let A have mutually orthogonal rows. Then $\|A\| \leq 1$, $\|A^T\| = 1$. Also, for $\alpha > \max_j D_{jj}$, the spectral radius of N (defined above) is less than 1.

Proof: Since the rows of A are mutually orthogonal, there is a matrix B such that $P = \begin{bmatrix} A \\ B \end{bmatrix}$ is $n \times n$ and is orthonormal (B can be obtained by

Gram-Schmidt orthogonalization process). Also, $P^{-1} = P^T$ and $\|Px\| = \|x\|$ for every x. Thus

$$1 = \|P\| = \max_{\|x\|=1}\|Px\| = \max_{\|x\|=1}\left\|\begin{bmatrix} Ax \\ Bx \end{bmatrix}\right\| \geq \max_{\|x\|=1}\|Ax\| = \|A\|$$

and, to see the result of $\|A^T\|$, note that

$$\max_{\|x\|=1}\|A^T x\| = \max_{\|x\|=1}\|A^T x + B^T 0\| = \max_{\|x\|=1}\|(x,0)^T\| = 1.$$

Thus,

$$\begin{aligned} \|N\| &\leq \|A\|\|I - \frac{1}{\alpha}D\|\|A^T\| \\ &\leq \max_j |1 - \frac{1}{\alpha}D_{jj}| \\ &< 1 \end{aligned}$$

and we are done. ∎

Since the data matrix A of a given linear program almost never has the property that its rows are mutually orthogonal, the data almost never satisfies the conditions of the above theorem. We now show how LQ factorization of A can be used to transform the data A such that rows of the resulting matrix are mutually orthogonal.

Let $A = LQ$ be the LQ factorization of A. (This was introduced in Section 2.3.16). Then, the rows of $Q = L^{-1}A$ are mutually orthogonal. Given the linear program

$$\begin{aligned} \text{minimize} \quad & c^T x \\ Ax &= b \\ x &\geq 0 \end{aligned}$$

we can transform it in to an equivalent problem

$$\begin{aligned} \text{minimize} \quad & c^T x \\ L^{-1}Ax &= L^{-1}b \\ x &\geq 0 \end{aligned}$$

and data matrix $L^{-1}A = Q$ satisfies the required property. In an application, this orthogonalization is performed only once on the data. (Compare this to the factorization (6.7) used in the conjugate gradient method).

Infinitely Summable Series Implementation

We have seen three factorizations of the system (6.4), namely (6.7) in the use of conjugate gradient method, (6.10) in the use of partial Cholesky method and (6.12) in the use of LQ factorization. System (6.4) is transformed into an equivalent system
$$(I - N)z = u \tag{6.13}$$
where N is a positive semi-definite matrix with spectral radius less than 1. For such a matrix, using Theorem 2.3.4, we can write the inverse of $(I - N)$ as the expansion
$$(I - N)^{-1} = I + N + N^2 + \cdots.$$
We now develop a methodology to use the above power expansion in the solution of (6.13). Note that we can find the solution by summing the series
$$z = u + Nu + N^2 u + \cdots.$$
The methodology we now develop is based on, iteratively, summing the above infinite series. For each $k = 1, 2, \cdots$, define
$$\begin{aligned} s^k &= u + Nu + \cdots + N^k u \\ &= u + N s^{k-1} \\ t^k &= N^{k+1} u + N^{k+2} u + \cdots. \end{aligned}$$
Here s^k is the sum of the first k terms and t^k is the sum of the remaining terms of the series, i.e., its tail. We could declare s^k as an approximate solution, but it is generally a poor approximation. On the other hand if the spectral decomposition $N = Q \Lambda Q^T$ is known, where Λ is the diagonal matrix of eigenvalues, and Q is the matrix of eigenvectors with $QQ^T = Q^T Q = I$ (see Theorem 2.2.22), then
$$\begin{aligned} t^k &= Q(\Lambda^{k+1} + \Lambda^{k+2} + \cdots) Q^T \\ &= Q \Lambda^{k+1} (I - \Lambda)^{-1} Q^T \end{aligned}$$
and we have a formula for the tail. Unfortunately obtaining the spectral decomposition of a matrix is computationally as expensive as solving the system of equations, and thus this does not give us a meaningful solution to our problem. We will now develop a linear model which will estimate the coefficients of the characteristic polynomial of N, and use this estimate to compute an estimate $\hat{t}^k$ of t^k and we will declare
$$z^k = s^k + \hat{t}^k$$

as an estimate of the solution z. We will also show tnat the "error" of this estimate is the same as the residual, when the linear model is solved as a least squares problem.

The model is based on the fact that, for large k, the contribution to t^k of small eigenvalues of N is minimal, and is of the same order for eigenvalues that are close to each other. The model uses functions of the type $(I-N)^p N^q$ where the contribution to the tail of eigenvalues close to 1 and 0 is minimal for p and q, as well as functions of the type $(\alpha I-N)^o(I-N)^p N^q$ for $0 < \alpha < 1$ which reduce the contribution of eigenvalues close to 0, α, and 1. We generate these functions by using finite differencing on the vectors $u, Nu, \cdots, N^r u$. We now introduce this operator.

6.2.11 Finite Difference Operators

Given an ordered sequence of vectors $x^1, x^2, \cdots, x^r$, we define the operator ∂ by the following rule: for $j < r$

$$\partial x^j = x^j - x^{j+1}.$$

Thus,

$$\partial^{k+1} x^j = \partial(\partial^k x^j) = \partial^k x^j - \partial^k x^{j+1}.$$

Also, given a real number α, we define the operator ∂_α by the following rule:

$$\partial_\alpha x^j = \alpha x^j - x^{j+1}.$$

Then,

$$\partial_\alpha^{k+1} x^j = \partial_\alpha(\partial_\alpha^k x^j) = \alpha \partial_\alpha^k x^j - \partial_\alpha^k x^{j+1}.$$

The following two results relate to these operators:

6.2.12 Lemma

Given any sequence of vectors $x^1, x^2, \cdots, x^r$ and any two real numbers α and β, $\partial_\alpha \partial_\beta x^j = \partial_\beta \partial_\alpha x^j$.

Proof: The following identities follow from definitions.

$$\begin{aligned}
\partial_\alpha \partial_\beta x^j &= \alpha \partial_\beta x^j - \partial_\beta x^{j+1} \\
&= \alpha(\beta x^j - x^{j+1}) - (\beta x^{j+1} - x^{j+2}) \\
&= \beta(\alpha x^j - x^{j+1}) - (\alpha x^{j+1} - x^{j+2}) \\
&= \beta \partial_\alpha x^j - \partial_\alpha x^{j+1} \\
&= \partial_\beta \partial_\alpha x^j
\end{aligned}$$

and we are done. ∎

6.2 IMPLEMENTATION OF INTERIOR POINT METHODS

6.2.13 Theorem Given $u, Nu, \cdots, N^r u$, $\partial^p(N^q u)$ and $\partial^o_\alpha \partial^p(N^q u)$ are well defined for all $o + p + q \leq r$, with

$$\partial^p(N^q u) = (I - N)^p N^q u$$

and

$$\partial^p \partial^o_\alpha(N^q u) = (\alpha I - N)^o (I - N)^p N^q u.$$

Proof: We will show this for the first result. The second result can be proved by a similar argument. This is clearly true for $p = 1$. Assume it true for some $p \geq 1$ and consider $\partial^{p+1}(N^q u)$, $p + q + 1 \leq r$. But this is, by definition, $\partial^p(N^q u) - \partial^p(N^{q+1} u)$ with both terms well defined. Using the induction hypothesis, we see that $\partial^{p+1}(N^q u) = (I - N)^p N^q u - (I - N)^p N^{q+1} u = (I - N)^{p+1} N^q u$ and we are done. ∎

Given $o + p + q \leq r$, generating $(\alpha I - N)^o (I - N)^p N^q u$ will require vector additions and vector-scalar multiplications. The procedure to do this is straight forward and so we omit the details.

6.2.14 The Model We now develop a linear model to obtain an estimate of the tail t^k. Before we do this, given the spectral decomposition of $N = Q \Lambda Q^T$, we establish the following straightforward lemma:

6.2.15 Lemma Let $N = Q \Lambda Q^T$. Then, there is a matrix $\bar{Q}$ such that for every $k = 1, 2, \cdots$, $N^k u = \bar{Q} \Lambda^k e$, where e is a vector of all ones.

Proof: Let $\bar{u} = Q^T u$ and $\bar{Q}_j = \bar{u}_j Q_j$ where Q_j is the jth column of Q. Since Q is orthonormal, $N^k u = Q \Lambda^k Q^T u = \bar{Q} \Lambda^k e$, and we are done. ∎

We note that, from the above lemma, we can write

$$Nu = \sum_{j=1}^m \lambda_j \bar{Q}_j.$$

Let the sets $S_1, \cdots, S_l$ be the indices of multiple eigenvalues of N, i.e., for each $i = 1, \cdots, l$, $S_i \subset \{1, \cdots, m\}$ and for some $\bar{\lambda}_i$, $\lambda_j = \bar{\lambda}_i$ for every $j \in S_i$. Then

$$Nu = \sum_{i=1}^l \sum_{j \in S_i} \bar{\lambda}_i \bar{Q}_j = \sum_{i=1}^l \bar{\lambda}_i \bar{P}_i.$$

In addition

$$N^k u = \sum_{i=1}^l \bar{\lambda}_i^k \bar{P}_i.$$

We are now ready to develop the linear model that will enable the estimation of the tail $\hat{t}^k$. Assume that at iteration k, the sequence $u, Nu, \cdots, N^k u$ has been generated. Also, assume the eigenvalues fall into classes $S_1, S_2, \cdots, S_l$ (which form a disjoint partition of $\{1, \cdots, m\}$); and for each class, assume that the functional from $(\alpha_i I - N)^{o_i}(I - N)^{p_i} N^q u$, $o_i \geq 0$, $s_i \geq 1$ and $o_i + p_i + q \leq k$, distinguishes the class S_i from the others. Based on this information, we now develop the linear model.

Let $o_i + p_i + t \leq k$, and define

$$\begin{aligned}\Delta_t^{(i)} &= \partial_{\alpha_i}^{o_i} \partial^{p_i}(N^t u) \\ &= (\alpha_i I - N)^{o_i}(I - N)^{p_i} N^t u \\ &= \sum_{j=1}^{l} \bar{Q}_j (\alpha_i I - \Lambda_j)^{o_i}(I - \Lambda_j)^{p_i} \Lambda_j^t e\end{aligned}$$

Here $\bar{Q}_j$, Λ_j are the corresponding matrices associated with S_j, and, as mentioned earlier, $\Delta_t^{(i)}$ is known, and can be generated from the original sequence by finite differencing.

Under the assumption that o_i, s_i and α_i distinguishes the eigenvalues of S_i from the others, we can write:

$$\Delta_t^{(i)} = \bar{Q}_i (\alpha_i I - \Lambda_i)^{o_i}(I - \Lambda_i)^{p_i} \Lambda_i^t e + \varepsilon_t^{(1)} \quad (6.14)$$

where we treat $\varepsilon_t^{(1)}$ as the "error" of ignoring the terms pertaining to the eigenvalues not in S_i.

For each i, let $o_i + p_i + t_i \leq k$. t_i is the number of eigenvalues in S_i. Define the $m \times t_i$ matrix:

$$\Delta^{(i)} = (\Delta_{k-o_i-p_i-t_i+1}^{(i)}, \cdots, \Delta_{k-o_i-s_i}^{(i)}).$$

Using (6.14), it is seen that

$$\Delta^{(i)} = \bar{Q}_i (\alpha_i I - \Lambda_i)^{o_i}(I - \Lambda_i)^{p_i} \Lambda_i^{k-o_i-p_i-t_i+1} V_i + \Upsilon^{(1)}$$

where

$$V_i = \begin{bmatrix} 1 & \lambda_1^{(i)} & (\lambda_1^{(i)})^2 & \cdots & (\lambda_1^{(i)})^{t_i-1} \\ 1 & \lambda_2^{(i)} & (\lambda_2^{(i)})^2 & \cdots & (\lambda_2^{(i)})^{t_i-1} \\ \cdot & \cdot & \cdot & \cdots & \cdot \\ 1 & \lambda_{t_i}^{(i)} & (\lambda_{t_i}^{(i)})^2 & \cdots & (\lambda_{t_i}^{(i)})^{t_i-1} \end{bmatrix}$$

is a $t_i \times t_i$ Vandermonde's matrix. Here, $\lambda_j^{(i)}$ are the eigenvalues of N indexed in S_i. One would expect $\Upsilon^{(1)}$ to be small only when such distinguishing o_i, p_i and α_i exist.

6.2 IMPLEMENTATION OF INTERIOR POINT METHODS

Thus, for each i, we can write

$$\bar{Q}_i = \Delta^{(i)} V_i^{-1} \Lambda_i^{-(k-o_i-p_i-t_i+1)} (I - \Lambda_i)^{-p_i} (\alpha_i I - \Lambda_i)^{-o_i} + \Upsilon^{(2)}. \quad (6.15)$$

Now, $N^k u = \bar{Q} \Lambda^k e = \sum_{i=1}^{l-1} \bar{Q}_i \Lambda_i^k e + \bar{Q}_p \Lambda_p^k e$ where we will assume that S_p contains the indices of all the 'small' eigenvalues. Substituting (6.15) into the above, we get

$$N^k u = \sum_{i=1}^{l-1} \Delta^{(i)} V_i^{-1} \Lambda_i^{o_i+s_i+t_i-1} (I - \Lambda_i)^{-p_i} (\alpha_i I - \Lambda_i)^{-o_i} e + \varepsilon^{(2)}.$$

Defining, for each $i = 1, \cdots, l-1$, $\alpha^{(i)} = V_i^{-1} \Lambda_i^{o_i+p_i+t_i-1} (I - \Lambda_i)^{-s_i} (\alpha_i I - \Lambda_i)^{-o_i} e$, and $\Delta^{(0)} = N^k u$, we obtain the linear model

$$\Delta^{(0)} = \sum_{i=1}^{l-1} \Delta^{(i)} \alpha^{(i)} + \varepsilon^{(2)} \quad (6.16)$$

where $\varepsilon^{(2)}$ is the 'deterministic error', and $\alpha^{(i)}$ are the parameters to be estimated.

A comment on this model is in order here. The parameters $\alpha^{(i)}$ are complicated functions of the eigenvalues of N, but by a simple analysis, can be shown to generate, through a linear transformation, coefficients of a polynomial which has $\lambda_j^{(i)}$ as roots (besides others). Fortunately, we do not need to compute $\lambda_j^{(i)}$, and need only estimates of $\alpha^{(i)}$ (see the next section). Also, the data, $\Delta^{(i)}$ for $i = 0, \cdots, l-1$ in the model is known, and is generated by finite differencing of the original data.

In pedagogical terms, this model sets up a linear relationship between the most recently generated data $N^k u$ and the past data $u, Nu, \cdots, N^{k-1} u$. In estimation of the tail, this linear relationship is assumed to hold for all future data (to be estimated) and the past data (assumed known).

This model now allows us to estimate the parameters by the "least squares" procedure, requiring the solution of the normal equation:

$$\Delta^T \Delta \alpha = \Delta^T \Delta^{(0)}. \quad (6.17)$$

where $\Delta = (\Delta^{(1)}, \cdots, \Delta^{(l-1)})$, $\alpha = (\alpha^{(1)}, \cdots, \alpha^{(l-1)})^T$, and is a $q \times q$ system of linear equations, with $q = \sum_{i=1}^{l-1} t_i$.

6.2.16 Estimating the Tail and the Solution

Let $\hat{\alpha}$ be the estimate of the coefficient α, and be obtained by solving the model (6.16) by the method of least squares, i.e., by solving (6.17). We now show how this estimate can be used to compute, $\hat{t}^{k-1}$, an estimate of the tail

$$t^{k-1} = \sum_{j=0}^{\infty} N^{k+j} u.$$

Note that the "known" term $N^k u$ has been included in the tail and the reason for this inclusion will become clear later.

Using Lemma 6.2.15, we can write

$$\begin{aligned} t^{k-1} &= \sum_{j=0}^{\infty} \sum_{i=1}^{l-1} \bar{Q}_i \Lambda_i^{k+j} e + \varepsilon^{(3)} \\ &= \sum_{i=1}^{l-1} \bar{Q}_i \Lambda_i^k (I - \Lambda_i)^{-1} e + \varepsilon^{(3)} \end{aligned}$$

where $\varepsilon^{(3)}$ is the error vector.

Our approach now is to obtain an expression for $\bar{Q}_i$ in terms of Λ_i, such that its substitution into the above expression can be readily identified as some function of $\alpha^{(i)}$. We could use formula (6.15), but it is unsatisfactory (the final terms involve both $\alpha^{(i)}$ and Λ_i in the simplification). We now desire an expression for $\bar{Q}_i$ that does the trick.

Define

$$\bar{\Delta}_t^{(i)} = \partial_{\alpha_i}^{o_i} \partial^{p_i - 1} (N^t u)$$

Using the above expression, for $t = k - 0_i - p_i - t_i + 1, \cdots, k - o_i - p_i$, we can define

$$\bar{\Delta}^{(i)} = \bar{Q}_i \Lambda_i^{k-o_i-p_i-t_i+1} (\alpha_i I - \Lambda_i)^{o_i} (I - \Lambda_i)^{p_i - 1} V_i + \Upsilon^{(3)}.$$

As before, $\bar{\Delta}^{(i)}$ is obtained from the original data by finite differencing, and is known. It is generated in the steps that generates $\Delta^{(i)}$.

Defining $\bar{Q}_i$, as in (6.15), and substituting in above, we get

$$\begin{aligned} t^{k-1} &= \sum_{i=1}^{l-1} \bar{\Delta}^{(i)} V_i^{-1} \Lambda_i^{o_i + p_i + t_i - 1} (\alpha_i I - \Lambda_i)^{o_i} (I - \Lambda_i)^{-p_i} e + \varepsilon^{(4)} \\ &= \sum_{i=1}^{l-1} \bar{\Delta}^{(i)} \alpha^{(i)} + \varepsilon^{(4)} \end{aligned}$$

and we can use

$$\hat{t}^{k-1} = \bar{\Delta} \hat{\alpha}$$

6.2 IMPLEMENTATION OF INTERIOR POINT METHODS

(where $\bar{\Delta} = (\bar{\Delta}^{(1)}, \cdots, \bar{\Delta}^{(l-1)})$, and $\hat{\alpha}^T = (\hat{\alpha}^{(1)}, \cdots, \hat{\alpha}^{(l-1)})$ is the "least squares" estimate of α in (6.16)) as an estimate of t^{k-1}. There is clearly a relationship between $\bar{\Delta}$ and Δ.

Using the estimate $\hat{t}^{k-1}$ we set

$$\begin{aligned} \hat{z}^k &= s^{k-1} + \hat{t}^{k-1} \\ &= s^{k-1} + \bar{\Delta}\hat{\alpha} \end{aligned}$$

and use this as an estimate of the solution to (6.13). Of interest is the error in the equation (6.13) when this estimate is used. Define the vector

$$\epsilon^k = u - (I - N)\hat{z}^k.$$

Indeed, a good estimate of the solution to (6.13), results in a smaller error ϵ^k, and a good model (6.16) would generate a lower residual error in the least squares setting. The next theorem asserts that these two errors, are the same, and also justifies the inclusion of $N^k u$ as a part of the tail.

6.2.17 Theorem Let $\hat{z}^k$ be the estimate of the solution to (6.13) as computed above using $\hat{\alpha}$, the least squares solution to (6.16). Then

$$\epsilon^k = u - (I - N)\hat{z}^k = N^k u - \sum_{i=1}^{p-1} \Delta^{(i)} \hat{\alpha}^{(i)}.$$

Proof:

$$\begin{aligned} u - (I - N)\hat{z}^k &= u - (I - N)(s^{k-1} + \bar{\Delta}\hat{\alpha}) \\ &= u - (I - N)(u + Nu + \cdots + N^{k-1}u + \bar{\Delta}\hat{\alpha}) \\ &= N^k u - (I - N)\bar{\Delta}\hat{\alpha}. \end{aligned}$$

But $(I - N)\bar{\Delta}_t^{(i)} = \partial(\partial_{\alpha_i}^{o_i} \partial^{p_i-1}(N^t u)) = \Delta_t^{(i)}$ and we have our result. ∎

This is an important theorem, since it connects the error in the solution of (6.13) with the residual vector of the least-squares solution. Thus, if the residual is large, the resulting error in the equation will also be large. Hence a large residual would indicate a model modification which is defined by the number l of sets S_i eigenvalues are partitioned into, the number, of eigenvalues, t_i, in each set, the "distinguishing" powers o_i and p_i, and the reals $0 < \alpha_i < 1$.

6.2.18 Validation of the Model We will apply model (6.16) to the system generated by a partial Cholesky factorization of the matrix encountered when solving an assignment problem. This problem has an optimum solution which is highly degenerate, and can thus present special problems for interior point methods. In several examples we tested this model on, the spectral radius of N approaches 1, as the algorithm proceeds to the solution (see Tables A.1-A.6 in the Appendix A), thus, making the summation of the series very hard. This happens when the optimum solution is unique, and thus a highly degenerate vertex of the assignment polytope. In the case when the algorithm converges to an interior solution, our model performs exceptionally well (see Tables A.1-A.6 in the Appendix A). These computational results use the dual affine scaling method of Section 5.8.

We now derive the matrix N used for the validation. An assignment problem is the following linear program: given n^2 positive real numbers $c_{i,j}$, $i = 1, \cdots, n$, $j = 1, \cdots, n$; find $x_{i,j}$ that solve the following linear program:

$$\begin{aligned}
\min \quad & \sum_{i=1}^{n} \sum_{j=1}^{n} c_{i,j} x_{i,j} \\
& \sum_{j=1}^{n} x_{i,j} = 1 && i = 1, \cdots, n \\
& \sum_{i=1}^{n} x_{i,j} = 1 && j = 1, \cdots, n \\
& x_{i,j} \geq 0 && i = 1, \cdots n \, , j = 1, \cdots, n.
\end{aligned}$$

For this problem, the vertex optimum solutions are known to be highly degenerate. The dual to this problem is the linear program:

$$\begin{aligned}
\max \quad & \sum_{i=1}^{n} u_i + \sum_{j=1}^{n} v_j \\
& u_i + v_j + s_{i,j} = c_{i,j} && i = 1, \cdots n \, , j = 1, \cdots, n. \\
& s_{i,j} \geq 0 && i = 1, \cdots n \, , j = 1, \cdots, n.
\end{aligned}$$

where u_i, v_j are the dual variables and $s_{i,j}$ are the dual slacks.

We apply the dual affine scaling method, starting with the interior dual solution $u_i = 0$, $v_j = 0$ and $s_{i,j} = c_{i,j} (> 0)$ for each i and j. In this case the diagonal matrix, D, is a $n^2 \times n^2$ matrix with the diagonal entries, $D_{i,j} = 1/s_{i,j}^2$ for $i = 1, \cdots, n$, $j = 1, \cdots, n$; and the matrix A has the partition (6.9) with

$$G = \begin{bmatrix} e^T & 0 & \cdots & 0 \\ 0 & e^T & \cdots & 0 \\ . & . & & . \\ 0 & 0 & \cdots & e^T \end{bmatrix} ; H = (I, I, \cdots, I),$$

where e is a n-vector of all 1's, and I is a $n \times n$ identity matrix. Thus, GDG^T and HDH^T of lemma 6.2.7 are diagonal matrices with positive diagonal entries. L_1 and L_2 are also diagonal matrices, with $Q = L_1^{-1} HDG^T$. The

6.2 IMPLEMENTATION OF INTERIOR POINT METHODS

matrix $N = L_2^{-1}QQ^T L_2^{-1}$, is a $n \times n$ positive and positive semi-definite matrix with spectral radius less than 1, see, Theorem 6.2.8.

Computational testing of the model is performed on three sets of randomly generated assignment problems, i.e, the positive constants $c_{i,j}$ are determined randomly. Each set consists of two similarly behaving problems, one of size 200×200 and the other 300×300. These problems are solved by the dual affine scaling method adopting adapting partial Cholesky factorization presented above, and present increasing degree of difficulty to the affine scaling and model. On the first set of problems, the method converges to a solution on the boundary of the primal polyhedron with a very few positive variables at value 1; on the second set of problems set the method converges to almost a vertex, with about 75% of positive variables at value 1; and, on the third set of problems the method converges to a vertex, with all positive variables at value 1. The "step-size" in the affine scaling method is given the following values: during the first 20 iterations the values, respectively, are 0.5, 0.5, 0.5, 0.5, 0.5, 0.5, 0.5, 0.5, 0.5, 0.5, 0.5, 0.5, 0.9, 0,90, 0,93, 0.93, 0.94, 0.95, 0.95, 0.97 and remains at 0.97 for the 21st and higher iterations.

The matrix N for testing the model is generated by the affine scaling algorithm. The estimated solution $\hat{y}$ generated by the model is not used in the method, but (6.13) is solved by the technique presented in the partial Cholesky factorization section. This is done to guarantee that the same matrix N is used in testing of different parameters. During each iteration of the affine scaling method, the model is tested on the resulting N with $l = 2$ and $l \leq 14$. In the later case, the iterative method automatically increases l as more entries $N^j u$ of the series are generated. The values $o_1 = 0$, $s_1 = 1$ and $t_1 = 5$ are used with $l = 2$. For $l \leq 14, o_i = 0$ for every i, with s_i and t_i as given in the table below:

i	1	2	3	4	5	6	7	8	9	10	11	12	13
t_i	3	2	2	2	2	2	2	2	2	2	2	2	2
s_i	1	4	6	8	10	12	14	16	18	20	22	24	26

The results of this test are presented in Tables A.1-A.6 in the Appendix A. At each iteration of the dual affine scaling method the matrix N is used to recursively generate $Nu, \cdots, N^K u$. This sequence is then used in the model to estimate the solution to the system (6.13). This system is considered adequately solved when either the error "ERROR" (reported in Tables A.1-

A.6 in the Appendix A) of this estimate satisfies

$$\text{ERROR} \leq \min\{10^{-5}, ||u||_2 \times 10^{-4}\}.$$

or $N^{500}u$ has been generated.

As can be verified from the Tables A.1-A.6 in the Appendix A, the problem complexity greatly effects the model performance. The first set of problems (converging to the interior of a face) is solved quickly; i.e., a good estimate of the solution is obtained for small K. Solving (6.13), for example requires, (at iteration 14) 44 or 23 matrix multiplications for the 200×200 and 44 or 24 for 300×300 case, instead of the inversion of $I - N$. From the tables, it appears that the second class of problems is easier to solve than the third; and for these problems, the model performs well at the earlier and the later iterations. It can be verified from Tables A.1-A.6 in the Appendix A, that the spectral radius of the matrix N approaches 1, specially for the harder problems, and that the model performance improves in the later iterations. This may be because the model performance is effected by the number of classes of "similar" eigenvalues, and not necessarily by the magnitude of the largest one.

6.2.19 Acceleration by Chebychev Polynomials

These polynomials, provide an efficient mechanism for making a best lower order polynomial approximation, over a range of values of the variables, to a given polynomial. In this section we show how these polynomials can be used to accelerate the convergence of the techniques developed in this paper. Before we do this, we present a brief introduction.

The Chebychev polynomial of degree n is defined as

$$T_n(x) = a_0^{(n)} + a_1^{(n)}x + \cdots + a_n^{(n)}x^n \qquad (6.18)$$

for all $x \in [-1, 1]$; and, the relationship between these polynomials is defined by the difference equation

$$\begin{aligned} T_0(x) &= 1 \\ T_1(x) &= x \\ T_{n+1}(x) &= 2xT_n(x) - T_{n-1}(x). \end{aligned}$$

This difference equation can be used to generate coefficients of T_n, and the formulii for these coefficients can be found in Rivlin [183], and other references.

Given a polynomial $p(x)$ of degree n, a lower order approximation $q(x)$ of degree m, $m < n$, is generated by first expressing $p(x)$ in terms of $T_j(x)$;

6.2 IMPLEMENTATION OF INTERIOR POINT METHODS

and then dropping the terms involving the higher order Chebychev polynomials. Thus, to get a m degree approximation $q(x)$ to $p(x)$, we first define $A_0, \cdots, A_n$ such that

$$p(x) = A_0 T_0(x) + A_1 T_1(x) + \cdots + A_n T_n(x) \tag{6.19}$$

and then define

$$q(x) = A_0 T_0(x) + A_1 T_1(x) + \cdots + A_m T_m(x). \tag{6.20}$$

The following is well known:

6.2.20 Theorem $|p(x) - q(x)| \leq \sum_{j=m+1}^{n} |A_j|$ for all $x \in [-1, 1]$.

In our application, we will apply Chebychev approximations to the series

$$p(N) = I + N + N^2 + \cdots + N^{k-1} \tag{6.21}$$

of matrices N by the lower order series

$$q(N) = A_0 + A_1 T_1(N) + \cdots + A_{k'} T_{k'}(N) \tag{6.22}$$

where $k' < k - 1$. A caution is in order here. For the ease of exposition, we have abused notation, and used the same symbols, p, q and T, having the set of reals (as in (6.18) - (6.20)) or the set of matrices (as in (6.21) and (6.22)), as their domain and range. We can show that

6.2.21 Theorem For every k

$$\|T_k(N)\|_2 \leq \max_i |T_k(\lambda_i)| \leq 1.$$

Proof: Using the fact that $N = Q\Lambda Q^T$ for some orthonormal matrix Q, we see that

$$T_k(N) = Q(a_0^{(k)} I + a_1^{(k)} \Lambda + \cdots + a_k^{(k)} \Lambda^k) Q^T$$

and thus

$$\begin{aligned}
\|T_k(N)\|_2 &\leq \|a_0^{(k)} I + a_1^{(k)} \Lambda + \cdots + a_k^{(k)} \Lambda^k\| \\
&= \max_i |(a_0^{(k)} + a_1^{(k)} \lambda_i + \cdots + a_k^{(k)} \lambda_i^k| \\
&= \max_i |T_k(\lambda_i)| \\
&\leq 1
\end{aligned}$$

and we are done. ∎

296   CHAPTER 6: IMPLEMENTATON

In the estimation of the solution to (6.13) we will use

$$\hat{s}^{k-1} = q(N)u$$

instead of s^{k-1}.

Since p and q are polynomials in N, the error calculation of Theorem 6.2.20 will involve the eigenvalues λ_i of N. This follows from the fact that $\|p(N) - q(N)\| \leq \sum_{k'+1}^{k-1} |A_j| \|T_j(N)\|_2$ and, from Theorem 6.2.21, $\|T_j(N)\|_2 \leq \max_i |T_j(\lambda_i)|$. Since $\lambda_i \in [0,1]$, we need to modify the definition of the polynomials so that the corresponding domain of interest is $[0,1]$. This is readily done by a change of variables to obtain

$$\begin{aligned} T_0(x) &= 1 \\ T_1(x) &= 2x - 1 \\ T_{n+1}(x) &= 2(2x-1)T_n(x) - T_{n-1}(x). \end{aligned} \qquad (6.23)$$

We now compute the coefficients of $T_n(x)$.

6.2.22 Theorem Let T_j be defined by (6.23). Then for each $n \geq 1$ and $j \leq n$

$$a_j^{(n)} = \frac{(-1)^{n+j} n}{n+j} \binom{n+j}{2j} 4^j.$$

Proof: Using (6.23), we can readily establish that for each $n \geq 1$:

$$\begin{aligned} a_{n+1}^{(n+1)} &= 4a_n^{(n)} \\ a_n^{(n+1)} &= 4a_{n-1}^{(n)} - 2a_n^{(n)} \\ a_j^{(n+1)} &= -2a_j^{(n)} + 4a_{j-1}^{(n)} - a_j^{(n-1)}, 0 < j < n \\ a_0^{(n+1)} &= -2a_0^{(n)} - a_0^{(n-1)}. \end{aligned}$$

The result now follows by a simple induction hypothesis using $a_0^{(0)} = 1$, and $a_1^{(1)} = 2$, $a_0^{(1)} = -1$. ∎

To generate the approximation (6.22) we need to determine the coefficients, $A_j^{(k-1)}$, $j = 0, \cdots, k-1$, such that

$$p(N) = \sum_{j=0}^{k-1} A_j^{(k-1)} T_j(N). \qquad (6.24)$$

6.2 IMPLEMENTATION OF INTERIOR POINT METHODS

It can be readily established, using the form (6.18) of T_j, that $A_j^{(k-1)}$ solve the following triangular system of equations:

$$\begin{bmatrix} a_0^{(0)} & a_0^{(1)} & \cdots & a_0^{(k-1)} \\ & a_1^{(1)} & \cdots & a_1^{(k-1)} \\ & & \ddots & \vdots \\ & & & a_{k-1}^{(k-1)} \end{bmatrix} \begin{bmatrix} A_0^{(k-1)} \\ A_1^{(k-1)} \\ \vdots \\ A_{k-1}^{(k-1)} \end{bmatrix} = \begin{bmatrix} 1 \\ 1 \\ \vdots \\ 1 \end{bmatrix}. \tag{6.25}$$

6.2.23 Theorem The solution to the system (6.25) is: For $j \geq 1$,

$$A_j^{(k-1)} = \sum_{i=0}^{k-j-1} \binom{2(j+i)}{i} 4^{-j-i+1}, \tag{6.26}$$

with

$$A_0^{(k-1)} = 1 + \sum_{j=1}^{k-1} (-1)^{j+1} A_j^{(k-1)}. \tag{6.27}$$

Proof: The (6.25) can be rewritten as

$$\begin{bmatrix} a_0^{(0)} & a_0 \\ & L \end{bmatrix} \begin{bmatrix} A_0 \\ \bar{A}_0 \end{bmatrix} = \begin{bmatrix} 1 \\ e \end{bmatrix}$$

for the appropriate upper triangular matrix L, vectors $\bar{A}_0$, a_0 and $e = (1, 1, \cdots, 1)^T$. After substituting the result of Theorem 6.2.22, we can obtain the solution of (6.26) by solving

$$8L\bar{A}_0 = r \tag{6.28}$$

where

$$L = \begin{bmatrix} 1 & -\binom{2}{1} & \binom{3}{1} & \cdots & (-1)^{k-1}\binom{k-1}{1} \\ & 1 & -\binom{4}{3} & \cdots & (-1)^k \binom{k}{3} \\ & & 1 & \cdots & (-1)^{k+1}\binom{k+1}{5} \\ & & & \ddots & \vdots \\ & & & & 1 \end{bmatrix},$$

$$\bar{A}_0 = \begin{bmatrix} A_1^{(k-1)} \\ 2A_2^{(k-1)} \\ \vdots \\ (k-1)A_{k-1}^{(k-1)} \end{bmatrix}, r = \begin{bmatrix} 1 \\ 2(4)^{-1} \\ \vdots \\ (k-1)4^{-k+2} \end{bmatrix}.$$

Using the standard identity

$$\binom{r}{j} - \binom{r-1}{j} = \binom{r-1}{j-1},$$

it can be readily shown that

$$L^{-1} = M_1^2 M_2^2 \cdots M_{k-2}^2$$

where

$$M_j = \begin{bmatrix} 1 & & & & & \\ & 1 & & & & \\ & & 1 & 1 & & \\ & & & 1 & 1 & \\ & & & & \ddots & \vdots \\ & & & & & 1 & 1 \end{bmatrix} \leftarrow \text{row } j$$

which specifies that columns, starting with the jth, be successively added. Since L^{-1} is triangular, we can define

$$L^{-1} = \begin{bmatrix} \hat{L} & u \\ & v \end{bmatrix}$$

and; if we define

$$r = \begin{bmatrix} \hat{r} \\ 0 \end{bmatrix} + r_{k-1} u_{k-1}$$

where $u_{k-1} = (0, 0, \cdots, 1)^T$ (the $(k-1)st$ unit vector),

$$L^{-1}r = \begin{bmatrix} \hat{L}^{-1}\hat{r} \\ 0 \end{bmatrix} + r_{k-1} L^{-1} u_{k-1}$$

$$= \begin{bmatrix} \hat{L}^{-1}\hat{r} + r_{k-1} u \\ r_{k-1} v \end{bmatrix}.$$

Using the last identity, the result of the theorem can be established by a straightforward induction on k. ∎

6.2 IMPLEMENTATION OF INTERIOR POINT METHODS

Theorem 6.2.23 can now be used to derive the coefficients of the approximating polynomial (6.22). If
$$q(N) = A'_0 + A'_1 N + \cdots + A'_{k'} N^{k'}$$
then
$$A' = L'A$$
where $A' = (A'_0, \cdots, A'_{k'})^T$, $A = (A_0, \cdots, A_{k'})^T$ and

$$L' = \begin{bmatrix} a_0^{(0)} & a_0^{(1)} & \cdots & a_0^{(k')} \\ & a_1^{(1)} & \cdots & a_1^{(k')} \\ & & \ddots & \\ & & & a_{k'}^{(k')} \end{bmatrix}.$$

Implementation using Schur Complements

Matrices of the form $A - UD^{-1}V$ are called Schur complements. Such matrices were encountered during Cholesky factorization. This factorization generates positive definite Schur complements. We consider the inverse of a matrix which has the form $A - EE^T$, and, its application to linear programming. Here A and $A - EE^T$ are $m \times m$ symmetric and positive definite matrices. We assume E is an $m \times r$ matrix. When $r = 1$, the inverse can be obtained by the Sherman-Morrison-Woodbury formula:

$$(A - uu^T)^{-1} = A^{-1} + vDv^T$$

where $Av = u$ and $D = (1 - u^T A^{-1} u)^{-1}$. The formula we derive is

$$(A - EE^T)^{-1} = A^{-1} + GDG^T \qquad (6.29)$$

where E and G are $m \times r$ matrices and D is a diagonal matrix with positive diagonal elements.

6.2.24 The Inductive Scheme   We now prove that there are matrices G and D which satisfy formula (6.29). For this purpose we introduce the following: Define

$$B_{k+1} = \begin{cases} A & \text{if } k = 0 \\ B_k - E_k E_k^T & \text{if } 1 \leq k \leq r \end{cases}$$

and note that $B_{r+1} = A - EE^T$. Also, for each $k = 1, \cdots, r$, define

$$B_k E_k^{(k)} = E_k.$$

We can then prove:

6.2.25 Theorem There exist matrices G and D such that D is diagonal with positive diagonal entries, and
$$(A - EE^T)^{-1} = A^{-1} + GDG^T$$
where, for each $k = 1, \cdots, r$, $G_k = E_k^{(k)}$ and $D_{kk} = (1- <E_k^{(k)}, E_k>)^{-1}$.

Proof: Since $A - EE^T$ is positive definite, for each $k = 1, \cdots, r$ B_k is positive definite. Also, $\det(B_{k+1}) = \det(B_k)(1- <B_k^{-1}E_k, E_k>)$, and so for each $k = 1, \cdots, r$, $D_{kk} > 0$. From Sherman-Morrison-Woodbury formula, the theorem holds for $r = 1$. Thus, assume that the theorem holds for some $r = l - 1$, $l \geq 2$. We now show the result for $r = l$, and thus the theorem follows by induction. Now
$$B_{l+1} = B_l - E_l E_l^T$$
thus from the Sherman-Morrison-Woodbury formula
$$B_{l+1}^{-1} = B_l^{-1} + \frac{1}{1- <E_l, B_l^{-1}E_l>} B_l^{-1} E_l (B_l^{-1}E_l)^T.$$
From the induction hypothesis,
$$B_l^{-1} = A^{-1} + \sum_{j=1}^{l-1} \frac{1}{1- <E_l, E_j^{(j)}>} E_j^{(j)} E_j^{(j)T}$$
and our result follows. ∎

The inductive procedure used in the proof of the Theorem 6.2.25 may not exist when the matrix $A - EE^T$ is indefinite. As an example consider $I - uu^T - vv^T$ where $u^T u = 1$ and $v^T v = 1$ with $u^T v \neq 0$. In this case both the matrices $I - uu^T$ and $I - vv^T$ are singular, but the matrix after two rank 1 updates is non-singular and thus has an inverse. It may have the form of the inverse stipulated by formula (6.29), but it cannot be generated by the procedure developed in this section.

6.2.26 Updates of Identity Matrix A case of special interest is when the inverse of a matrix with several rank one updates of the identity, i.e., the inverse of
$$I - EE^T$$
is required, and E is $m \times r$. In case $r \leq m$, it is computationally more efficient to use the formula given by (6.29). Otherwise, one can use the following standard identity,
$$(I - EE^T)^{-1} = I + E(I - E^T E)^{-1} E^T,$$

6.2 IMPLEMENTATION OF INTERIOR POINT METHODS

and, since $I - EE^T$ is positive definite if and only if $I - E^T E$ is, the result of Theorem 6.2.25 gives the existence of G and D such that

$$(I - E^T E)^{-1} = I + G^T DG$$

where G is an $m \times r$ matrix. Thus we obtain the following alternative formula for the inverse:

$$(I - EE^T)^{-1} = I + EE^T + EG^T DGE^T. \tag{6.30}$$

6.2.27 Application to Boundary and Interior Methods See the section on partial Cholesky factorization for application to interior point methods. We show how formula (6.29) can be used to implement the simplex method. This may provide a framework for unifying implementation of these two methods.

Let L be a lower triangular matrix which is the Cholesky factor of AA^T, i.e.,

$$AA^T = LL^T.$$

Then, for the given basis matrix B, we have

$$BB^T = AA^T - \sum_{j \in N} A_j A_j^T. \tag{6.31}$$

where A_j is the jth column of the matrix A. Thus,

$$BB^T = L(I - \sum_{j \in N} \bar{A}_j \bar{A}_j^T) L^T$$

where $L\bar{A}_j = A_j$ for each $j \in N$. Defining

$$Q = I - \sum_{j \in N} \bar{A}_j \bar{A}_j^T \tag{6.32}$$

we can write

$$BB^T = LQL^T. \tag{6.33}$$

Our aim is to use the decomposition (6.33) during the steps of the simplex method, and to update this decomposition during the iterations of the method. This involves updating Q when one column in B is replaced by one indexed in N.

During the application of the simplex method, two linear systems, the primal system

$$B\bar{a} = a$$

for some give vector a, and the dual system

$$B^T y = c_B$$

where c_B is the cost vector of the basic variables, are solved. In addition, the reduced costs for each nonbasic index $j \in N$

$$\bar{c}_j = c_j - y^T A_j$$

are also computed. It is readily confirmed that, using the decomposition (6.33), primal system is solved by the following sequence of steps:

Step 1 Compute $\hat{a}$ such that
$$Q\hat{a} = L^{-1}a$$

Step 2 Compute $\tilde{a}$ such that
$$\tilde{a} = L^{-T}\hat{a}$$

Step 3 Compute the solution $\bar{a}$ by
$$\bar{a} = B^T \tilde{a}.$$

Also, the dual system is solved by the following:

Step 1 Compute $\bar{y}$ such that
$$Q\bar{y} = L^{-1}Bc_B$$

Step 2 Compute the solution y by
$$y = L^{-T}\bar{y}.$$

And, we note that

$$\bar{c}_j = c_j - y^T A_j = c_j - \bar{y}^T L^{-1} A_j = c_j - \bar{y}^T \bar{A}_j,$$

and step 2 of the dual system can be eliminated.

We note that at Step 1 of both the primal and dual systems, the inverse of matrix Q is required. In that sense, we will consider this implementation as an application of formula (6.29).

6.2 IMPLEMENTATION OF INTERIOR POINT METHODS

6.2.28 Updating Q^{-1} We now show how the formula for Q^{-1} can be updated when the column A_{j_v} leaves the basic set, and the column A_{j_s} enters. That is, the new non-basic index set is $\bar{N} = \{j_1, \cdots, j_{s-1}, j_{s+1}, \cdots, j_r, j_v\}$. Thus, if the updated matrix

$$\bar{Q} = I - \bar{E}\bar{E}^T$$

then

$$\bar{E}_k = \begin{cases} E_k & 1 \leq k \leq s-1 \\ E_{k+1} & s \leq k \leq r-1 \\ E_v & k = r \end{cases}$$

and

$$\bar{B}_k = \begin{cases} B_k & 1 \leq k \leq s-1 \\ B_{k+1} + E_s E_s^T & s \leq k \leq r \end{cases}$$

with

$$\bar{B}_{r+1} = B_{r+1} + E_s E_s^T - E_v E_v^T.$$

From the above definitions, and section 2, we readily see that for each $1 \leq k \leq s-1$

$$\bar{B}_k^{-1} = B_k^{-1} = I + G^{(k)} F^{(k)} G^{(k)T}.$$

Now, consider $s \leq k \leq r-1$. Defining $\bar{B}_k \bar{E}_k^{(k)} = \bar{E}_k$ we have

$$(B_{k+1} + E_s E_s^T) \bar{E}_k^{(k)} = E_{k+1}$$

or

$$\bar{E}_k^{(k)} = E_{k+1}^{(k+1)} - \frac{<E_s, E_{k+1}^{(k+1)}>}{1+ <E_s, E_s^{(k+1)}>} E_s^{(k+1)}$$

We now show how this can be obtained. Define $a = G^T E_s$, which has been obtained during the 'column updating' phase of the simplex method, and let $a^{(k)} = (a_1, \cdots, a_k)^T$. Then

$$E_s^{(k+1)} = E_s + G^{(k)} F^{(k)} a^{(k)}$$

and

$$\begin{aligned} <E_s, E_s^{(k+1)}> &= <E_s, E_s> + E_s^T G^{(k)} F^{(k)} G^{(k)T} E_s \\ &= \|E_s\|^2 + \sum_{i=1}^{k} F_{ii} a_i^2 \end{aligned}$$

and thus
$$\bar{E}_k^{(k)} = E_{k+1}^{(k+1)} - \frac{a_{k+1}}{1 + \|E_s\|^2 + \sum_{i=1}^k F_{ii} a_i^2} E_s + G^{(k)} F^{(k)} a^{(k)}$$

and
$$\begin{aligned}\bar{F}_{kk}^{-1} &= 1 - <\bar{E}_k, \bar{E}_k^{(k)}> \\ &= 1 - F_{k+1,k+1}^{-1} + \frac{a_{k+1}}{1 + \|E_s\|^2 + \sum_{i=1}^k F_{ii} a_i^2}(<E_{k+1}, E_s> \\ &\quad + E_{k+1}^T G^{(k)} F^{(k)} a^{(k)}).\end{aligned}$$

For the case $k = r$, we obtain
$$\bar{B}_r \bar{E}_r^{(r)} = \bar{E}_r$$

or
$$(B_{r+1} + E_s E_s^T)\bar{E}_r^{(r)} = E_v$$

and thus
$$\bar{E}_r^{(r)} = E_v^{r+1} - \frac{<E_s, E_v^{r+1}>}{1 + <E_s, E_s^{r+1}>} E_s^{r+1}$$

where
$$E_v^{r+1} = E_v + GFG^T E_v$$

and the formula for
$$\bar{F}_{rr} = 1 - <\bar{E}_r, \bar{E}_r^{(r)}>$$

can be readily obtained.

The above formulae have been developed for updating in a parallel computing environment, where each processor, for $k \geq s$ works on generating $\bar{G}_k$ and $\bar{F}_{kk}$. For computing on a single processor, using the above results, a simpler and inductive formula can be developed, i.e., $\bar{E}_{k+1}^{(k+1)}$ is generated after $\bar{E}_k^{(k)}$ has been generated. In case the number of non-basic indices is larger than m, the above representation is not suitable, and we should use the formula (6.30). The updating formula for this case is also generated in a similar manner.

6.3 Notes

Section 6.1 deals with the implementation of boundary methods. More on this can be found in Dantzig [36] and Murty [170].

6.3 NOTES

Section 6.2 presents the basic ideas of implementing an interior point method. Most implementations use the sparse Cholesky factorization heuristics developed by George and Liu [78], and then use the conjugate gradient method during the asymptotic phase of the iterations. The infinitely summable series implementation is based on the work of Saigal [207].

Appendix A

Tables

200 × 200 ASSIGNMENT PROBLEM					
Iter no	Interior Facet				Max Eigenvalue
	$l = 2$		$l \leq 14$		
	Error	K	K	Error	
1	.23E-06	4	4	.23E-06	.9901447238672399
2	.66E-06	4	4	.66E-06	.9900961443084537
3	.22E-05	4	4	.22E-05	.9900607937971756
4	.17E-06	5	5	.17E-06	.9899965061222104
5	.30E-05	5	5	.30E-05	.9899036721613828
6	.32E-05	6	6	.34E-05	.9898362574474263
7	.99E-06	8	9	.21E-06	.9899194893548973
8	.26E-05	9	9	.48E-05	.9902630967647999
9	.36E-05	11	11	.22E-05	.9909347495951451
10	.31E-05	14	13	.11E-05	.991698350999346
11	.42E-05	16	14	.32E-05	.9923520122869073
12	.16E-05	20	16	.10E-05	.9930303653401724
13	.11E-05	25	18	.64E-06	.9935758356117154
14	.78E-07	44	23	.12E-06	.9952617651221247
15	.72E-08	19	15	.70E-08	.9932782887906138
16	.95E-09	60	29	.10E-08	.9965906303594346
17	.33E-10	18	15	.18E-10	.9931044028591927
18	.57E-11	81	39	.57E-11	.9971093266767113
19	.19E-12	18	15	.11E-12	.9922950323648496
20	.22E-13	115	52	.36E-13	.9973853729691724
21	.13E-14	22	17	.81E-15	.9936907023585453
22	.23E-15	215	95	.23E-15	.9980711062108224
23	.60E-17	18	15	.43E-17	.9911968175530437
24	.85E-18	77	36	.11E-17	.9969672707273868

Table A.1: Last column is maximum eigenvalue of N

	200 × 200 ASSIGNMENT PROBLEM				
Iter no	Almost Vertex				Max Eigenvalue
	$l = 2$		$l \leq 14$		
	Error	K	K	Error	
1	.21E-07	5	5	.21E-07	.9896131723699559
2	.87E-07	5	5	.87E-07	.9899056464756422
3	.66E-06	5	5	.66E-06	.9901720102742565
4	.85E-05	5	5	.85E-05	.9905098393465249
5	.25E-05	6	6	.28E-05	.9909246054695797
6	.62E-05	7	8	.58E-05	.9914143785352492
7	.57E-05	9	10	.14E-05	.9920528292478605
8	.53E-05	12	12	.20E-05	.9927454269019744
9	.38E-05	17	14	.87E-05	.9934552173991646
10	.92E-05	25	18	.71E-05	.9942917996965845
11	.76E-05	30	20	.73E-05	.9950296766226088
12	.87E-05	43	23	.91E-05	.9957786450015449
13	.91E-05	58	29	.98E-05	.9963645109429915
14	.66E-05	127	61	.89E-05	.99770385857701
15	.99E-05	268	136	.10E-04	.9982106973541132
16	.19E-05	282	147	.18E-05	.9986350889093472
17	.17E-05	500	500	.67E-06	.9993053726008107
18	.30E-05	500	500	.12E-05	.9998011882172851
19	.85E-06	500	500	.28E-05	.9999726712124946
20	.62E-07	500	500	.92E-06	.9999978542185801
21	.20E-08	500	500	.20E-08	.9999998914042579
22	.81E-11	500	500	.81E-11	.9999999841440135
23	.12E-11	177	79	.11E-11	.9999999997658635
24	.23E-12	500	319	.87E-13	.9999999999165852
25	.60E-14	187	86	.60E-14	.9999999999985193
26	.59E-15	500	276	.41E-15	.999999999999502
27	.30E-16	239	110	.29E-16	.9999999999999911
28	.24E-17	410	210	.25E-17	.9999999999999983

Table A.2: Last column is maximum eigenvalue of N

200 × 200 ASSIGNMENT PROBLEM					
Iter no	Vertex				Max Eigenvalue
	$l = 2$		$l \leq 14$		
	Error	K	K	Error	
1	.38E-05	7	7	.38E-05	.9902704011386622
2	.15E-07	8	8	.15E-07	.9901099097716491
3	.84E-06	8	8	.84E-06	.9899756389655071
4	.10E-05	10	10	.10E-05	.9898290242692684
5	.10E-05	11	11	.10E-05	.9896816509502613
6	.44E-06	16	16	.44E-06	.9895654883729561
7	.41E-05	19	19	.41E-05	.9895965563950285
8	.85E-05	26	21	.35E-05	.9900471983168925
9	.71E-05	41	33	.50E-05	.9910334510805776
10	.95E-05	65	47	.12E-05	.9924207814025372
11	.92E-05	97	70	.72E-05	.9937359111635224
12	.59E-05	117	79	.50E-05	.9949046663890485
13	.82E-05	183	110	.75E-05	.9960325015350879
14	.91E-05	475	367	.86E-05	.9986825570584397
15	.20E-02	500	500	.32E-03	.9994308944679383
16	.66E+00	500	500	.52E-02	.9998158719261547
17	.13E+00	500	500	.25E-01	.9999507392517608
18	.37E-01	500	500	.15E-01	.9999914233005139
19	.40E-02	500	500	.29E-02	.9999975689126962
20	.90E-04	500	500	.72E-04	.999999824948421
21	.31E-05	500	500	.20E-05	.9999999816879248
22	.29E-07	3	3	.29E-07	.9999999995858695
23	.42E-09	2	2	.42E-09	.9999999999067528
24	.16E-11	2	2	.16E-11	.9999999999997641
25	.68E-14	2	2	.68E-14	.9999999999998697
26	.31E-16	2	2	.31E-16	.9999999999999836
27	.20E-18	2	2	.20E-18	.9999999999999946
28	.62E-21	2	2	.62E-21	.9999999999999996

Table A.3: Last column is maximum eigenvalue of N

300 × 300 ASSIGNMENT PROBLEM				
Iter no	Interior Facet			Max Eigenvalue
	l = 2		l ≤ 14	
	Error	K	K	Error	
1	.16E-06	4	4	.16E-06	.9934978175014585
2	.44E-06	4	4	.44E-06	.9934186912268195
3	.20E-05	4	4	.20E-05	.9933559054478737
4	.11E-06	5	5	.11E-06	.9932791318740413
5	.24E-05	5	5	.24E-05	.9931941062181071
6	.11E-05	6	6	.12E-05	.9931267666279024
7	.86E-06	7	7	.69E-05	.9930952597987705
8	.42E-05	8	9	.98E-06	.9931056428988306
9	.49E-05	10	11	.58E-06	.9938442214108793
10	.52E-05	12	11	.93E-05	.9933668666248594
11	.44E-05	14	13	.24E-05	.9936250698922473
12	.35E-05	16	14	.33E-05	.9939487465725621
13	.20E-05	23	17	.16E-05	.9943471698211603
14	.42E-06	44	24	.31E-06	.9951934721853056
15	.81E-08	24	17	.90E-08	.9945002638745208
16	.29E-08	100	36	.26E-08	.9958084656045198
17	.47E-10	16	14	.33E-10	.9942191630970714
18	.14E-10	150	50	.16E-10	.9966531473552512
19	.39E-12	21	17	.12E-12	.9948769102561883
20	.97E-13	138	57	.91E-13	.9967373895081607
21	.20E-14	20	16	.15E-14	.9935361068659556
22	.40E-15	116	57	.35E-15	.9969089383222424
23	.24E-16	41	23	.23E-16	.9957296613353846
24	.21E-17	125	53	.22E-17	.9966921636270423
25	.25E-18	67	28	.24E-18	.9968005968116425

Table A.4: Last column is maximum eigenvalue of N

300 × 300 ASSIGNMENT PROBLEM					
Iter no	Almost Vertex				Max Eigenvalue
	l = 2		l ≤ 14		
	Error	K	K	Error	
1	.44E-07	5	5	.44E-07	.9932347666942302
2	.49E-07	5	5	.49E-07	.9933083337386703
3	.43E-06	5	5	.43E-06	.9933864437617831
4	.61E-05	5	5	.61E-05	.9935104212676482
5	.10E-05	6	6	.11E-05	.9936881369782937
6	.67E-06	7	7	.60E-05	.9939091182799246
7	.67E-05	8	9	.97E-06	.9941600282091113
8	.51E-05	11	11	.21E-05	.9944891617919949
9	.91E-05	15	14	.30E-05	.9949007307038975
10	.58E-05	21	17	.20E-05	.9953666836700457
11	.69E-05	26	19	.43E-05	.9957957479586064
12	.99E-05	33	22	.54E-05	.9961745224941985
13	.81E-05	51	27	.67E-05	.9966200644015995
14	.82E-05	125	63	.90E-05	.9978513542039164
15	.77E-05	213	119	.96E-05	.9986067117779875
16	.39E-05	488	288	.38E-05	.9990509284216897
17	.37E-05	500	500	.25E-05	.9993768099938902
18	.52E-05	500	500	.43E-05	.999656820722268
19	.18E-05	500	500	.15E-05	.99980850749491
20	.21E-06	500	500	.57E-05	.999939258173573
21	.16E-07	500	500	.27E-06	.9999894909179856
22	.39E-10	124	31	.38E-10	.9999985974184675
23	.30E-11	77	28	.38E-11	.9999999503334166
24	.20E-12	41	21	.13E-12	.9999999881180519
25	.22E-13	69	27	.21E-13	.9999999999413154
26	.11E-14	77	29	.71E-15	.9999999999677954
27	.93E-16	49	23	.27E-16	.9999999999964501
28	.51E-17	250	40	.11E-17	.9999999999998771

Table A.5: Last column is maximum eigenvalue of N

300 × 300 ASSIGNMENT PROBLEM					
Iter no	Vertex				Max Eigenvalue
	l = 2		l ≤ 14		
	Error	K	K	Error	
1	.43E-05	6	6	.43E-05	.9934762989524124
2	.26E-06	7	7	.26E-06	.9933895689884312
3	.89E-07	9	9	.89E-07	.9933091899889467
4	.66E-08	11	10	.44E-06	.9932136733431952
5	.23E-05	10	10	.23E-05	.9931259144676824
6	.39E-05	12	12	.39E-05	.9930849464235598
7	.47E-05	16	16	.47E-05	.9931239002123157
8	.38E-05	24	24	.38E-05	.9932731057716612
9	.35E-05	37	37	.35E-05	.9935938447387953
10	.81E-05	58	42	.20E-05	.9941748293727437
11	.98E-05	84	84	.98E-05	.9949332859291869
12	.69E-05	139	67	.38E-05	.9957201695856718
13	.83E-05	185	107	.87E-05	.9964116190494283
14	.99E-05	354	226	.91E-05	.9982736953381484
15	.51E-03	500	500	.38E-04	.9987384902708429
16	.12E+00	500	500	.46E-02	.9993068865577555
17	.14E+00	500	500	.73E-01	.9996908839712081
18	.89E-01	500	500	.56E-01	.9998352646075588
19	.32E-01	500	500	.22E-01	.9998816294842967
20	.79E-02	500	500	.52E-02	.9999699825813522
21	.27E-02	500	500	.15E-02	.9999984634373612
22	.20E-03	500	500	.25E-03	.999999833594602
23	.11E-04	500	500	.67E-05	.9999999907302287
24	.88E-07	500	500	.34E-07	.9999999991419738
25	.27E-08	32	17	.24E-08	.9999999999731211
26	.19E-10	53	17	.14E-10	.9999999999951627
27	.55E-11	39	17	.23E-12	.9999999999999893
28	.83E-12	61	19	.99E-13	.9999999999999925

Table A.6: Last column is maximum eigenvalue of N

Bibliography

[1] I. Adler, M. G. C. Resende, G. Veiga, and N. K. Karmarkar. An implementation of Karmarkar's algorithm for linear programming. *Mathematical Programming*, 44:297–335, 1989.

[2] I. Adler, N. K. Karmarkar, M. G. C. Resende, and G. Veiga. Data structures and programming techniques for the implementation of Karmarkar's algorithm. *ORSA Journal on Computing*, 1:84–106, 1990.

[3] I. Adler and R. D. C. Monteiro. Limiting behavior of the affine scaling continuous trajectories for linear programming problems. *Mathematical Programming*, 50:29–51, 1991.

[4] A. V. Aho, J. K. Hopcroft, and J. D. Ullman. *The Design and Analysis of Computer Algorithms*. Addison-Wesley, Reading, MA, 1976.

[5] F. Alizadeh. *Combinatorial Optimization with Interior Point Methods and Semi-Definite Matrices*. PhD thesis, University of Minnesota, October 1991.

[6] E. L. Allgower and K. Georg. *Numerical Continuation Methods*. Springer - Verlag, New York, 1990.

[7] K. M. Anstreicher. A monotonic projective algorithm for fractional linear programming. *Algorithmica*, 1:483–498, 1986.

[8] K. M. Anstreicher. A combined phase 1-phase 2 projective algorithm for linear programming. *Mathematical Programming*, 43:209–223, 1989.

[9] K. M. Anstreicher. A worst-case step in Karmarkar's algorithm. *Mathematics of Operations Research*, 14:294–302, 1989.

[10] K. M. Anstreicher. A standard form variant, and safeguarded linesearch for the modified Karmarkar algorithm. *Mathematical Programming*, 47:337–351, 1990.

[11] K. M. Anstreicher and R. A. Bosch. Long steps in an $O(n^3 L)$ algorithm for linear programming. *Mathematical Programming*, 54:251–265, 1992.

[12] O. Bahn, J.-L Goffin, J.-Ph. Vial, and O. Du Merle. Implementation and behavior of an interior point cutting plane algorithm for convex programming: an application to geometric programming. Working Paper, Universite de Geneve, Geneve, Switzerland, to appear in *Discrete Applied Mathematics*, 1991.

[13] E. R. Barnes. A variation on Karmarkar's algorithm for solving linear programming problems. *Mathematical Programming*, 36:174–182, 1986.

[14] R. H. Bartels and G. H. Golub. The simplex method of linear programming using LU-decomposition. *Communications of the ACM*, 12:266–268 and 275–278, 1969.

[15] D. A. Bayer and J. C. Lagarias. The nonlinear geometry of linear programming I. affine and projective scaling trajectories II. Legendre transform coordinates and central trajectories III. projective Legendre transform coordinates and Hilbert geometry. *Transactions of the American Mathematical Society 314*, 2:499–581, 1989.

[16] D. A. Bayer and J. C. Lagarias. Karmarkar's algorithm and Newton's method. *Mathematical Programming*, 50:291–330, 1991.

[17] M. S. Bazaraa, J. J. Jarvis, and H. D. Sherali. *Linear Programming and Network Flows*. John Wiley & Sons, New York, 1990.

[18] E. M. L. Beale. Cycling in the dual simplex algorithm. *Naval Research Logistics Quarterly*, 2:269–276, 1955.

[19] M. J. Best and K. Ritter. *Linear Programming Active Set Analysis and Computer Programs*. Prentice-Hall, Inc.., Englewood Cliffs, New Jersey, 1985.

[20] R. G. Bland. New finite pivoting rules for the simplex method. *Mathematics of Operations Research*, 2:103–107, 1977.

[21] P. T. Boggs, P. D. Domich, J. R. Donaldson, and C. Witzgall. Algorithmic enhancements to the method of centers for linear programming problems. *ORSA Journal on Computing*, 1:159–171, 1990.

[22] G. Boole. *Calculus of Finite Differences*. Chelsea Publishing Company, New York, 1970.

[23] K. H. Borgwardt. *The Simplex Method: A Probabilistic Analysis*. Springer-Verlag, New York, 1987.

[24] R. P. Brent. Some efficient algorithms for solving systems of non-linear equations. *SIAM Journal on Numerical Analysis*, 10:327–344, 1973.

[25] C. Caratheodory. Uber den variabilitatsbereich der koeffizienten von potenzreihen, die gegebene werte nicht annehmen. *Mathematische Annalen*, 64:95–115, 1907.

[26] V. Chandru and B. S. Kochar. A class of algorithms for linear programming. Research Memorandum 85–14, School of Industrial Engineering, Purdue University, West Lafayette, Indiana, 1986.

[27] A. Charnes. Optimality and degeneracy in linear programming. *Econometrica*, 20:160–170, 1952.

[28] A. Charnes and W. W. Cooper. *Management Models and Industrial Applications of Linear Programming*. John Wiley & Sons, New York, 1961.

[29] Y.-C. Cheng, D. J. Houck, J. -M Liu, M. S. Meketon, L. Slutsman, R. J. Vanderbei, and P. Wang. The AT&T KORBX system. *AT&T Technical Journal*, 68:7–19, 1989.

[30] I. C. Choi, C. L. Monma, and D. F. Shanno. Further development of primal-dual interior point methods. Report 60-88, Rutgers University, New Brunswick, New Jersey, 1990.

[31] I. C. Choi, C. L. Monma, and D. F. Shanno. Further development of a primal-dual interior point method. *ORSA Journal on Computing 2*, 4:304–311, 1990.

[32] R. W. Cottle. Manifestations of Shur complements. *Linear Algebra and Its Applications*, 8:189–211, 1974.

[33] G. B. Dantzig. Programming in a linear structure. *Econometrica*, 17:73–74, 1949.

[34] G. B. Dantzig. Maximization of a linear function of variables subject to linear inequalities. In T. C. Koopmans, editor, *Activity Analysis of Production and Allocation*, pages 339–347, John Wiley & Sons, New York, 1951.

[35] G. B. Dantzig. Programming of interdependent activities, ii, mathematical model. In T. C. Koopmans, editor, *Activity Analysis of Production and Allocation*, John Wiley & Sons, New York, 1951.

[36] G. B. Dantzig. *Linear Programming and Extensions*. Princeton University Press, Princeton, New Jersey, 1963.

[37] G. B. Dantzig and A. Orden. Duality Theorems. Technical Report No. 1265, The RAND Corporation, Santa Monica, Calif., 1953.

[38] G. B. Dantzig, L. R. Ford, and D. R. Fulkerson. A primal-dual algorithm for linear programs. In H. W. Kuhn and A. W. Tucker, editors, *Linear Inequalities and Related Systems*, Annals of Mathematics Study No. 30, pages 171–181, Princeton University Press, Princeton, New Jersey, 1956.

[39] G. B. Dantzig and P. Wolfe. Decomposition principle for linear programs. *Operations Research*, 8:101–111, 1960.

[40] G. B. Dantzig and Y. Ye. A build up interior method for linear programming: Affine scaling form. Technical Report SOL 90-4, Stanford University, Stanford, California, 1990.

[41] D. Davidenko. On a new method of numerical solution of systems of nonlinear equations (in Russian). *Dokl. Akad. Nauk USSR*, 88:601–602, 1953.

[42] D. Davidenko. On approximate solution of systems of nonlinear equations. *Ukraine Math. Ž.*, 5:196–206, 1953.

[43] G. De Ghellinck and J.-Ph. Vial. A polynomial Newton method for linear programming. *Algorithmica*, 1:425–453, 1986.

[44] G. De Ghellinck and J.-Ph. Vial. An extension of Karmarkar's algorithm for problems in standard form. *Mathematical Programming*, 37:81–90, 1987.

[45] D. Den Hertog. *Interior Point Approach to Linear, Quadratic and Convex Programming-Algorithms and Complexity*. PhD thesis, Technishe Universiteit, Delft, The Netherlands, September 1992.

[46] D. Den Hertog. *Interior Point Approach to Linear, Quadratic and Convex Programming*. Kluwer Academic Publishers, Boston, 1994.

[47] D. Den Hertog and C. Roos. A survey of search directions in interior point methods for linear programming. *Mathematical Programming*, 52:481–509, 1991.

[48] D. Den Hertog, C. Roos, and T. Terlaky. A polynomial method of weighted centers for convex quadratic programming. *Journal of Information & Optimization Sciences 12*, 2:187–205, 1991.

[49] D. Den Hertog, C. Roos, and T. Terlaky. A build-up variant of the path-following method for LP. *Operations Research Letters*, 12:181–186, 1992.

[50] D. Den Hertog, C. Roos, and T. Terlaky. Inverse barrier methods for linear programming. *Recherce Opèrationelle*, 28:135 – 163, 1994.

[51] D. Den Hertog, C. Roos, and J.-Ph. Vial. A complexity reduction for the long-step path-following algorithm for linear programming. *SIAM Journal on Optimization*, 2:71–87, 1992.

[52] J. E. Dennis and R. B. Schnabel. *Numerical Methods for Unconstrained Minimization*. Prentice-Hall, Inc., Englewood Cliffs, New Jersey, 1983.

[53] I. I. Dikin. Iterative solution of problems of linear and quadratic programming. *Doklady Akademiia Nauk SSSR 174*, 747–748. Translated into English in *Soviet Mathematics Doklady* 8:674–675, 1967.

[54] I. I. Dikin. On the convergence of an iterative process. *Upravlyaemye Sistemi*, 12:54–60, 1974.

[55] I. I. Dikin. The convergence of dual variables. Technical Report, Siberian Energy Institute, Irkutsk, Russia, 1990.

[56] I. I. Dikin. Determination of interior point of one system of linear inequalities. *Kibernetica and system analysis*, 1:74–96, 1992.

[57] I. I. Dikin and V. I. Zorkaltsev. *Iterative solution of mathematical programming problems: Algorithms for the method of interior points*. Nauka, Novosibirsk, USSR , 1980.

[58] P. D. Domich, P. T. Boggs, J. E. Rogers, and C. Witzgall. Optimizing over three-dimensional subspaces in an interior-point method for linear programming. *Linear Algebra and its Applications*, 152:315–342, 1991.

[59] B. C. Eaves. Homotopies for the computation of fixed points. *Mathematical Programming*, 3:1–22, 1972.

[60] B. C. Eaves and R. Saigal. Homotopies for computation of fixed points on unbounded regions. *Mathematical Programming*, 3:225–237, 1972.

[61] H. G. Eggleston. *Convexity*. Cambridge University press, Cambridge, 1958.

[62] S.-C. Fang. An Unconstrained Convex Programming View of Linear Programming. *Zeischrift fur Operations Research-Theory*, 36:149–161, 1992.

[63] S.-C. Fang and S. Puthenpura. *Linear Optimization and Extensions*. Prentice Hall, Englewood Cliffs, New Jersey, 1993.

[64] S.-C. Fang and H.-S. J. Tsao. Linear Programming with Entropic Perturbation. *Zeischrift fur Operations Research*, 37:171–186, 1993.

[65] J. Farkas. Uber die Theorie der einfachen Ungleichungen. *Journal fur die reine und angewandte Mathematic*, 124:1–27, 1901.

[66] A. V. Fiacco and G. P. McCormick. *Nonlinear Programming, Sequential Unconstrained Minimization Techniques*. John Wiley & Sons, New York, 1968.

[67] R. M. Freund. Theoretical efficiency of a shifted barrier function algorithm for linear programming. *Linear Algebra and its Applications*, 152:19–41, 1991.

[68] R. M. Freund. A potential reduction algorithm with user-specified Phase I-Phase II Balance, for solving a linear program from an infeasible warm start. *Mathematical Programming*, 52:441–466, 1991.

[69] R. M. Freund. Polynomial-time algorithms for linear programming based only on primal scaling and projected gradients of a potential function. *Mathematical Programming*, 51:203–222, 1991.

[70] R. M. Freund. An infeasible-Start algorithm for linear programming whose complexity depends on the distance from the starting point to the optimal solution. Working Paper, Sloan School of Management, Massachusetts Institute of Technology, USA, 1993.

[71] R. M. Freund. Projective Transformation for Interior-Point Algorithms, and a Superlinearly Convergent Algorithm for the W-Center Problem. *Mathematical Programming*, 58:385–414, 1993.

[72] K. R. Frisch. The logarithmic potential method for solving linear programming problems. Memorandum, Institute of Economics, Oslo, Norway, 1955.

[73] K. R. Frisch. La resolution des problemes de programme lineaire par la methode du potential logarithmique. *Cahiers du Seminaire D'Econometrie*, 4:7–20, 1956.

[74] D. Gale. The basic theorems of real linear equations, inequalities, linear programming and game theory. *Naval Research Logistics Quarterly*, 3:193–200, 1956.

[75] D. Gale. *The Theory of Linear Economic Models*. McGraw-Hill Book Company, New York, 1960.

[76] D. Gale, H. W. Kuhn, and A. W. Tucker. Linear Programming and the Theory of Games, In T. C. Koopmans, editor, *Activity Analysis of Production and Allocation*, John Wiley & Sons, New York, 1951.

[77] D. M. Gay. A variant of Karmarkar's linear programming algorithm for problems in standard form. *Mathematical Programming*, 37:81–90, 1987.

[78] A. George and J. W–H. Liu. *Computer Solution of Large Sparse Positive Definite Systems*. Prentice–Hall, Inc., New Jersey, 1981.

[79] P. E. Gill, W. Murray, and M. H. Wright. *Practical optimization*. Academic Press, Inc., San Diego, California, 1988.

[80] P. E. Gill, W. Murray, M. A. Saunders, J. A. Tomlin, and M. H. Wright. On projected Newton barrier methods for linear programming and an equivalence to Karmarkar's projective method. *Mathematical Programming*, 36:183–209, 1986.

[81] J.-L. Goffin and J.-Ph. Vial. Cutting planes and column generation techniques with the projective algorithm. *Journal of Optimization Theory and Applications*, 65:409–429, 1990.

[82] J.-L. Goffin and J.-Ph. Vial. Short steps with Karmarkar's projective algorithm for linear programming. Gerad, McGill University, Montreal, Canada, 1990.

[83] D. Goldfarb and S. Liu. An $O(n^3L)$ primal interior point algorithm for convex quadratic programming. *Mathematical Programming*, 49:325–340, 1991.

[84] D. Goldfarb and M. J. Todd. Linear programming. In A.H.G. Rinnooy Kan, G.L. Nemhauser, and M. J. Todd, editors, *Handbooks in Operations Research and Management Science, Optimization, 1*, page 141–170. North-Holland, Amsterdam, 1989.

[85] G. H. Golub and C. F. Van Loan. *Matrix Computations*. The Johns Hopkins Press Ltd., London, 1989.

[86] C. C. Gonzaga. An algorithm for solving linear programming problems in $O(nL)$ operations. In N. Megiddo, editor, *Progress in Mathematical Programming: Interior-Point and Related Methods*, pages 1–28. Springer-Verlag, New York, 1987.

[87] C. C. Gonzaga. Polynomial affine algorithms for linear programming. Report ES-139/88, COPPE Federal University of Rio de Janeiro, Rio de Janeiro, Brasil, 1988.

[88] C. C. Gonzaga. An algorithm for solving linear programming problems in $O(n^3L)$ operations. In N. Megiddo, editor, *Progress in Mathematical Programming, Interior Point and Related Methods, 1- 28*. Springer Verlag, New York, 1989.

[89] C. C. Gonzaga. Conical projection algorithms for linear programming. *Mathematical Programming*, 43:151–173, 1989.

[90] C. C. Gonzaga. Polynomial affine algorithms for linear programming. *Mathematical Programming*, 49:7–21, 1990.

[91] C. C. Gonzaga. Interior point algorithms for linear programming problems with inequality constraints. *Mathematical Programming*, 52:209–225, 1991.

[92] C. C. Gonzaga. Large-steps path-following methods for linear programming, Part I: Barrier function method. *SIAM Jorrnal on Optimization*, 1:268–279, 1991.

[93] C. C. Gonzaga. Large-steps path-following methods for linear programming, Part II: Potential reduction method. *SIAM Journal on Optimization*, 1:280–292, 1991.

[94] C. C. Gonzaga. Search directions for interior linear programming methods. *Algorithmica*, 6:153–181, 1991.

[95] O. Güler, D. Den Hertog, C. Roos, T. Terlaky, and T. Tsuchiya. Degeneracy in interior point methods for linear programming. Technical Report, Faculty of Mathematics and Computer Science, Delft University of Technology, Delft, Netherlands, 1991.

[96] W. W. Hager. Updating the inverse of a matrix. *SIAM Review*, 31:221–239, 1989.

[97] L. A. Hall, and R. J. Vanderbei. Two-thirds is sharp for affine scaling. *O. R. Letters*, 13:197–201, 1993.

[98] M. Hamala. Quasibarrier methods for convex programming. In *Survey of Mathematical Programming 1, 465–477*, Budapest, 1976.

[99] C.-G. Han, P.M. Pardalos, and Y. Ye. On interior-point algorithms for some entropy optimization problems. Working Paper, The Pennsylvania State University, University Park, Pennsylvania, 1991.

[100] F. L. Hitchcock. Distribution of a product from several sources to numerous localities. *Journal of Mathematical Physics*, 20:224–230, 1941.

[101] A. J. Hoffman. Cycling in the simplex algorithm. Technical Report No. 2974, National Bureau of Standards, Washington, D.C., 1953.

[102] H. Imai. On the convexity of the multiplicative version of Karmarkar's potential function. *Mathematical Programming*, 40:29–32, 1988.

[103] M. Iri. A proof of the polynomiality of the Iri-Imai method. Report metro 91-08, University of Tokyo, Tokyo, Japan, 1991.

[104] M. Iri and H. Imai. A mutiplicative barrier function method for linear programming. *Algorithmica*, 1:455–482, 1986.

[105] G.-W. Jan and S.-C. Fang. A Variant of Affine Scaling Algorithm for Linear Programming, *Optimization*, 22:681–715, 1991.

[106] F. Jarre. On the method of analytic centers for solving smooth convex programs. In S. Dolecki, editor, *Lecture Notes in Mathematics 1405, Optimization*, pages 69–85. Springer-Verlag, New York, 1989.

[107] F. Jarre. On the convergence of the method of analytic centers when applied to convex quadratic programs. *Mathematical Programming*, 49:341–358, 1991.

[108] F. Jarre. Interior-point methods for convex programming. *Applied Mathematics & Optimization*, 26:287–311, 1992.

[109] J.A. Kaliski and Y. Ye. A decomposition variant of the potential reduction algorithm for linear programming. Working Paper 91-11, The University of Iowa, Iowa City, Iowa, 1991.

[110] L. V. Kantorovich. *Mathematical Methods in the Organization and Planning of Production*. Publication House of the Leningrad State University, 1939.

[111] L. V. Kantorovich. Functional analysis and applied mathematics. *Uspehi. Mat. Nauk.*, 3:39–185, 1948.

[112] S. Kapoor and P. M. Vaidya. Fast algorithms for convex quadratic programming and multicommodity flows. *Proc. of the 18th Annual ACM Symposium on Theory of Computing*, 1986.

[113] N. K. Karmarkar. A new polynomial-time algorithm for linear programming. *Combinatorica*, 4:373–395, 1984.

[114] N. K. Karmarkar and K. Ramakrishnan. Computational results of an interior point algorithm for large scale linear programming. *Mathematical Programming* , 52:555–586, 1991.

[115] L. G. Khachian. A polynomial algorithm in linear programming. *Doklady Akademiia Nauk SSSR 244, 1093-1096,* Translated into English in Soviet Mathematics Doklady 20:191–194, 1979.

[116] L. G Khachian, S.P. Tarasov, and A.I. Erlich. The inscribed ellipsoid method. *Dokl. Akad. Nauk SSSR*, 298:(Engl. trans. Soviet Math. Dokl.), 1988.

[117] V. Klee and G. J. Minty. How good is the simplex algorithm? In O. Shisha, editor, *Inequalities III*, pages 159–175, Academic Press, New York, 1972.

[118] M. Kojima, N. Megiddo, and A. Yoshise. A primal-dual interior point algorithm for linear programming. In N. Megiddo, editor, *Progress in Mathematical Programming: Interior-Point and Related Methods*, pages 29–48. Springer-Verlag, New York, 1989.

[119] M. Kojima, S. Mizuno, and A. Yoshise. A polynomial time algorithm for a class of linear complementarity problems. *Mathematical Programming*, 44:1–26, 1989.

[120] M. Kojima, S. Mizuno, and A. Yoshise. An $O(\sqrt{n}L)$ iteration potential reduction algorithm for linear complementarity problems. *Mathematical Programming*, 50:331–342, 1991.

[121] M. Kojima, N. Megiddo, and S. Mizuno. A primal-dual, infeasible-interior-point algorithm for linear programming. *Mathematical Programming*, 61:263–280, 1993.

[122] M. Kojima, N. Megiddo, T. Noma, and A. Yoshise. *A Unified Approach to Interior Point Algorithms for Linear Complementarity Problems*. Lecture Notes in Computer Science 538, Springer Verlag, New York, 1991.

[123] T. C. Koopmans. Optimum utilization of the transportation system. *Econometrica*, 17:3–4, 1949.

[124] T. C. Koopmans. *Activity Analysis of Production and Allocation*. John Wiley & Sons, New York, 1951.

[125] K. O. Kortanek and H. No. A second order affine scaling algorithm for the geometric programming dual with logarithmic barrier. *Optimization*, 23:501–507, 1992.

[126] K. O. Kortanek and J. Zhu. New purification algorithms for linear programming. *Naval Research Logistics Quarterly*, 35:571–583, 1988.

[127] H. W. Kuhn and A. W. Tucker. Nonlinear programming. In J. Neyman, editor, *Proceedings of the Second Berkeley Symposium on Mathematical Statistics and Probability*, pages 481–492, University of California Press, Berkeley, Calif, 1950.

[128] H. W. Kuhn and A. W. Tucker. *Linear Inequalities and Related Systems*. Annals of Mathematics Study No. 38, Princeton University Press, Princeton, New Jersey, 1956.

[129] C. E. Lemke. The dual method of solving the linear programming problem. *Naval Research Logistics Quarterly*, 1:36–47, 1954.

[130] W. W. Leontief. *The Structure of the American Economy, 1919–1939*. Oxford University Press, New York, 1951.

[131] J. Leray and J. Schauder. Topologie et èquations functionelles. *Ann. Sci. École Norm. Sup.*, 51:45–78, 1934.

[132] F. A. Lootsma. *Boundary Properties of Penalty Functions for Constrained Minimization*. PhD thesis, Eindhoven, The Netherlands, 1970.

[133] I. J. Lustig. Feasibility issues in an interior point method for linear programming. Technical Report SOR 89-9, Princeton University, Princeton, New Jersey, 1989.

[134] I. J. Lustig. Feasibility issues in a primal-dual interior-point method for linear programming. *Mathematical Programming*, 49:145–162, 1990.

[135] I. J. Lustig, R. E. Marsten, and D. F. Shanno. Computational experience with a primal-dual interior point method for linear programming. *Linear Algebra and Its Applications*, 152:191–222, 1991.

[136] I. J. Lustig, R. E. Marsten, and D. F. Shanno. On implementing Mehrotra's predictor-corrector interior point method for linear programming. *SIAM Journal on Optimization*, 2:435–449, 1992.

[137] I. J. Lustig, J. M. Mulvey, and T. J. Carpenter. The formulation of stochastic programs for interior methods. Technical Report SOR-89-16, Princeton University, Princeton, New Jersey, 1989.

[138] O. L. Mangasarian. *Non-Linear Programming*. McGraw-Hill, New York, 1969.

[139] K. T. Marshall, and J. W. Suurballe. A note on cycling in the simplex method. *Naval Research Logistics Quarterly*, 16:121–137, 1969.

[140] R. E. Marsten, M. J. Saltzman, D. F. Shanno, G. S. Pierce, and J. F. Ballintijn. Implementation of a dual affine interior point algorithm for linear programming. *ORSA Journal on Computing*, 1:287–297, 1990.

[141] W. F. Mascarenhas. The affine scaling algorithm fails for $\lambda = 0.999$. Technical Report, Universidade Estadual de Campinas, Campinas S. P., Brazil, 1993.

[142] K. A. McShane, C. L. Monma, and D. F. Shanno. An implementation of a primal-dual interior point method for linear programming. *ORSA Journal on Computing*, 1:70–83, 1989.

[143] N. Megiddo. Pathways to the optimal set in linear programming. In N. Megiddo, editor, *Progress in Mathematical Programming, Interior Point and Related Methods*, pages 131–158. Springer Verlag, New York, 1989.

[144] N. Megiddo and M. Shub. Boundary behavior of interior point algorithms in linear programming. *Mathematics of Operations Research 14*, 1:97–146, 1989.

[145] S. Mehrotra. On finding a vertex solution using interior point methods. Technical Report 89-17, Northwestern University, Evanston, IL, 1989.

[146] S. Mehrotra. On the implementation of a (primal dual) interior point method. *SIAM Journal on Optimization*, 2:575–601, 1992.

[147] H. Minkowski. *Geometry der Zahlen*. Teubner, Leipig, 1910.

[148] M. Minoux. *Mathematical Programming, Theory and Algorithms*. John Wiley & Sons, New York, 1986.

[149] J. E. Mitchell. *Karmarkar's Algorithm and Combinatorial Optimization Problems*. PhD thesis, Cornell University, Ithaca, NY, 1988.

[150] J. E. Mitchell and M. J. Todd. On the relationship between the search directions in the affine and projective variants of Karmarkar's linear programming algorithm. In *Contributions to Operations Research and Economics, 237-250*, Cambridge, MA, 1989.

[151] S. Mizuno. Polynomiality of infeasible-interior-point algorithm for linear programming. *Mathematical Programming*, 67:109–120, 1994.

[152] S. Mizuno and A. Nagasawa. A primal-dual affine scaling potential reduction algorithm for linear programming. *Mathematical Programming*, 62:119–131, 1993.

[153] S. Mizuno, R. Saigal, and J. B. Orlin. Determination of optimal vertices from feasible solutions in unimodular linear programming. *Mathematical Programming*, 59:23 – 31, 1993.

[154] S. Mizuno and M. J. Todd. An $O(n^3L)$ adaptive path following algorithm for a linear complementarity problem. *Mathematical Programming*, 52:587–592, 1991.

[155] S. Mizuno, M. J. Todd, and Y. Ye. Anticipated behavior of path-following algorithms for linear programming. Technical Report No. 878, Cornell University, Ithaca, NY, 1989.

[156] S. Mizuno, M. J. Todd, and Y. Ye. Anticipated behavior of long-step algorithms for linear programming. Research Report 24, Tokyo Institute of Technology, Tokyo, Japan, 1990.

[157] S. Mizuno, M. J. Todd, and Y. Ye. On adaptive-step primal-dual interior-point algorithms for linear programming. *Mathematics of Operations Research*, 18:964–981, 1993.

[158] C. L. Monma and A. J. Morton. Computational experience with a dual affine variant of Karmarkar's method for linear programming. *Operations Research Letters*, 6:261–267, 1987.

[159] R. D. C. Monteiro and I. Adler. An $O(n^3 L)$ primal-dual interior point algorithm for linear programming. Report ORC 87-4, University of California, Berkeley, CA, 1987.

[160] R. D. C. Monteiro and I. Adler. Interior path following primal dual algorithms, Part I: Linear programming. *Mathematical Programming*, 44:27–41, 1989.

[161] R. D. C. Monteiro and I. Adler. Interior path-following primal-dual algorithms. Part II: Convex quadratic programming. *Mathematical Programming*, 44:43–66, 1989.

[162] R. D. C. Monteiro, I. Adler, and M. G. C. Resende. A polynomial time primal-dual affine scaling algorithm for linear and convex quadratic programming and its power series extension. *Mathematics of Operations Research*, 15:191–214, 1990.

[163] R. D. C. Monteiro, T. Tsuchiya, and Y. Wang. A simplified global convergence proof of the affine scaling algorithm. *Annals of Operations Research*, 47:443 – 482, 1993.

[164] T. L. Morin and T. B. Trafalis. A polynomial time algorithm for finding an efficient face of a polyhedron. Technical Report, Purdue University, West Lafayette, Indiana, 1989.

[165] T. S. Motzkin. *Beitrage zur Theorie der Linearen Ungleichungen*. PhD thesis, University of Zurich, Switzerland, 1936.

[166] M. Muramatsu. *Convergence analysis of affine scaling method for linear programming*. PhD thesis, Graduate School of Advanced Studies, Tokyo, Japan, 1994.

[167] M. Muramatsu and T. Tsuchiya. An affine scaling method with an infeasible starting point. Research Memorandum No. 490, The Institute of Statistical Mathematics, Tokyo, Japan, 1993.

[168] M.Muramatsu and T. Tsuchiya. A convergence analysis of a long-step variant of the projective scaling algorithm. Research Memorandum No. 454, The Institute of Statistical Mathematics, Tokyo, Japan, 1992.

[169] W. Murray and M. H. Wright. Line search procedures for the logarithmic barrier function. Numerical Analysis Manuscript 92-01, AT&T Bell Laboratories, Murray Hill, New Jersey, 1992.

[170] K. G. Murty. *Linear Programming*. John Wiley & Sons, New York, 1986.

[171] J. L. Nazareth. Homotopy techniques in linear programming. *Algorithmica*, pages 529–535, 1986.

[172] J. L. Nazareth. Pricing criteria in linear programming. In N. Megiddo, editor, *Progress in Mathematical Programming: Interior-Point and Related Methods*, pages 105–129. Springer-Verlag, New York, 1989.

[173] A. S. Nemirovsky. On an algorithm of Karmarkar's type. *Izvestija AN SSSR*, Tekhnitcheskaya Kibernetika, 1:105–118 (In Russian.), 1987.

[174] Yu. E. Nesterov. Polynomial time iterative methods in linear and quadratic programming. *Voprosy Kibernetiki*, Moscow, 3:102–125 (In Russian.), 1988.

[175] Yu. E. Nesterov. Dual polynomial time algorithms for linear programming. *Kibernetika*, 1:34–54 (In Russian.), 1989.

[176] Yu. E. Nesterov and A. Nemirovskii. *Interior-Point Polynomial Algorithms in Convex Programming*. Society for Industrial and Applied Mathematics, Philadelphia, 1994.

[177] J. M. Ortega and W. C. Rheinboldt. *Iterative Solution of Nonlinear Equations in Several Variables*. Academic Press, New York, 1970.

[178] H. Poincarè. *Sur les courbes definè par un èquation differentielle. I - IV*. Oeuvers I. Gauthier-Villars, Paris, 1881-1886.

[179] F. A. Potra. An infeasible interior-point predictor-corrector algorithm for linear programming. Report No. 26, The University of Iowa, Department of Mathematics, Iowa City, Iowa, 1992.

[180] S. Puthenpura, L. Sinha, R. Saigal, and S. C. Fang. Solving stochastic programming problems via Kalman filter and affine scaling. Technical Report 94-24, Department of Industrial and Operations Engineering, University of Michigan, Ann Arbor, Michigan 48109, to appear in *European Journal of Operational Research*, 1994.

[181] J. Renegar. A polynomial-time algorithm, based on Newton's method, for linear programming. *Mathematical Programming*, 40:59–93, 1988.

[182] J. Renegar and M. Shub. Simplified complexity analysis for Newton LP methods. Report No. 807, Cornell University, Ithaca, NY, 1988.

[183] T. J. Rivlin. *The Chebychev Polynomial*. John Wiley & Sons, New York, 1974.

[184] R. T. Rockafellar. *Convex Analysis*. Princeton University Press, Princeton, New Jersey, 1970.

[185] C. Roos. New trajectory-following polynomial time algorithm for linear programming problems. *Journal of Optimization Theory and Applications*, 63:433–458, 1989.

[186] C. Roos. A projective variant of the approximate center method for linear programming. Report No. 90-83, Faculty of Mathematics and Informatics/Computer Science: Delft University of Technology, Delft, Holland, 1990.

[187] C. Roos and J.-Ph. Vial. Analytic centers in linear programming. Report No. 88-74, Delft University of Technology, Delft, Holland, 1988.

[188] C. Roos and J.-Ph. Vial. Long steps with the logarithmic penalty barrier function in linear programming. In J.-F. Richard J. Gabszevwicz and L. Wolsey, editors, *Economic Decision-Making: Games, Economics and Optimization*, pages 433–441. Elsevier Science Publisher B.V., 1990.

[189] C. Roos and J.-Ph. Vial. A polynomial method of approximate centers for linear programming. *Mathematical Programming*, 54:295–305, 1992.

[190] W. Rudin. *Principles of Mathematical Analysis*. McGraw-Hill Book Company, New York, 1964.

[191] D. G. Saari and R. Saigal. Some generic properties of paths generated by fixed point algorithms. In S. M. Robinson, editor, *Analysis and computation of fixed points*, Academic Press, New York, 1980.

[192] R. Saigal. Block Triangularization of Multistage Linear Programs. ORC Technical Report 66-5, University of California, Berkeley, 1966.

[193] R. Saigal. Multicommodity flows in directed networks. ORC Report 67-38, University of California, Berkeley, Calif., 1967.

[194] R. Saigal. A constrained shortest route problem. *Operations Research*, 16:205–209, 1968.

[195] R. Saigal. A proof of the Hirsch coNew Jerseyecture on the polyhedron of the shortest route problem. *SIAM Journal on Applied Mathematics*, 17:1232–1238, 1969.

[196] R. Saigal. On the modularity of a matrix. *Linear Algebra and Its Applications*, 5:39–48, 1972.

[197] R. Saigal. On paths generated by fixed point algorithms. *Mathematics of Operations Research*, 1:359–380, 1976.

[198] R. Saigal. On the convergence rate of algorithms for solving equations that are based on the methods of complementary pivoting. *Mathematics of Operations Research*, 2:108–124, 1977.

[199] R. Saigal. Class Notes. Northwestern University, Evanston, IL, 1980.

[200] R. Saigal. Tracing homotopy paths in polynomial time. Talk given at the ORSA/TIMS Joint National Meeting in Washington, D.C., Department of Industrial and Operations Engineering, University of Michigan, Ann Arbor, MI 48109, 1988.

[201] R. Saigal. Matrix partitioning methods for interior point methods. Technical Report No. 92-39, Department of Industrial and Operations Engineering, University of Michigan, Ann Arbor, Michigan 48109, 1992.

[202] R. Saigal. A simple proof of the primal affine scaling method. Technical Report No. 92-60, Department of Industrial and Operations Engineering, University of Michigan, Ann Arbor, MI 48109, to appear in *Annals of Operations Research*, 1992.

[203] R. Saigal. On the inverse of a matrix. Technical Report No. 93-41, Department of Industrial and Operations Engineering, University of Michigan, Ann Arbor, Michigan 48109, 1993.

[204] R. Saigal. A three step quadratically convergent implementation of the primal affine scaling method. Technical Report No. 93-9, Department of Industrial and Operations Engineering, University of Michigan, Ann Arbor, MI 48109, 1993.

[205] R. Saigal. The primal power affine scaling method. Technical Report No. 93-21, Department of Industrial and Operations Engineering, University of Michigan, Ann Arbor, MI 48109, to appear in *Annals of Operations Research*, 1993.

[206] R. Saigal. On the primal-dual affine scaling method. *Opsearch*, 31:261–278, 1994.

[207] R. Saigal. An infinitely summable series implementation of interior point methods. *Mathematics of Operations Research*, 20, 1995.

[208] R. Saigal and M. J. Todd. Efficient acceleration techniques for fixed point algorithms. *SIAM Journal on Numerical Analysis*, 2:997–1007, 1978.

[209] M. Sakarovitch and R. Saigal. An extension of the generalized upper bounding techniques for structured linear programs. *SIAM Journal on Applied Mathematics*, 15:906-914, 1967.

[210] A. Schrijver. *Theory of Linear and Integer Programming*. John Wiley & Sons, Inc., New York, 1986.

[211] J. Sherman and W. J. Morrison. Adjustment of an inverse of matrix corresponding to changes in elements of a given column or a given row of the original matrix. *Ann. Math. Statis.*, 20:621, 1949.

[212] R. L. Sheu and S. C. Fang. On the generalized path-following methods for linear programming. O. R. Technical Report No. 261, North Carolina State University, Raleigh, NC 27695, to appear in *Optimization*, 1992.

BIBLIOGRAPHY

[213] M. A. Simonnard. *Linear Programming*. Prentice-Hall, Englewood Cliffs, New Jersey, 1966.

[214] G. Sonnevend. An analytic center for polyhedrons and new classes of global algorithms for linear (smooth, convex) programming. In Budapest; Lecture Notes in Control, Hungary and Vol. 84 Inform. Sci., *Proc. 12th IFIP Conf. on System Modelling and Optimization*. Springer-Verlag, New York, 1985.

[215] G. Sonnevend. On the complexity of following the central path of linear programs by linear extrapolation II. Manuscript, University of Wurtzburg, Germany, 1990.

[216] G. Sonnevend, J. Stoer, and G. Zhao. On the complexity of following the central path of linear programs by linear extrapolation. *Mathematics of Operations Research*, 63:19–31, 1989.

[217] G. Strang. *Linear Algebra and its Applications*. Academic Press, New York, 1980.

[218] J. Stoer and C. Witzgall. *Convexity and Optimization in Finite Dimensions, I*. Springer Verlag, New York, 1970.

[219] K. Tanabe. Centered Newton method for mathematical programming, system modelling and optimization. In *Proceedings of the 13th IFIP Conference, 197–206*, Tokyo, Japan, 1987.

[220] K. Tanabe. Centered Newton method for linear programming: Interior and 'exterior' point method (in Japanese). In K. Tone, editor, *New Methods for Linear Programming 3*, pages 98–100, The Institute of Statistical Mathematics, Tokyo, Japan, 1990.

[221] M. J. Todd. Improved bounds and containing ellipsoids in Karmarkar's linear programming algorithm. *Mathematics of Operations Research*, 13:650–659, 1988.

[222] M. J. Todd. Anticipated behavior of Karmarkar's algorithm. Technical Report No. 879, Cornell University, Ithaca, NY, 1989.

[223] M. J. Todd. Recent developments and new directions in linear programming. In M. Iri and K. Tanabe, editors, *Mathematical Programming: Recent Developments and Applications*, pages 109–157. Kluwer Academic Press, Dordrecht, Holland, 1989.

[224] M. J. Todd. A Dantzig-Wolfe-like variant of Karmarkar's interior-point linear programming algorithm. *Operations Research*, 38:1006-1018, 1990.

[225] M. J. Todd. On Anstreicher's combined phase I-phase II projective algorithm for linear programming. *Mathematical Programming*, 55:1–15, 1992.

[226] M. J. Todd and B. P. Burrell. An extension of Karmarkar's algorithm for linear programming using dual variables. *Algorithmica*, 1:409–424, 1986.

[227] M. J. Todd and Y. Wang. On combined phase I phase II projective methods for linear programming. Technical Report No. 877, Cornell University, Ithaca, NY, 1989.

[228] M. J. Todd and Y. Ye. A centered projective algorithm for linear programming. *Mathematics of Operations Research*, 15:508–529, 1990.

[229] J. A. Tomlin. An experimental approach to Karmarkar's projective method for linear programming. *Mathematical Programming Study*, 31:175–191, 1987.

[230] K. Tone. An active-set strategy in interior point method for linear programming. Working Paper, Saitama University, Urawa, Saitama 338, Japan, 1991.

[231] B. Trong Lieu and P. Huard. La methode des centres dans un espace topologique. *Numer. Mat.*, 8:56–67, 1966.

[232] P. Tseng and Z. Q. Luo. On the convergence of the affine-scaling algorithm. *Mathematical Programming*, 56:301–319, 1992.

[233] T. Tsuchiya. Global convergence of the affine scaling methods for degenerate linear programming problems. *Mathematical Programming*, 52:377–404, 1991.

[234] T. Tsuchiya. Quadratic convergence of Iri and Imai's algorithm for degenerate linear programming problems. Research Memorandum No. 412, The Institute of Statistical Mathematics, Tokyo, Japan, to appear in *Journal of Optimization Theory and Applications*, 1991.

[235] T. Tsuchiya. Global convergence property of the affine scaling methods for primal degenerate linear programming problems. *Mathematics of Operations Research*, 17:527–557, 1992.

[236] T. Tsuchiya and R. D. C. Monteiro. Superlinear convergence of the affine scaling algorithm. Research memorandum, 1992.

[237] T. Tsuchiya and M. Muramatsu. Global convergence of a long-step affine scaling algorithm for degenerate linear programming problems. Research memorandum no. 423, The Institute of Statistical Mathematics, Tokyo, Japan, to appear in *SIAM Journal on Optimization*, 1992.

[238] P. M. Vaidya. A new algorithm for minimizing convex functions over convex sets. Technical Report, AT&T Bell Laboratories, Murray Hill, New Jersey, 1989.

[239] P. M. Vaidya. An algorithm for linear programming which requires $0(((m + n)n^2 + (m + n)^{1.5}n)l)$ arithmetic operations. *Mathematical Programming*, 47:175–201, 1990.

[240] F. A. Valentine. *Convex Sets*. McGraw–Hill, New York, 1964.

[241] R. J. Vanderbei and J. C. Lagarias. I. I. Dikin's convergence result for the affine-scaling algorithm. In J. C. Lagarias and M. J. Todd, editors, *Mathematical Developments Arising from Linear Programming*, pages 109–119. Contemporary Mathematics 114, 1990.

[242] R. J. Vanderbei, M. S. Meketon, and B. A. Freedman. A modification of Karmarkar's linear programming algorithm. *Algorithmica*, 1:395–407, 1986.

[243] J. Von Neumann. Uber ein okonomisches gleichungssystem und ein verallgemeinerung des Boouwerschen fixpunktsatzes. *Ergebnisse eines Mathematischen Kollguims*, 8, 1937.

[244] J. Von Neumann and O. Morgenstern. *Theory of Games and Economic Behavior*. Princeton University Press, Princeton, New Jersey, 1944.

[245] J. H. Wilkinson. *The Eigenvalue Problem*. Oxford University Press, New York, 1965.

[246] A. C. Williams. Complementary theorems for linear programming. *SIAM Review*, 12:135–137, 1970.

[247] P. Wolfe. A technique for resolving degeneracy in linear programming. *Journal of SIAM*, 11:205–211, 1963.

[248] M. Woodbury. Inverting modified matrices. Memorandum Report 42, Statistical Research Group, Princeton University, Princeton, New Jersey, 1950.

[249] D. Xiao and D. Goldfarb. A path-following projective interior point method for linear programming. Working Paper, Columbia University, New York, NY, 1990.

[250] H. Yamashita. A polynomially and quadratically convergent method for linear programming. Working Paper, Mathematical Systems Institute, Inc., Tokyo, Japan, 1986.

[251] H. Yamashita. A class of primal-dual method for constrained optimization. Working Paper, Mathematical Systems Institute, Inc., Tokyo, Japan, 1991.

[252] Y. Ye. Further development on the interior algorithm for convex quadratic programming. Technical Report, Stanford University, Stanford, California, 1987.

[253] Y. Ye. *Interior Algorithms for Linear, Quadratic and Linearly Constrained Convex Programming*. PhD thesis, Stanford University, Stanford, CA, 1987.

[254] Y. Ye. The "build down" scheme for linear programming. *Mathematical Programming*, 46:61–72, 1990.

[255] Y. Ye. A class of projective transformations for linear programming. *SIAM Journal on Computing*, 19:457–466, 1990.

[256] Y. Ye. Line search in potential reduction algorithms for linear programming. manuscript, The University of Iowa, Iowa City, IA, 1990.

[257] Y. Ye. An $O(n^3 L)$ potential reduction algorithm for linear programming. *Mathematical Programming*, 50:239–258, 1991.

[258] Y. Ye. A further result on the potential reduction algorithm for the P-matrix linear complementarity problem. In P. M. Pardalos, editor, *Advances in Optimization and Parallel Computing*, pages 310–316. North-Holland, Amsterdam, 1992.

[259] Y. Ye. A potential reduction algorithm allowing column generation. *SIAM Journal on Optimization*, 2:7–20, 1992.

BIBLIOGRAPHY

[260] Y. Ye and M. Kojima. Recovering optimal dual solutions in Karmarkar's polynomial algorithm for linear programming. *Mathematical Programming*, 39:305–317, 1987.

[261] Y. Ye and P. Pardalos. A class of linear complementarity problems solvable in polynomial time. *Linear Algebra and Its Applications*, 152:1–9, 1991.

[262] S. Zhang. On the convergence property of Iri-Imai's method for linear programming. Report 8917/A, Erasmus University, Rotterdam, The Netherlands, 1989.

[263] Y. I. Zhang. On the convergence of a class of infeasible interior-point algorithms for the horizontal linear complementarity problem. *SIAM Journal on Optimization*, 4:208–227, 1994.

[264] Y. I. Zhang, R. A. Tapia, and J. E. Dennis. On the superlinear and quadratic convergence of primal–dual interior point linear programming algorithms. *SIAM Journal on Optimization*, 3:413–422, 1993.

[265] W. I. Zangwill. *Nonlinear Programming A Unified Approach*. Prentice-Hall, Englewood Cliffs, New Jersey, 1969.

Index

accelerated affine scaling method 148
adjacent vertices 41
Adler 263
affine combination 37
affine hull 37
affine scaling step 151, 158, 172, 174, 176, 181, 190, 192, 211
affine space 37
Allgower 65
analytic center 139, 143, 146, 148, 150, 155, 156, 165, 222, 279
arc length 58, 224
Arden 83
assignment problem 292
asynchronous 274, 275

backward solve 24, 267, 273
band method 273
basic feasible solution 34
basis 14
basis inverse 92
Bazaraa 5
Beale 110
big M method 95, 198, 232
bordering method 271, 274
boundary 10
bounded variable affine scaling 203
bounded variable simplex method 98

Burrell 264

central path 222
Chandru 263
characteristic polynomial 19, 285
Charnes 110
Chebychev polynomial 294
Cholesky factor 25, 270, 271, 272, 278
Cholesky factorization 25, 270, 277
closed set 10
closure 10
column space 16, 151
compact set 10
complementary pair 76
complementary slackness theorem 75
conjugate gradient method 270, 276, 277, 278, 285
continuation method 55
continuous function 10
continuously differentiable function 12
convex combination 28
convex hull 29
convex polyhedral set 36
convex set 27
corrector step 158, 190, 229
cost minimization model 4
critical value 56

Dantzig 5, 83, 110, 304

INDEX

degeneracy resolution 139, 181
Diet problem 3
differentiable function 12
Dikin 262
dimension 15
dual affine scaling method 203, 292
dual of other forms 68
dual problem 67
dual simplex method 103

edge 39
efficiency of convergence 197
Eggleston 65
eigenvalue 19, 285, 286, 287, 288, 289, 294, 296
eigenvector 19
elementary matrix 23, 266, 267
ellipsoidal approximating problem 112
ellipsoid 112
embedding method 55
envelope method 273
Euclidean space 7, 15

face 38
facet 39
Farkas 65
Farkas' lemma 35
feasible solution 33
Fiacco 264
fill-in 272, 273
finite difference Newton's method 49
finite difference operator 286
Ford 110
forward solve 24, 267, 273
Fulkerson 110

Gale 65, 83
Gaussian elimination 22, 266, 267, 268

Georg 65
George 305
Givens matrix 269
global existence theorem 56
global Newton method 55
gradient 12
Gram-Schmidt orthogonalization 276, 284

half affine space 38
halfspace 28
Hessian 12
Hitchcock 5
Hoffman 110
homogeneous solution 40
homotopy method 55, 86, 111, 222
Householder matrix 268
hyperplane 27
hypersphere 113, 147, 246

infeasible start path following method 236
initial value problem 58
inner product method 272
integral data 43
interior feasible solution 33, 112
interior 10

Jacobian 12
Jarvis 5

Kantorovich 5, 65
Kantorovich's theorem 45
Karmarkar 263, 264
Karush-Kuhn-Tucker conditions 144, 149, 150, 154
Khachian 2
Klee 110
Kochar 263
Kojima 264
Koopmans 5

Lagarias 262
Lagrange multipliers 124, 144, 155
Lagrangian 123
large step affine method 118, 129, 155, 278
large step power method 166
least squares 119, 286, 289
left null space 16
Lemke 110
Leontief 5
Leray Schauder Theorem 57
Leray 65
lexico-graphic ordering 94
lexico-graphic min ratio test 94
linear combination 14
linear convergence rate 130, 186
linear independence 14
linear model 285, 287, 288, 289
linear space 13
linear system of equations 20
line 27
Lipschitz constant 11, 45, 50
Lipschitz continuous 11, 12, 45
Liu 305
local potential function 140, 181
LQ factor 26, 284
LU factorization 25, 266
Lustig 264

Marshall 110
Marsten 264
Mascarenhas 139, 262
McCormick 264
Megiddo 264
merit function 183
min $\bar{c}_j$ rule 89
min ratio test 90
minimum degree method 273
Minkowski 65
Minty 110

Mizuno 263
Monteiro 263
Motzkin 65
Motzkin's Theorem 36
Muramatsu 262
Murty 5, 304

Nagasawa 263
Neumann 83
Newton direction 155, 156
Newton step 151, 153, 158, 159, 172, 181
Newton's method 45, 148, 150, 161, 211, 218
non-degenerate:
    assumption, 87, 171, 209
    face, 41
    problem, 41, 137
    uniqueness, 82
non-singular matrix, 21
norm:
    definition, 7
    l_1, 8
    l_∞, 8
    equivalent, 8
    Euclidean, 7
    general, 7
    matrix, 18, 21
    vector, 19
null space 16, 20, 32, 115

open set 10
orthogonal:
    definition, 18, 20, 59, 283
    complement, 17, 124
    matrix, 267
orthonormal:
    basis, 20, 124, 127
    matrix, 20, 287, 295
    vectors, 26, 283

INDEX 341

outer product method 271, 280

parallel chord method 53
partial Cholesky factorization 280, 301
path following method 111, 222
permutation matrix 23, 266, 269, 271, 272, 273
pivot operation 24
Poincarè 65
polynomial time convergence 230, 233, 243, 252
polytope 39
positive definite 20, 25, 26, 27, 201, 210, 265, 270, 281, 299
positive semi-definite 285
potential function 139, 150
power center 174, 176, 185
predictor step 58, 157, 158, 190, 228
predictor-corrector method 57
pricing 89
primal affine scaling method 112
primal power affine scaling method 166
primal simplex method 87
primal-dual path following method 228
primal-dual affine scaling method 211
primal-dual method 107
principal rearrangement 23
profit maximization model 3
projection 17
projective transformation method 244
projective transformation 244, 245

QR factorization 267, 270
QR factor 26, 267

quadratic convergence 47, 148, 156, 165

rank 16
ray 27
re-inversion 267, 270
regular value 56
Resende 263
resolution theorem 42
revised simplex method 91
Rivlin 294
Rockafellar 65
row space 16
Rudin 64

Saigal 263, 305
Schauder 65
Schur complements 299
sequence:
    bounded, 9
    Cauchy, 9, 47, 170, 216, 217
    convergence, 9
    definition, 8
    subsequence, 9
Shanno 264
Sherman-Morrison-Woodbury formula 281, 282, 299, 300
slack variables 73
solution set 20
sparse Cholesky factorization 272
spectral decomposition 285
spectral radius 281, 283, 285, 292, 293, 294
Stoer 65
Strang 64
strict complementarity theorem 77
strict complementary pair 76
strict separation theorem 30
strictly complementary solution 137

strong duality theorem 71
submatrix 16
subspace 15
superlinear convergence rate 196
Suurballe 110
symbolic factorization 273
symmetric matrix 20

Tanabe 263
Todd 264
Transportation problem 2
triangular matrix 24
Tsuchiya 262
two phase method 96, 199

uniqueness theorem 82
Unit Ball 8
unrestricted variables and affine scaling 203
upper Hessenburg matrix 269

Valentine 65
Vanderbei 262
Vandermonde's matrix 288
vector-parallel environments 273, 282
Veiga 263
vertex 39

weak duality theorem 70
weak separation theorem 29
Wilkinson 64
Williams 83
Witzgall 65

Yoshise 264